Chicago Public Library
W9-AWQ-953
Linking science and technology to s

NATIONAL FORUM ON SCIENCE AND TECHNOLOGY GOALS

Linking Science and Technology to Society's Environmental Goals

Policy Division
National Research Council

NATIONAL ACADEMY PRESS
Washington D.C. 1996

NATIONAL ACADEMY PRESS • 2101 Constitution Ave., N.W. • Washington, DC 20418

NOTICE: This volume was produced as part of a project approved by the Governing Board of the National Research Council, whose members are drawn from the councils of the National Academy of Sciences, the National Academy of Engineering, and the Institute of Medicine. The members of the expert committee were chosen for their special competences and with regard for appropriate balance. This report has been reviewed by a group other than the authors according to procedures approved by the NRC and the Report Review Committee.

The project resulting in this report was supported by the Carnegie Corporation of New York and the Kellogg Endowment Fund of the National Research Council.

Library of Congress Cataloging-in-Publication Data

National Research Council (U.S.). Policy Division.
Linking science and technology to society's environmental goals / National Research Council, Policy Division.
p. cm. — (National forum on science and technology goals)
Includes bibliographical references and index.
ISBN 0-309-05578-4
1. Environmental risk assessment—Social aspects—United States. 2. Environmental policy—United States. 3. Environmental management—Government policy—United States. I. Title. II. Series.
GE150.N38 1996
363.7—dc21

96-47176
CIP

Internet Access: This report is available on the National Academy of Sciences' Internet host. It may be accessed via World Wide Web at http://www.nas.edu.

Printed in the United States of America

POLICY DIVISION

COMMITTEE ON THE NATIONAL FORUM ON SCIENCE AND TECHNOLOGY GOALS: ENVIRONMENT

JOHN F. AHEARNE (*Co-chair*), Lecturer in Public Policy, Duke University, and Director, Sigma Xi Center
H. GUYFORD STEVER (*Co-chair*), Science Consultant, Gaithersburg, Maryland
ALVIN L. ALM,[*] Assistant Secretary for Environmental Management, US Department of Energy
JAN E. BEYEA, Consultant, National Audubon Society
BARBARA L. BENTLEY, Professor, Department of Ecology and Evolution, State University of New York, Stony Brook
HARVEY BROOKS,[**] Professor of Technology and Public Issues, Emeritus, John F. Kennedy School of Government, Harvard University
PATRICIA A. BUFFLER, Dean, School of Health, University of California, Berkeley
JOHN B. CARBERRY, Director, Environmental Technology, DuPont Research and Development
EMILIO Q. DADDARIO, Former member of Congress, Washington, D.C.
PERRY L. McCARTY, Silas H. Palmer Professor of Civil Engineering, Stanford University
RODNEY W. NICHOLS, President and Chief Executive Officer, New York Academy of Sciences
PAUL R. PORTNEY, President, Resources for the Future
F. SHERWOOD ROWLAND, Donald Bren Research Professor, Department of Chemistry, University of California, Irvine
ROBERT M. WHITE, President, Washington Advisory Group

Principal Project Staff

LAWRENCE E. McCRAY, Director, Policy Division
DEBORAH D. STINE, Study Director
PATRICK P. SEVCIK, Program Assistant
NORMAN GROSSBLATT, Editor

---

[*]Dr. Alm resigned from the committee on May 8, 1996, to become Assistant Secretary for Environmental Management at the US Department of Energy

[**]Dr. Brooks was unable to participate in the public forum or in the committee meeting immediately following the forum.

The **National Academy of Sciences** (NAS) is a private, nonprofit, self-perpetuating society of distinguished scholars engaged in scientific and engineering research, dedicated to the furtherance of science and technology and to their use for the general welfare. Under the authority of the charter granted to it by Congress in 1863, the Academy has a working mandate that calls on it to advise the federal government on scientific and technical matters. Dr. Bruce M. Alberts is president of the NAS.

The **National Academy of Engineering** (NAE) was established in 1964, under the charter of the NAS, as a parallel organization of distinguished engineers. It is autonomous in its administration and in the selection of members, sharing with the NAS its responsibilities for advising the federal government. The National Academy of Engineering also sponsors engineering programs aimed at meeting national needs, encourages education and research, and recognizes the superior achievements of engineers. Dr. William A. Wulf is interim president of the NAE.

The **Institute of Medicine** (IOM) was established in 1970 by the NAS to secure the services of eminent members of appropriate professions in the examination of policy matters pertaining to the health of the public. The Institute acts under the responsibility given to the NAS in its congressional charter to be an adviser to the federal government and, on its own initiative, to identify issues of medical care, research, and education. Dr. Kenneth I. Shine is president of the IOM.

The **National Research Council** (NRC) was organized by the NAS in 1916 to associate the broad community of science and technology with the Academy's purposes of furthering knowledge and advising the federal government. Functioning in accordance with general policies determined by the Academy, the Council has become the principal operating agency of both the NAS and the NAE in providing services to the government, the public, and the scientific and engineering communities. The Council is administered jointly by both Academies and the IOM. Dr. Bruce M. Alberts is chairman and Dr. William A. Wulf is interim vice-chairman of the NRC.

Preface

OBJECTIVES

This study, first in a series of National Forums on Science and Technology Goals, is the first response to a recommendation in a September 1992 report of the Carnegie Commission on Science, Technology, and Government entitled *Enabling the Future: Linking Science and Technology to Societal Goals*. That report recommended that: "A nongovernmental National Forum on Science and Technology Goals should be established to facilitate the process of defining, debating, focusing, and articulating science and technology goals in the context of federal, national, and international goals, and to monitor the development and implementation of policies to achieve them." As envisioned, the forum would convene persons from industry, academe, nongovernment organizations, and the interested public to explore and seek consensus on long-term science and technology goals as they related to other social objectives and the potential contribution of scientific and engineering advances to societal goals.

PROJECT HISTORY

After extended discussions between the National Research Council and representatives of the Carnegie Corporation of New York, agreement was reached that the National Research Council would conduct two pilot forum studies. A grant from the Carnegie Corporation of New York set these experiments in motion. This is the first of those experiments. If an evaluation of this experimental effort and the following one show that the forums are successful, the forums will continue.

In response to a request from the White House Office of Science and Technology Policy (OSTP), the NRC chose environment as the topic for the first forum. A committee was then appointed that had not only broad environmental expertise, but expertise in science and technology policy as well. This selection was also reflected in the co-chairs for the forum. In addition, to maintain a linkage to the original Carnegie Commission report, several members of this committee were also members of the originating Carnegie Commission Committee. A Steering committee then met via conference call to plan the first forum.

FORUM HISTORY

To prepare for the forum and committee meetings, a number of activities were undertaken. Questions were sent to various organizations and persons (the respondents are listed in Appendix E), and 128 replies were received; these are summarized in Appendix D. Part II provides eight commissioned papers on several specific subjects. And, finally, Part III lists the 10 presentations made at the forum by leaders of government agencies. Throughout this document, boxes highlight comments made to the committee.

The centerpiece of the study was the August 20-30, 1995, meeting at the Arnold and Mabel Beckman Center in which a forum activity was joined to a committee meeting. The agenda for the forum is provided in Appendix B. Participants in the forum are listed in Appendix C. Comments and questions after the presentations involved the forum attendees, but the most effective way to involve them was to mix them with committee members in breakout sessions, where more intimate exchanges of views took place. A summary of the breakout sessions is provided in Appendix F.

After the interchange with forum participants, the committee, whose members are listed in Appendix A, met for several days to select and enlarge on a number of the important points to present as results of the forum process. Several principles guided the committee's work. First, the committee was to study the direction of science and technology to be helpful for reaching long-term societal goals 15-25 years in the future. Second, the committee was to depend on the forum results (in oral form and in written form via the call for comments) and the debate within the committee and among the forum participants. This constraint is the principal reason that this report is different from other NRC reports.

SELECTION OF REPORT TOPICS

The diversity of the ideas from so many sources provided the committee with an abundance of input for its deliberations. The committee focused on science and technology topics that it concluded merited increased attention and resources—realizing that this focus might be at the expense of other environmental issues.

> *We need a definition of environmental goals that people can understand and become passionate about pursuing.*
> —Forum Participant Comment*

For example, two major global environmental problems have not been addressed in this report. They include climate change and biodiversity. For both of these issues, there are now international treaties that have been signed, goals and plans have been developed, and substantial international research resources are being devoted to them. It was the belief of the committee that these two very important environmental issues were being adequately treated.

The committee instead chose to concentrate on six critical environmental subjects:

- Economics and risk assessment.
- Environmental monitoring and ecology.
- Chemicals in the environment.
- The energy system.
- Industrial ecology.
- Population.

Almost all have been the subject of continuing science and technology work—some of it substantial—but additional research beyond this is needed if societal goals are to be met.

Other issues raised but considered to be outside the purview of this forum or to have lower priority at this time include the environmental/dimensions of weapons cleanup, human health, international relations, complex systems, agriculture, and transportation.

AN INNOVATIVE APPROACH

This project is the first in what might, if the initial experiment is successful, become a series of projects with the mission of establishing science and technology goals to address major national (and international) problems. The basic concept is to involve a broad group of knowledgeable and concerned people to develop these goals. The process is to include committee surveys, commissioned background papers, and a forum. The NRC would serve as a convener for the forum and establish a group that would distill and synthesize the results.

This experimental forum where the public and the scientific and technological community interact has suggested a number of ways for the science and tech-

*Throughout this report, boxes present comments received by the committee from forum participants and respondents to the committee's call for comments. These do not necessarily represent the committee's views but are presented here to provide the flavor of the forum process. In accordance with the committee's agreement with the authors, the sources of these quotations are not identified.

nology community to focus its activities, but the approach taken was an innovative one that differs from other NRC reports.

A key difference between this report and others that have discussed the issue of the environment, is that this document is not based on an analysis of the literature, empirical analysis of the cost and benefits of various actions, etc., but, as befits its charge, is based on the judgments of the committee after hearing comments in the forum, via the call for comments, and on its own deliberations. It therefore does not refer heavily to the scientific literature or specific information. Instead, our "references" are the comments from the public that are shown in boxes throughout this report. Ideally, of course, the committee would have both published and public "references," but that was not possible in the few days allocated to this activity, nor would it necessarily be better relative to its "public" objectives.

Several cautions are in order in reading this report:

- This forum was the first. Later efforts will benefit from the lessons learned from it, and many of the problems evident here can be corrected.
- The concept envisioned more extensive preparation, which was reduced because of financial limitations, and periodic followup meetings, to reach consensus on the goals developed in the first forum. It has not been decided whether such meetings will be held, but this report should be read as a first draft of the goals. It would be only after several such iterative meetings (probably over 1-2 years) that a true consensus would be reached on a fully developed set of science and technology goals for the environment.

The next step of this activity is to encourage the science and technology community to undertake activities in and discussion of these areas.

ACKNOWLEDGMENTS

The committee acknowledges the invaluable information and opinions received from individuals and organizations who responded to our call for comments and who participated in the forum. The committee would also like to thank those who wrote the papers commissioned by the committee as shown in Part II. Each paper, on its own, is a valuable source of information and guidance.

The project was aided by the invaluable help of its professional staff: Deborah Stine, who directed the project, and Lawrence McCray, who oversaw the project as director of the NRC's Policy Division. Patrick Sevcik ably assisted the project. The report was improved by its diligent editor, Norman Grossblatt.

John F. Ahearne
H. Guyford Stever
Co-Chairs

Contents

PART I: COMMITTEE REPORT

PART II: COMMISSIONED PAPERS

PART III: KEYNOTE ADDRESSES AND PRESENTATIONS

Keynote Addresses

Presentations

PART IV: APPENDIXES

Linking Science and Technology to Society's Environmental Goals

PART I

COMMITTEE REPORT

Summary

The general desire on the part of societies to improve their standard of living has resulted in greater consumption of water, food, mineral, and energy services and expansion of our manufacturing capability. Although wastes and other products that result from such consumption produce a greater burden on the environment, it is also generally true that the higher the living standard, the greater the resources that are devoted to protecting human health and the environment.

Thus, industrially advanced societies tend to have safer food, water, and air than do less-advanced societies. Nevertheless, there remains a conflict between the desire for more goods and services, particularly in a growing population, and the desire for a healthy environment.

Following the recommendations of the Carnegie Commission report *Enabling the Future: Linking Science and Technology to Societal Goals*, the National Research Council convened the first National Forum on Science and Technology Goals, which addressed goals related to the environment. The central question was, "How can science and technology contribute most effectively to meeting societal environmental goals?"

The National Forum on Science and Technology Goals Environment Committee developed conclusions and recommendations based on four sources:

- Responses from 128 persons and organizations to a questionnaire sent to a broad array of national, state, and local governments, citizen groups, environmental groups, industry and academe.
- Eight commissioned papers, summaries of which were presented at the forum.
- Discussions by forum participants, especially in small breakout groups.
- The committee members' own special knowledge.

Using that information as a basis, the committee adopted the following process for selecting subjects for this report:

1. Via the call for comments and forum discussion, the committee developed a list of some 30-odd topics that could possibly be discussed in the report.
2. The committee then selected the topics that they believed to be the most important for the scientific and engineering community to focus on, given the current intellectual and financial resources devoted to each. Eight topics were selected.
3. Of the eight topics, six eventually emerged as subjects for the chapters of this report (several topics were merged).

The six topics are

- Economics and risk assessment.
- Environmental monitoring and ecology.
- Chemicals in the environment.
- The energy system.
- Industrial ecology.
- Population.

The chapter titles incorporate those topics but have been revised to reflect the content of the chapters.

Many other topics are also important, but the committee believes that they are already receiving considerable attention elsewhere. For example, two major global environmental problems have not been addressed in this report: climate change and biodiversity. For both, international treaties have been signed, goals and plans have been developed, and substantial international research resources are being applied. The committee believes that these two very important environmental issues needed no further committee discussion.

The committee's conclusions and recommendations regarding the chosen topics follow, with some discussion of the nature of goals and the environment.

USE SOCIAL SCIENCE AND RISK ASSESSMENT TO MAKE BETTER SOCIETAL CHOICES

In recent years, many segments of society have become concerned about the economic impact of environmental regulation. Business, government, and individuals together spend an estimated $150-180 billion of their revenues each year to meet environmental regulations.

Present regulatory strategies do not sufficiently differentiate between minor and major risks. Furthermore, the costs incurred to reduce risks often do not bear a consistent relation to the magnitude of the risks and the number of people potentially affected. The nation's existing environmental goals could be met less expensively or faster by substituting incentive-based approaches to environmental regulation for command-and-control approaches. Incentive approaches provide a more flexible and cost-effective regulatory environment for industry, business, and government while maintaining or perhaps even improving environmental quality much less expensively.

The tools needed to implement incentive-based approaches—specifically quantitative risk assessment (QRA) and cost-benefit analysis (CBA) (including some attention to distributional effects)—are not as well developed as they need to be if they are to be reliable aids in decision-making. Nor are these tools widely understood or accepted by decision-makers or other interested parties and the general public.

Recommendations

1. Research to improve the analytical tools available to decision-makers should be expanded. Several specific questions need attention by researchers

- With respect to cost-benefit analysis, what values do people attach to the services provided by ecosystems? To the protection of endangered species? To the extension of human life? To aesthetics and the quality of life? How do those values differ for different states of ecosystem and human health?
- Also, with respect to benefit-cost analysis, how can models better estimate the costs of regulatory proposals? How can models predict human behavior, as opposed to relying only on the more reliable predictive value of current technologies and practices? How should models account for lost opportunities, even when no out-of-pocket expenditures are made, and for continuing technological changes that bring down the costs of regulatory compliance? The distribution of risks and benefits involves issues that are inherently political, and cost-benefit analysis generally is silent on these distributional matters. In light of that, how can models more realistically take into account the distribution of risks, costs, and benefits among all affected stakeholders?
- What are the best ways to assess noncancer health risks, such as neurological and reproductive disorders? Disproportionate attention is paid today to collecting information on cancer risks compared with, for example, risks of neurological and reproductive disorders. Future research on quantitative risk assessment should be directed toward correcting this imbalance of emphasis.
- Can quantitative risk assessment and cost-benefit analysis be integrated so that the health or ecological end points that risk assessors predict are the ones that the public understands and cares about? To what extent could and should probabilistic techniques be used with worst-case assumptions? In the face of uncertainty, what degree of conservatism should be used in connection with probabilistic techniques?

2. Research should be increased and demonstration projects launched to expand the application of incentive-based approaches to environmental protection

(these include pollution taxes, systems of marketable discharge permits, and deposit-refund schemes). The research should include the evaluation of institutional impediments that often reduce the savings associated with incentive-based systems. Experimental economics—in which a group of participants act out the decisions that they would make in given regulatory or other scenarios—is a particularly promising avenue for research. As more data on incentives become available, decision-makers can have a better intellectual rationale for choice between incentive-based approaches to environmental protection and command-control regulations.

3. Social-science research on comparative risk assessment (or risk ranking) can be helpful at all levels of government to help to establish regulatory and legislative priorities. Comparative risk-assessment activities should involve elected and appointed officials; members of business, environmental, and civic organizations; and lay persons. In addition, natural and social scientists who can provide information on the magnitude of various risks to health and the environment, the likely costs of mitigating these risks, and the uncertainties associated with both should be part of the process.

FOCUS ON MONITORING TO BUILD BETTER UNDERSTANDING OF OUR ECOLOGICAL SYSTEMS

Science and technology provide quantitative data to characterize the state of the environment. Scientists analyze and interpret this information to provide society with a deeper understanding of its relation to the state of the environment. As part of this analysis, they compare past conditions with projections to determine the potential for achieving a desired state of environmental quality at a feasible rate of change.

One particular concern relative to achieving a desired state of environmental quality is our understanding of the ecological system in which plants, animals, and humans live. Current ecological data and understanding are inadequate to

- Detect, monitor, and characterize environmental changes.
- Evaluate the consequences of human activities.
- Provide an information base for sustainable management (i.e., "no loss") of both natural and ecological human-designed systems.

Therefore, it is difficult to conduct the comparative analysis of past, current, and future ecological states (as described earlier) to determine what actions are needed to achieve a desired end of environmental quality. Furthermore, current programs do not address these issues in a sufficiently coherent and comprehensive manner on a national basis. Indicators that are needed to measure the current status of ecological systems, to gauge the likelihood of meeting society's environmental goals, or to anticipate problems resulting from economic growth are not available. We are spending much money to collect data that are neither complete nor always relevant to the decisions that society needs to make about land use, transportation, industrial activity, agriculture, and other human activities. The system of monitoring the state of the environment needs to be improved to make it more relevant to decision-makers, and there is need for a more-sophisticated and better-informed discussion of what needs to be measured and why.

Recommendations

1. The White House Office of Science and Technology Policy should review and evaluate the quality of existing measurement and monitoring systems for relevance to and usefulness in meeting environmental goals. That would include establishing a system-design process to complete and maintain the monitoring system.

2. Congress should assign an existing or new federal research organization the mission of working with the scientific community to identify key subject areas for ecological research and ensuring that this research is being pursued adequately somewhere in the overall environmental research system which includes not only the Environmental Protection Agency and the Department of the Interior, but also the National Science Foundation, National Aeronautics and Space Administration, National Oceanic and Atmospheric Administration, Department of Defense, and Department of Energy.

3. Research should aim at identifying and developing reliable indicators of the health and sustainability of the environment and ecosystems. Such factors might include chemical nutrient systems, habitat fragmentation, and changes in biodiversity.

4. New systems of monitoring that meet society's decision-making needs should be identified and implemented. This includes provision of access to data sets via the Internet, integration of data from various scales (such as integrating satellite data with land-based measurements), and finding sources of historical data (such as glacial samples that have preserved the earth's history).

5. The nation's existing monitoring system needs to be reviewed and evaluated for its relevance to indicators of environmental progress and identification of emerging issues. This evaluation procedure is a first step toward improving the overall system.

REDUCE THE ADVERSE IMPACTS OF CHEMICALS IN THE ENVIRONMENT

Concerns about chemicals in the environment have focused major attention on the possible consequence for humans, animals, and whole ecosystems. Substantial progress has been made, and some contaminated bodies of water have been restored to use. However, we still lack basic knowledge and procedures for evaluating the potential impacts of chemicals, compound mixtures, or artificial concentrations of natural substances that have an adverse effect on human health and the environment. Such knowledge will be essential for developing products with adequate safeguards against unwanted side effects.

Reasonably good methods are available for testing the potential carcinogenic effects of chemicals on surrogate species for humans, particularly rodents. The correlation between these surrogates and humans is far from proved and is hotly debated. However, on the basis of experience and buttressed by significant testing, use of surrogate species seems to have been most helpful in reducing exposure to many suspect carcinogens.

However, there is a need for better tests to assess ecological damage potentially caused by single compound chemicals, the byproducts of various waste-treatment processes, and the degradation products of intentional products or unintentional process emissions that find their way into the environment. Better understanding of the basic biochemical processes occurring in the environment is necessary to decide where to look, what to look for, what to measure, and how to measure it.

Recommendations

1. Better test methods should be developed to evaluate, model, and monitor the potential long-term environmental impacts of single compounds emitted as a result of new products or processes. Emphasis should be placed on compounds that degrade only very slowly.

2. Better test methods should be developed to define and ultimately to model and predict the byproducts and degradation products associated with production and use of materials.

3. Basic studies of biochemical effects and of the impact of various chemicals and other adverse effects on the biochemistry of sensitive plant and animal species should be strongly supported. It is from such studies and the monitoring program that the most-effective hypotheses about items of greatest concern and about the continual development of testing will arise.

4. Strong support should be given to innovative ideas for modeling and tests on lower-order surrogate species that help to reduce the cost of tests for potential adverse environmental health effects on humans or shorten the response time needed to obtain that information.

5. International standardization of testing and international sharing of testing responsibilities should be promoted to reduce costs and speed the availability of reliable and reproducible assessments.

6. The emerging concept of developing experimental "miniecosystems"—focused on controlled-exposure environments for testing and for developing mathematical simulations of ecosystem impact based on limited, specific tests—should be supported.

DEVELOP ENVIRONMENTAL OPTIONS FOR THE ENERGY SYSTEM

Energy production and use underlie the growth of modern industrial society, but production and use are often replete with environmental problems. Of particular concern is the use of fossil fuels that lead to environmental problems, such as urban air pollution, acid rain, resource extraction, and global warming.

Responses to these concerns can be actions on either the supply side or the demand side. In both cases, knowledge, technical, and social barriers need to be overcome before these actions can be implemented. The barriers can be overcome by additional research and development.

Energy research and development should create options for an uncertain future of energy availability and the environmental impact of that energy. Development of cleaner and economically viable efficiency and production alternatives will be key to preserving options against a number of contingencies. One contingency is scarcity caused by rising energy demand and limited supplies. Another contingency would arise from new knowledge that indicated severe environmental impacts of CO_2, radionuclides, or other emissions from conventional energy sources than now expected. If either of these contingencies arises, alternative energy sources and end-use technologies will be critical.

Recommendations

1. The committee recommends sustained research and development that will lead to more options for energy generation and use, less emission of carbon into the atmosphere, and more-efficient use of natural resources. In particular, the following topics should be explored:

- *Electricity*. Thus while the United States is still the largest national consumer of primary energy, its relative contribution to pollution from energy use in the world is declining. Per-capita electricity use remains strongly correlated with development as electricity continues to replace other forms of energy because of its environmental and other advantages. Further increases in electrification should be accompanied by research in non-fossil-fuel sources (described further below) load management, and in other conservation approaches.
- *Renewable energy sources*. Solar energy, especially as used in photovoltaic cells, and biomass are the leading options for renewable energy sources. Research efforts should focus on making these more economical. Transition to widespread use will not occur until the cost of electricity from these sources is so low that large public-sector subsidies are no longer required to make them cost-competitive.
- *Coal*. The United States, as well as Russia and China, has vast reserves of coal. About 60% of U.S. electricity comes from coal plants. Coal is also a major energy source in other countries. Thus, even though efforts should be made to

defossilize our energy sources, research and development to improve the efficiency and reduce the emissions of coal plants (clean-coal technology) can help the U.S. environment and be a major factor in the Asian market, where India and China will burn increasingly large amounts of coal.

- *Nuclear fission.* A major source of U.S. electricity (21%), nuclear plants do not emit carbon or other pollutants. However, the U.S. nuclear industry has been crippled by the high cost of plants, by the government's inability to solve the problem of safe and reliable disposal of nuclear waste, and by the resulting disenchantment of the public and investors. Nuclear research should focus on these problems, including designs with improved safety. Until such problems are solved, further expansion of installed nuclear-power capacity is unlikely, at least in the United States.
- *Nuclear fusion.* For more than 35 years, researchers in the United States, Russia (previously the Soviet Union), Japan, and the European Community have sought to use the energy potential of fusion to generate electricity. The potential fuel source is vast; yet the scientific and technological problems remain daunting. Although it might offer substantial advantages over fission plants in waste, fusion is unlikely to be a major energy source within the next 30 years. Basic and applied research should be continued.
- *Hydrogen.* The committee recommends long term R&D to investigate the feasibility of hydrogen-energy cycles, because of their potential as an efficient and clean carrier for distribution of energy to users.
- *Transportation.* The largest present primary source of energy in the United States is petroleum products; more than 50% of energy from petroleum products is used in transportation (DOE/EIA 1995). The ubiquitous automobile influences our choice of jobs, where we live, and how we spend our leisure time. There has been enormous progress (including improvements in fuel efficiency) in the reduction of automobile emissions implicated in urban air pollution; but the automobile is still the major source of such pollution in part because growth in automobile ownership and in driving per vehicle has nearly offset this progress. Continued research to improve the fuel efficiency of automobiles will help, but further major improvements will almost certainly require switching from vehicles powered by the internal-combustion engine to electric (or possibly hybrid-electric) or hydrogen-fueled cars that need such technologies as fuel cells, flywheels, and greatly improved batteries. Research is needed so that this transition can occur.

2. The United States should continue to address ways of making energy use more efficient, including pollution reduction. It should also help to conduct the R&D and policy analysis required to take account of the severe needs of the developing world, where primary energy use for economic development is less efficient, the growth in primary energy consumption is more rapid, and severe

environmental damage per unit of energy is much greater than in developed countries. This will persist until better methods of energy use are developed and political and economic incentives for achieving their adoption sufficiently rapidly on a large enough scale can be devised.

USE A SYSTEMS ENGINEERING AND ECOLOGICAL APPROACH TO REDUCE RESOURCE USE

In our current efforts to reduce the pollution generated by and the ecological impact of society's industrial activities, we most often use "end-of-the-pipe" controls. However, end-of-pipe treatment is increasingly less likely to be the most cost-effective or the most-desirable means of pollution control.

In recent years, a new way of thinking about how to reduce environmental impacts has been developed. It is called industrial ecology, and it is influencing the thinking of many major corporations in how they handle environmental issues. Industrial ecology takes a systems-engineering and ecological approach to integrate the producing and consuming segments of the design, production, and use of services and products to reduce environmental impacts.

A key component of industrial ecology is analyzing the environmental effects of all materials in manufacture, use, and disposal. Companies find that the use of the industrial-ecology approach in the design process provides them with more options for reducing the human health and ecological effects of their products and processes. However, its use is in its infancy. The information, planning, standards, and societal changes needed to implement this concept on a scale sufficiently extensive to have a large impact is still lacking. The problem of implementation is made more difficult by the fragmentation of industry and, to some extent, by the present trend toward decentralization and devolution of hitherto vertically integrated industries.

Some form of "societal vertical integration" among many institutions and economic entities "from cradle to grave" will be involved in a solution. How it can be achieved is an important topic for research and public-policy debate. One key challenge is to formulate effective economic incentives to create a market-driven industrial ecology. Another is to alleviate the liability and regulatory barriers that inhibit the full application of industrial ecology.

Furthermore, industrial ecology requires substantial recycling, including the use of one plant's waste stream as feed for another plant, and therefore requires coordination, planning, and perhaps proximity, all of which could make it more difficult for it to achieve widespread use. One key challenge is to formulate effective economic incentives for developing a market-driven industrial ecology. Another is to alleviate the liability and regulatory barriers that inhibit the full application of industrial ecology.

Recommendations

1. Design of products and processes for environmental compatibility should make use of such mechanisms as life-cycle analysis, alternative manufacturing processes, and efficient separation technologies, which use energy to unmix materials that have been mixed.

2. Products and processes should be designed to accommodate recycling and reuse more readily.

3. Regulations introduced for other purposes often create barriers to the use of economic incentives for promoting the adoption of the principles of industrial ecology. Research should be aimed at identifying and eventually removing such barriers.

4. The use of industrial-ecology approaches should be expanded to many industries through dialogue among an ever-widening circle of corporations, governments, academic institutions and environmental and citizen organizations.

5. Research should be conducted to develop methods for chemical species-specific separations that yield streams that are economically recoverable or dischargeable to the environment.

6. Research on new or improved catalytic systems that offer improved yields and improved specificity from more-benign chemicals should be promoted.

IMPROVE UNDERSTANDING OF THE RELATIONSHIP BETWEEN POPULATION AND CONSUMPTION AS A MEANS TO REDUCING THE ENVIRONMENTAL IMPACTS OF POPULATION GROWTH

The current and potential future threats to environmental quality, of which there are many, are the results of the character and magnitude of today's economic activity and human population growth. As the experience of the United States, other industrialized nations, and developing countries indicates, birth rates and economic development are closely linked. Although the extent to which threats to environmental quality and ecological resources will be intensified by future population growth is debated, there is agreement that continued population growth has the effect of narrowing the options available for meeting these threats. The predicted addition of billions of people to the global population in the next few decades could overwhelm programs aimed at enhancing energy efficiency, global monitoring, and industrial ecology. Many elements of social science, such as demography and sociology, and of medical research address issues that affect population growth. Environmental engineering develops strategies and devices that can be used to decrease the impact of population growth on the needs of developing countries.

The political milieu makes it difficult for U.S. federal agencies to recognize publicly the interdependence of environmental quality and global population growth. The federal budget often omits support for studies related to population growth—studies of demography, ecology of population growth, and contraception—even in the U.S. Agency for International Development and the Department of Health and Human Services. This omission reflects a serious constriction in the horizons of current U.S. environmental policy. As noted in *The Sustainable Biosphere Initiative* (Lubchenco et al. 1991),

> The issues associated with population growth are broad, involving such factors as changes in per capita income and resource distribution; increasing pollution and environmental degradation; problems of health and poverty; the effects of urban, industrial, and agricultural expansion; and especially the integration of ecologic and socioeconomic considerations. Even those factors that are primarily economic will have substantial environmental effects.

Recommendations

1. The United States, in its efforts to cooperate with the world community, should recognize the linkages between birth rates, child survival, economic development, education, and the economic and social status of women in its environmental research efforts.

2. To cope with global population pressures, researchers should focus on ways to improve the potential for universal access to effective family-planning information, contraceptives, and health care.

3. U.S. policies, both domestic and foreign, need to provide support, through partnerships with developing countries, for the scientific and technological research needed by international population programs.

4. Interdisciplinary research should be conducted on the future environmental consequences of population growth, especially in vulnerable environments. This research should incorporate human biology, human behavior, epidemiology, and ecology to acquire a better understanding of all aspects of the population-environment interface. Research should be conducted on the possible adverse consequences of technology when introduced into a population (e.g., the possible adverse impacts of the use of artificial baby formula in developing countries).

SET ENVIRONMENTAL GOALS VIA RATES AND DIRECTIONS OF CHANGE

When speaking about environmental goals, people often focus on achieving a specific level of environmental quality by a specific time. However, such formulations might not be the best way to evaluate our rate of progress. Measuring progress requires a metric and a path to determine the direction in which progress is going.

Therefore, in setting environmental goals, rates and directions of change might be more important than end points, but it is endpoints that seem to be at the center of the public discussion of environmental policy.

Rates and directions of change should become the focused goals for current actions with end points as the tools for motivation (not just one-time goals). Both a desired end-point and a rate-of-change are needed in order for society to be successful in achieving its goals. Without an end point, it can be difficult to mobilize the political will to take action; without a predetermined rate-of-change, it can be difficult to establish the specific actions needed to achieve the desired end point.

Recommendation

1. Rather than stopping at the selected specific end points being discussed in the federal government and elsewhere, environmental goals should be formulated in terms of an adjustable strategy for continuous evolutionary improvement in environmental performance, including intermediate milestones.

CONCLUSION

The committee has had some difficulty in getting its hands around such a large, amorphous issue as the environment, but it hopes that it has made a credible effort to advance the discussion of the role of science and technology in defining and addressing society's environmental objectives. The major disadvantage of the effort is that the forum format, with its relatively limited time frame, at best permits only a first cut at these issues. Nevertheless, the committee believes that this report will be an important guidebook for both the scientific and policy communities and a starting point for further deliberations.

1

Society's Environmental Goals

This report focuses on science, technology, and the environment. As will be seen in discussion of some of the existing goal statements, the "environment" can be taken to cover an extremely broad range of topics. For the purposes of this report, the committee used the following definition:

The environment is all physical and biological features of the earth that can affect or be affected by human activities.

CHARGE TO THE COMMITTEE

As described in more depth in the preface, this committee had several charges.

- **Call for Comments.** First, to broaden the national awareness of long-term goals and to obtain the views of groups that will ultimately affect the attainment of goals, the committee was to conduct a national "call for comments" to obtain input before the summer study. These contributions were to be cataloged and synthesized (see Appendix D).
- **Commissioned Papers.** Second, the committee was to commission a series of papers that would discuss national science and technology goals related to domestic environmental policy. These papers were to become part of the final publication (see Part II).
- **National Forum.** Third, the committee was to hold a forum where there could be a dialogue among persons from industry, academe, nongovernment organizations, and the interested public as to what the long-term science and technology goals are to be to meet society's environmental goals. This dialogue took

place both in written form, via a "call for comments," and in a public forum, which took place on August 20-24, 1995, in Irvine, California. The forum would include presentations of the commissioned-paper authors and invited guests and discussion of the topics in plenary and breakout sessions in relation to several key questions that would form the framework of the report (see Parts II and III).

- **Report.** Fourth, the committee, which was to include a group of persons with broad relevant experience, was to convene, consider commissioned papers, discuss potential long-term goals, receive the views of selected representatives of those who are interested and affected by the policy in question, and attain consensus on recommended national goals in a published report. The short report would present the committee's consensus as to what should be the nation's science and technology goals for the environment, summarize the call for comments, and include the commissioned papers.

Questions that seemed to recur in the committee's deliberations were, Who is the audience for this report? and What, exactly, are science and technology goals? After some discussion, the committee concluded that the audience for its report was the science and engineering community—particularly, though not exclusively, those working in environmental research and development. In addition, the report provides guidance to government officials who fund research in some of these subjects as to which ones the committee believes need the most attention in response to societal needs. Furthermore, government agencies are a source of much of the information analyzed by scientists. The report suggests ways in which agencies can focus their efforts.

> *The environmental goals will never be achieved absolutely—this is an ongoing process.*
>
> —Forum Participant Comment*

Science and technology goals are more than a "research agenda" (i.e., more than a list of all the items in which we need research). Instead, science and technology goals identify research subjects on which there is insufficient focus by scientists and engineers relative to responding to societal goals.

FORUM PRESENTATIONS

Environmental goals were discussed with respect to industry, federal agencies, states, the nation, and the world.[1] These differ qualitatively and quantita-

*Throughout the report, the reader will find boxes that provide some of the comments from those who responded to the call for comments or who participated in the forum discussions. The purpose of these boxes is to describe some of the varied opinions on societal goals that influenced the committee's choices on topic selection. Inclusion of these boxes does not represent an endorsement of the opinion expressed by the authors. To be consistent with the ground rules that were established as part of this process, the committee does not identify the individual authors of the statement.

[1]One of the committee members noted that goals for the environment can be inferred in the preamble of the US Constitution, which calls on the federal government to "insure domestic tranquility, provide for the common defence, promote the general welfare . . . ."

tively. Recipients of the call for comments were asked to rank a set of current Environmental Protection Agency (EPA) goals. Clean air was ranked first by 50 of the 80 who offered a first-place ranking, and clean water was ranked second by 40 of 66. This forum focused on the future, several decades into the next century. Regarding future goals, there was no consensus in the responses to the call for comments. As one respondent said:

> *We have the answers to solve today's needs, if we think and act the right way.*
>
> —Forum Participant Comment

> Environmental goals are, by nature, multifaceted, and a detailed listing of all important issues is subject to preferences and priorities. To avoid these choices at this stage, we believe that stating the following overall goal is more productive and allows specifics to be developed later. The nation's environmental goal should be to achieve an economy built on the principles of Sustainable Development.

In the United States, much environmental interest has focused on the principal environmental agency, EPA. This agency was described as having "fashioned an environmental policy out of the sum of the parts," referring to the many statutes administered by the agency, each of which "contains some form of a goal statement." (see Truitt and Wise presentation in Part III). This approach was criticized by Morgenstern (see Part II): "Many of our major environmental statutes contain little more than hortatory phrases that offer scant guidance to the implementing agencies." In addition, he noted that concerns have arisen on "whether our legislative goals are really the right ones or provide sufficient direction for the present and/or the future." In his presentation to the forum, Department of Energy (DOE) Assistant Secretary for Environmental Management Thomas

> *A systems approach to achieving current environmental goals would include at least three steps: (1) identifying and characterizing problems, (2) formulating solutions, and (3) identifying end-states resulting from the application of specific solutions to specific problems. Science and technology (S&T) can make important contributions to all three steps. S&T are critical components of step (1) through elements such as basic research (e.g., increasing our knowledge of environmental systems), instrumentation (e.g., measuring/monitoring systems), and information technology (e.g., disseminating data and communicating the information provided by these data). S&T are critical components of step (2) through elements such as applied research and engineering. S&T are critical components of step (3) primarily through predictive studies (e.g., modeling) and retrospective studies (e.g., epidemiology). It is important to realize that S&T are a necessary but not sufficient part of all three steps. For example, S&T can describe and predict end state, but cannot determine whether a given end state is "desirable" or "worth achieving."*
>
> —Forum Participant Comment

Grumbly also questioned the adequacy of the legislative approach, stating that "we have been making environmental laws, not establishing environmental goals. We should consider what are we doing today that future generations will question."

Presented in Morgenstern's appendixes are three recent major efforts by the executive branch to develop environmental goals: EPA's *Proposed Environmental Goals for America with Milestones for 2005*, *Ten National Goals to Put the United States on a Path Toward Sustainable Development*, by the President's Council on Sustainable Development (PCSD), and the Committee on Environment and Natural Resources' (CENR) *Strategic Planning Document*. For example, the CENR presents five cross-cutting topics for integrated environmental research and development:

- Ecosystem research.
- Observations and data management.
- Social and economic dimensions of environmental change.
- Environmental technology.
- Science-policy tools: integrated assessments and characterizations of risks.

As background for the forum, Morgenstern analyzed those three major efforts. His paper is included in Part II. He notes that "the three goals reports address three fundamentally different sets of problems [and] . . . have different technical approaches . . ." Although each of the efforts has identifiable weaknesses, Morgenstern concludes, in an opinion shared by this committee, that "an overwhelming strength common to all three projects is the implicit recognition that our environmental management system is in need of significant reform."

> *We have yet to do the R&D that will enable us to address environmental issues seriously or even know what goals are appropriate and how to get there.*
>
> —Forum Participant Comment

Morgenstern compares the goals using such measures as scope, time frame, success, and inclusion of an implicit assessment of tradeoffs in goal choices. He finds that the three sets differ substantially in those measures. "CENR seeks to conduct relevant and useful research . . . the EPA aims to protect the environment . . . and the PCSD attempts to enhance the public welfare." He concludes that "none of the goal schemes prioritize individual goals or acknowledge the basic tradeoffs between desired outcomes." However, "the CENR strategy—which addresses research as opposed to policy goals—is clearly the most focused effort." He concludes that "the EPA project is oriented to implementation of current environmental statutes and treaties . . . the PCSD tries to fashion a vision for the next century . . . [and] the CENR project . . . is a research strategy rather than a gameplan for environmental policy."

EPA proposes 15 long-range environmental goals for the nation and provides quantitative milestones, usually for the year 2005. The PCSD interim report proposes eight "priority national goals" designed "to put the U.S. on a path to-

Broad, national environmental policies would incorporate science and technology as important elements in achieving the following scientifically driven outcomes:

- *Risk-based decision-making.*
- *Creation of human health effects knowledge that improves risk-based standards, which protect all sectors of the population.*
- *A knowledge base and ecosystem understanding that provides the scientific basis for risk-based decision-making by legislators and the public.*
- *Innovative technologies available or being developed to restore contaminated sites at home and abroad.*
- *U.S. legislators promoting strategies for assessing the risks of global environmental change and their impact on investment decisions and making the commitment to succeed in the assessment.*

—Forum Participant Comment

ward sustainable development." "The goals are oriented toward the basic objectives of promoting efficiency, protecting the environment, and ensuring equity." Morgenstern concludes that "the PCSD has clearly opted for breadth over specificity", which can be seen in that two of the goals are economic prosperity and sustainable communities. The CENR goals are presented in the context of five overall goals for science and technology.

Federal funding for the environment is spread across many agencies. As Albert Teich's paper in Part II indicates, EPA is not even the largest part of the federal environmental program.[2] Nevertheless, environmental goals tend to be interpreted on the basis of their relevance to EPA's programs. However, in addition to the federal agencies, states and industry are examining goals, as described in several of the commissioned papers.

In his paper, Richard Minard describes growing activity at the state level. He notes that "states appear to be setting up processes to identify and promote technological 'winners,' rather than focusing on the problems that most need to be solved." However, Minard perceives that "setting goals for the environment is useful if it helps people focus on problems and discover a shared commitment to solving them." He concludes that "almost every government effort to set goals will also suffer from the related dilemma of involvement: too few people will set the goals and too many people will have to pay in some way to achieve them." That is related to another Minard observation: "When real money is at stake and when decisions get close to home, the strength of the information-based, consensus-building process is put to the test."

[2]As shown in Teich's paper in Part II, the Department of Energy (DOE) Environmental Management (EM) program (i.e., the nuclear materials and weapon facilities clean up) is the largest. Of the $22.7 billion spent on environmental and natural-resource programs in FY 1995, $7 billion was spent on DOE's program, compared with about $5.5 billion for EPA's activities.

The paper by Konrad von Moltke provides an international context. He indicates that comparisons are difficult for several reasons, including the presence of different perspectives that stem from different histories, cultures, and educational systems. "At all levels of economic activity, however, social preferences for environmental quality may differ." He describes several possible approaches for international comparison, all of which have weaknesses. For example, "generally expressed in technical terms, standards appear to offer a comparable basis for evaluating environmental policies in different countries. However, two difficulties exist in comparing standards: variations in the definition of standards and in their application in practice." "These difficulties in comparing environmental policy have permitted public officials . . . everywhere to claim that their policies are the most advanced, the most stringent and the most effective . . . . By picking the times and the areas where a country has been active it is even possible to provide proof for these mutually exclusive statements." Moltke, not surprisingly, concludes that "attempts to 'harmonize' standards internationally have proven difficult."

Ehrenfeld and Howard, in their paper, discuss the approach taken by several major U.S. industries. They see industry as taking a market-driven, short-time view: "U.S. industry traditionally has a short time and myopic framework for setting goals." In their view, industry moves through a series of stages, beginning with seeing environmental issues as problems to be solved through compliance, emissions reduction, source reduction, and finally managing for the environment, "the Green Company." They also criticize the approach to environmental improvement through laws: "where specific targets have been written into laws, they have often been grossly over ambitious." Nevertheless, they acknowledge that sometimes the government has led industry to accomplish what it claimed it could not. They cite "the technology-forcing requirement to reduce auto exhaust emissions . . . the first instance of an environmental technology goal for indus-

The answer to the potential contribution of science and technology is intuitive in nature. Science and technology are the cornerstones to meeting any environmental goals that are developed. One can only learn new things by studying them. Research on environmental problems is rather new compared with other areas and is highly dependent upon technological advancements. Therefore, to insure that our national environmental goals are met, sound basic research is needed. My fear is that policies are being debated and enacted without the scientific support that is necessary to insure that they are sound. Some of this is the fault of the scientists by not publishing data in a timely manner and some of the fault is with the policy makers making hasty decisions (reactive instead of proactive) to please the public.

—Forum Participant Comment

The barriers to achieving environmental goals may be classified into the following major categories:

- *Economic—The current method of accounting for life-cycle costs, natural resource depletion or increase, capital depreciation, and other environmental expense does not provide the correct price signals to the owners, builders, engineers, taxpayers and other stakeholder groups.*
- *Legislative and Regulatory—This category includes a variety of barriers to innovation that either create prescriptive methods of addressing problems, increase the risk and liability for innovators, or increase the time/cost of changing current ways of doing business.*
- *Leadership and Political—There is no clear mandate from national and community leaders for innovation and insufficient numbers of "champions of innovation" in the industry.*
- *Knowledge—There is often no objective, credible data to provide demonstration or verification of the real-world cost and performance of environmental technologies with the potential for addressing the goals.*
- *Educational—The potential of positive change has not been sufficiently documented and communicated to leaders and the public at large.*

—Forum Participant Comment

try." The authors note that "many of the technological advantages of later Japanese automobiles that fueled their competitive onslaught in the American market were spawned by these earlier engineering approaches taken to meet environmental standards." The authors describe several approaches taken by industry, including life-cycle analysis, design for the environment, and the recommendations of the international group, the Business Council for Sustainable Development (BCSD). BCSD advocates "creating a new target for industrial performance—eco-efficiency. . . 'achieved only by profound changes in the goals and assumptions that drive corporate activities.'" Nevertheless, industry must keep a focus on the market and ensure that shareholders interests are protected. For the government, the shareholders are present and future generations. The government can focus on long-term programs that might not have any direct payoff to any specific industry; industry cannot.

The paper by Bowman addresses whether the environment issue is on the top of the public's agenda. If it were, there would be an additional urgency in establishing a clear set of national goals. According to polling data, it is not. "Today Americans remain committed to the goal of protecting and improving the environment, but they no longer see an urgent problem." This might conflict with an impression that some have. "Americans have been asked repeatedly in a wide variety of formulations to affirm a core value, in this case the importance of the environment. Each time, not surprisingly, they responded that a clean and healthful environment was important to them. These questions tell us little about what a

> *We need science to become a greater voice, and to use that voice to explain the "relevance" of scientific findings, helping the citizenry understand what implications might be, preventing panic or knee jerk reaction to incomplete information. Technology could be a greater boost to environment goals if there was a Science & Technology clearing house as a communication tool between it, industry, government, and community.*
>
> —Forum Participant Comment

society with many demands on it is willing to do to advance the value, what trade-offs the public is willing to make for it, or what happens when one important value clashes with another." Regarding the conflict often described as needing to choose between the environment and development, "in the 1990s, this belief that economic growth and a clean world are simultaneously obtainable has substantially broader support than previously." The public thinks that "growth versus the environment" is a false choice. Furthermore, in a belief shared by the committee, "Americans also believe that the United States is above average when compared with other nations in its efforts to protect the environment." "Viewed in isolation, the results to these questions seem to suggest enormous concern . . . . [However,] many polls like the ones above confirm the view that other problems are far more urgent than the environment for Americans today."

CALL FOR COMMENTS

A detailed overview of the responses to the call for comments is provided in Appendix D. Discussions at the forum and the comments indicate that there is no clear consensus on what should be the environmental goals for the next century, beyond "sustainable development" (which itself has many definitions), and that there is no agreed-on comprehensive set of baseline data to use as a foundation to measure progress (see paper by Phillip Ross et al. in Part II).

Although substantial progress has been made in the last 25 years in improving environmental quality, the country still lacks a unified national strategy. That is due partly to the fact that the current set of goal statements are not consistent and are in a constant state of change. However, they do provide a general framework aimed at improving the environment.

COMMITTEE APPROACH

The committee has chosen neither to develop a new set of goals nor to select a subset from the existing array on which to concentrate. Instead, the committee has focused on how science and technology can contribute substantially to improving the environment—the goal toward which the country is moving.

The commissioned papers in Part II provided a background for the public sessions. On the last day of the forum, participants met in small breakout groups to discuss the principal subjects on which the committee should focus (see Appendix F). The forum was held 25 years after the first Earth Day, and so it seemed appropriate to focus discussion on what we should do now to meet society's goals 25 years from now. For example, one of the breakout groups concluded that, looking back 25 years from now, people will ask, "Why did they not sort out some priorities?" This group noted that today *everything* seems to have high priority.

On the basis of those discussions, the responses to the call for comments, and their own experience, committee members discussed the critical subjects on which science and technology should be focused, looking ahead 20-25 years. Of immediate importance is a need to set priorities for the environmental R&D community.

Each member was asked to select his or her top three candidates. After discussion, a list of 38 possibilities was generated. The committee then selected those on which this report would concentrate: energy, in many aspects; monitoring, the structured and continued collection of data; environmental impacts; and population. In addition, four others were identified to be addressed: industrial ecology, ecological systems, economics, and risk analysis. The committee decided to develop those eight subjects and make recommendations on them. In writing the report, the committee combined several subjects and went from eight to six topics. The chapters of the report are based on that committee process. In summary, the following process was used:

1. Via the call for comments and forum discussion, the committee developed a list of some 30-plus topics that could possibly be discussed in the report.
2. The committee then voted on these topics, selecting the ones believed most important for the scientific and engineering community to focus on relative to the current intellectual and financial resources provided to each. Eight topics were selected.
3. Of the eight, six topics eventually emerged as the chapters for this report.

> *The nation should strive for sustainability, which means that this generation should be able to provide an acceptable standard of living for its people without affecting the ability of the next generation to provide for its own. The nation should strive to maintain the natural areas it has left by careful management of tourism and public use and restricting or eliminating the presence of private profit-makers on public lands. It should also do what it can to acquire new areas and improve those it already owns that are damaged. On the basis of sound risk assessment principles, individuals should have an environment that does not contribute to illness, both at home and at work.*
>
> —Forum Participant Comment

> *Quit copping the attitude that science and technology are all we need to problem-solve. Science and technology are only tools that humans can use to make decisions and to "manage" things. We need to balance our facts and figures with human dimensions of cooperation [and] attitudes.*
>
> —Forum Participant Comment

An issue related to setting priorities is funding. Given fiscal realities, the committee recognizes that recommending expanded effort, or even continued effort, in the face of declining resources implicitly selects subjects that should gain while others, unmentioned, lose. The committee recognizes that, especially in the near term, tradeoffs must be made; therefore, it suggests in Chapter 2, on societal choices, how such changes might be made. (In the long term, although tradeoffs still must be made, the tradeoffs are not necessarily those currently considered.[3])

Many topics are mentioned only in passing, both in the responses to the call for comments and in this report. That constitutes an implicit judgment that they will be or are of less importance to the public and the scientific and technological community. But several subjects, although not explicitly covered in this report, are extremely important. One is global climate change, on which there is an enormous federal effort, including that of National Aeronautics and Space Administration, National Science Foundation, National Oceanic and Atmospheric Administration, DOE, and EPA. Another is the largest federal environmental program, the management and cleanup of the wastes at DOE weapons facilities, currently estimated to cost at least $250 billion and to take more than 50 years. This program involves institutional and technical challenges as large as any other we discuss.

The committee has not had the material and time to evaluate how to implement its recommendations as financial and institutional arrangements change rapidly in the U.S. public sector's programs on environment and related R&D. Until it becomes clear to what degree environmental responsibility will devolve to state and local governments, for instance, reorganization of the federal apparatus is unlikely to be successful. At the same time, the committee believes that the national (i.e., the federal, state and local) environmental R&D effort needs more coherence. Alternatives for federal agencies have already been assessed and identified by the National Academy of Sciences and National Academy of Engineering, and the Carnegie Commission on Science, Technology, and Government (1992).

In spite of the above caveats, the committee in general agrees that the issues on which science and technology programs should especially focus for achieving the nation's environmental goals over the long term are as shown in the following chapters of this report.

---

[3]A good example of this is the original tradeoff between automobile emissions reduction and energy efficiency which proved to be less drastic than was originally expected after three-way catalysts were introduced.

STRUCTURE OF THIS REPORT

This report is divided into four parts. The first contains the result of the committee's discussions after receiving and hearing comments from forum and call-for-comments participants. As noted earlier, these discussions focused on the areas where the committee believes that the science and technology community should focus its efforts to be responsive to society's environmental goals. They led to chapters as follows:

- Use Social Science and Risk Assessment to Make Better Societal Choices.
- Focus on Monitoring to Build Better Understanding of Our Ecological Systems.
- Reduce the Adverse Impacts of Chemicals in the Environment.
- Develop Environmental Options for the Energy System.
- Use a Systems Engineering and Ecological Approach to Reduce Resource Use.
- Improve Understanding of the Relationship Between Population and Consumption as a Means to Reducing the Environmental Impacts of Population Growth.
- Set Environmental Goals via Rates and Directions of Change.

Although the chapters are of varied lengths, it should not be assumed that the subjects within shorter chapters are of less importance.

Part II of the report includes the papers commissioned by the committee. Each of these papers is interesting, useful, and well worth reading independently of the report. The titles of these papers are

- *National Environmental Goals: Implementing the Laws, Visions of the Future, and Research Priorities* (Richard D. Morgenstern).
- *Measurement of Environmental Quality in the United States* (N. Phillip Ross, Carroll Curtis, William Garetz, and Eleanor Leonard).
- *Attitudes Toward the Environment Twenty-Five Years After Earth Day* (Karlyn Bowman).
- *Environmental Goals and Science Policy: A Review of Selected Countries* (Konrad von Moltke).
- *Can States Make a Market for Environmental Goals?* (Richard A. Minard, Jr.).
- *Setting Environmental Goals: The View from Industry. A Review of Practices from the 1960s to the Present* (John R. Ehrenfeld and Jennifer Howard).
- *Status of Ecological Knowledge Related to Policy Decision-Making Needs in the Area of Biodiversity and Ecosystems in the United States* (Walter V. Reid).
- *The Federal Budget and Environmental Priorities* (Albert H. Teich).

Part III contains the keynote addresses and presentations made at the forum by

- D. James Baker, Under Secretary for Oceans and Atmosphere, National Oceanic and Atmospheric Administration.
- Thomas Grumbly, Assistant Secretary for Environmental Management, United States Department of Energy.
- Barry Gold, Chief, Scientific Planning and Coordination, National Biological Service, United States Department of the Interior.
- Harlan Watson, Staff Director, House of Representatives Committee on Science, Subcommittee on Energy and Environment.
- David Garman, Professional Staff Member, Senate Committee on Energy and Natural Resources.
- John Wise, Deputy Regional Administrator, Region 9, United States Environmental Protection Agency.
- Peter Truitt, Senior Analyst, Office of Policy, Planning, and Evaluation, and Manager, National Environmental Goals Project, United States Environmental Protection Agency.
- Judith Espinosa, Former Secretary of the Environment, New Mexico, and Member, President's Council on Sustainable Development.
- Peggy Duxbury, Coordinator for Principles, Goals, and Definitions Task Force, and Staff, President's Council on Sustainable Development.
- Gilbert Omenn, Dean, School of Public Health and Community Medicine, University of Washington, Seattle.

Part IV contains the appendixes, in which the committee provides the agenda for the forum, the participants in the forum and a call for comments, and a summary of the responses to the call for comments and of the forum breakout group discussions. Also included is biographical information on the committee members.

CONCLUSION

The committee has had some difficulty in getting its hands around such a large, amorphous issue as the environment, but it hopes that it has made a credible effort to advance the discussion of science and technology's role in defining and addressing society's environmental objectives. The major down-side of the effort is that the Forum format, with its relatively limited time frame, at best permits only a first cut at these issues. Nevertheless, the committee believes that this report will be an important guidebook for both the scientific and policy communities, and a starting point for further deliberations.

2

Use Social Science and Risk Assessment To Make Better Societal Choices

The world in which we legislate and regulate to protect our environment is constantly changing, as it has always been. Perhaps most importantly, it is becoming "smaller." That is, goods and services that once could be supplied only from the United States can now be produced in and offered from very distant places, often at prices below those of U.S. suppliers. Foreign competition has changed forever the economic situation facing the United States. Making matters more complicated, other countries have different "tastes" for environmental protection and, accordingly, different legislative and regulatory regimes, which often differ from those in the United States. That is particularly true of developing countries, where standards are sometimes nonexistent.

At the same time (and it is probably unrelated), there has developed in the United States a willingness to question the need for ever-tighter environmental standards, as well as the means chosen to meet those standards. One reason is that, after 25 years of pollution control in the United States, many of the easy control opportunities have long since been exploited. For example, the cost of preventing an additional ton of hydrocarbon from being released in the Los Angeles air basin has grown from a few dollars per ton years ago to as much as $25,000 per ton today. Throughout air, water, and solid-waste management, the marginal costs of added controls have grown dramatically, making people more willing than ever before to ask whether the benefits of the next ton of pollution control are worth its costs. The exact magnitude is unknown, but EPA estimates that $150 billion—about 2.4% of GDP—is spent annually by businesses, individuals, and governments to comply with federal environmental regulation (EPA 1990); this is only slightly less than the federal spending for Medicare. Although no efforts have ever been made to estimate the monetary value of the annual benefits of all environmental regulations, they are clearly important. On the basis of current knowledge, it appears that for many regulations, benefits exceed costs.

Furthermore, technology frequently improves, as does our ability to reduce pollution through process change and product redesign. For this reason, the cost of controlling pollution can often fall, thus making some programs that were heretofore prohibitively expensive now attractive as social investments.

Many also are questioning whether the environmental benefits resulting from new controls are as much as they were at the start of the environmental era. Some believe that the law of diminishing returns relative to long-regulated environmental pollutants has begun to set in and that, although added control will no doubt produce gains to society, these gains will sometimes be small in relation to their costs.

Those questions are being asked not only in corporate board rooms—although that is where the questioning might be loudest—but also in city halls, governors' offices, and even in the offices of the Secretaries of federal departments. That is because state and local governments and such federal agencies as the Departments of Interior, Defense, and Energy are also subject to federal environmental regulations. In fact, concerns about the cost of controlling environmental pollutants at the federal and state levels relative to the benefits received led to the recent passage of legislation that makes it much harder for the federal government to write regulations that impose costs on lower levels of government without appropriating funds to help the affected parties comply.

There is ample evidence that the concern is bipartisan. Presidents Ford, Carter, Reagan, and Clinton—two Republicans and two Democrats—issued executive orders requiring that federal regulatory agencies identify both the benefits and the adverse economic impacts of all major regulations and ensure, if relevant statutes permitted, that the benefits of a proposed regulation exceed the costs and the least-costly approach to meeting the environmental objective was chosen. In the 103rd Congress, before the Republican takeover of both houses, a measure to require even more cost-benefit analysis and more explicitness in risk assessment passed in both houses by substantial majorities. By mid-1995, environmental advocacy groups, business organizations, policy experts in academe and "think tanks," and the Clinton administration had all come to agree that environmental regulation needed to be rethought in important respects—even if there was much disagreement about just which respects. At the same time, agreement appears to have been reached on the need for a fair and careful balancing of environmental improvements with their associated costs.

ECONOMIC INCENTIVES VERSUS COMMAND AND CONTROL

It appears that the nation's existing environmental goals can be met for a good deal less money than the country is now spending if we substitute, as appropriate, what have come to be known as incentive-based approaches to environ-

> *This is obviously a large topic, with much commentary having been produced by previous panels that does not make sense to try to repeat off the top of my head. Of importance to me is recognition that social science has contributed and can continue to contribute to meeting environmental goals in a variety of ways (articulated best in the review by the NRC committee on the Human Dimensions of Global Change). Any research strategy should have a strong social science component that treats social science as more than merely implementing findings from physical and natural science. A fundamental challenge for any aspect of managing the environment is bringing about changes in human and organizational behavior; social scientists have an important role to play in helping to understand this.*
>
> —Forum Participant Comment

mental regulation for command-and-control approaches.[1] Incentive-based approaches consist of such things as taxes on pollution discharges, the use of marketable discharge permits (which allow firms to buy and sell the right to discharge specific quantities of pollutants), deposit-refund schemes, and even the provision to the public of information about the amounts of pollutants that sources discharge annually. The hallmark of these approaches is that they give regulated parties the flexibility that they need not only to decide how much they should reduce their air or water pollution emissions or the volume of solid or hazardous wastes that they generate, but also to determine how to go about accomplishing whatever reductions they do decide to make. Furthermore, such approaches have the potential of providing greater benefit for each dollar spent on environmental protection.

In contrast, command-and-control approaches have traditionally taken the form of outright bans on products, mandatory emissions reductions, and even requirements to install specific types of control equipment, such as electrostatic precipitators, stack-gas scrubbers, and water-filtration equipment. There is little doubt that command-and-control approaches played an important role in the environmental improvements that the United States has enjoyed over the last 25 years, there is growing recognition that they are increasingly inappropriate for many current environmental problems and there are better and less-expensive ways of achieving environmental protection.

At first, the evidence in support of incentive-based approaches was purely academic. It consisted of studies results showing how effluent taxes or marketable permits, by giving regulated parties the flexibility to meet their environmental requirements as inexpensively as they could, could make possible control-cost savings of 10-70% while meeting the same environmental goals. Those approaches have now begun to be tested in practical ways, and the evidence sug-

[1]See, for example, a joint study conducted by Amoco and EPA. (Amoco/EPA 1993).

Science and technology can contribute a lot to meeting current national environmental goals. In particular, science and technology can help define the costs and benefits associated with the national investment in environmental protection. The nation does not have an unlimited amount of resources to allocate to environmental protection, and we need scientifically and technically defensible information to help determine where we will allocate the limited resources. We should focus our environmental-protection activities where we get the biggest return on our investment in the environment. The current state of most science applied to the definition of environmental problems is that it uses very conservative assumptions whenever actual values are not known. Therefore, the less we know about a problem, the more we are likely to overestimate the risks associated with it and therefore overreact to it. Science can help overcome this problem (but it should also be noted that it was science that created the problem). Appropriate use of advanced technology can enable us to get a better return on the investment, in terms of environmental protection. There may be softer social sciences involved also. Not everything can be measured in common terms (e.g., dollars), and therefore we need to develop and apply better methods for the balancing environmental damages against costs.

—Forum Participant Comment

gests that, although they are more difficult to implement than was first recognized, they are as capable of producing substantial cost savings in practice as the theory and modeling suggested.

By far the best example to date concerns the SO_2 emission allowances created under the 1990 Clean Air Act amendments. When Congress decided, out of a concern about acid deposition, that nationwide emission of SO_2 should be reduced by nearly 50%, it directed EPA to take a very different approach. Instead of requiring a specific technology, Congress took a three-pronged performance-based-standard approach. It included capping annual SO_2 emission at a new and much-lower level, apportioning the initial emission reductions required of individual power plants, and directing EPA to establish an allowance-trading program so that a utility that found it too expensive to meet its initial emission-reduction requirement could contract with another utility (or even a nonutility source of SO_2) to go beyond the latter's required cutback as long as the combined reduction of the two parties would equal the sum of the initial reductions required of both. As of mid-1995, many such reallocations of control effort had taken place, most within individual companies but a large and growing number between companies.

The evidence to this point, still preliminary, suggests that that incentive-based approach will save the country $4-5 billion annually, inasmuch as implementation of the performance-based standard will cost the country $1-2 billion annually instead of the $5-7 billion for specified technology-based approach. In other words, the savings now being realized are in keeping with the most optimistic estimates from earlier studies. Other, admittedly smaller-scale incentive-based approaches (including taxes on chlorofluorocarbons [CFCs] and a trading program for the phaseout of lead in gasoline) also provide confirmatory evidence. Although the verdict is by no means in, incentive-based approaches have the potential to deliver more than their promise.

ANALYTICAL TOOLS: COST-BENEFIT ANALYSIS AND RISK ASSESSMENT

Analytical tools enable us both to predict the effects of regulatory interventions on human health and the environment and to value the effects and compare their benefits with their costs. At a time when much is being asked of these tools—most notably, quantitative risk assessment (QRA), the more-subjective comparative quantitative risk assessment (CRA), and cost-benefit analysis (CBA)—we are learning more and more about how the tools must be improved if they are to be reliable aids in decision-making.

Starting with QRA, we know far less than we would like to about, for example, the effects of multiple, potentially synergistic pollutants on human health; the link between environmental pollution and virtually all noncancer human health effects, including reproductive, neurological, and immune disorders; and the differential susceptibility of sensitive populations.

Regarding CRA, despite its potentially great usefulness, we know little about how to compare the value of prolonging the life of a 75-year-old by 3 years with, for instance, the value of ameliorating asthma for the entire lifetime of a newborn baby. That example is an illustration of why we need a way to deal with subjective judgments that are inherently part of CRA. Much work needs to be done to understand fully how to conduct CRA before it can be fully implemented.

Finally, with respect to CBA, there is great uncertainty even among experts about the value to be attached to the preservation of an endangered species, to the restoration of a contaminated aquatic ecosystem, to the enhancement of visibility in urban areas, or to the protection of remote natural areas in their pristine form. There is considerable uncertainty about the costs of environmental regulatory programs or about the rate at which future benefits and costs should be discounted (if at all) to make them commensurate with more-immediate effects.

> *I believe that the goals that are presently in place are adequate. However, having goals and achieving them are two different issues. Legislation needs to be strong enough to enforce these goals, and it has to be realistic. Although it would be ideal, zero pollution is not an option. Therefore, acceptable boundaries have to be developed, and science can provide the framework for this.*
>
> —Forum Participant Comment

There are other tools not treated here, such as negotiated conflict resolution, that are just beginning to be used more commonly and on which research is now being conducted. But, we need to know how to use our current tools—such as QRA, CRA, and CBA—better and how to develop even better tools that we can use in the future.

FINDINGS, CONCLUSION, AND RECOMMENDATIONS

Findings

One barrier to the implementation of incentive-based approaches is that some tools needed to implement them—specifically QRA and CBA—are not as mature as they need to be if they are to be reliable aids in decision-making. Nor are they widely understood or accepted.

There is an insufficient program of interdisciplinary research to improve and facilitate implementation of the analytical tools available to decision-makers—most notably QRA, CRA, and CBA—and to develop improved alternatives if possible. To achieve congressional objectives and ensure that the public has confidence in these tools, improved analysis is needed. Congress should not be simultaneously requiring the expanded use of these tools in decision-making and reducing the funds available to improve their deficiencies and gaps.

This is not the place to lay out a complete agenda for research, but among the subjects needing attention are the following:

- Benefit estimation, especially the tools and techniques for estimating the values that people place on ecosystem services (including so-called "existence values"), on the protection of endangered species, and on providing additional years of life to people in various health states.
- Cost estimation, including an improved understanding of how technological change and market discipline can combine to reduce regulatory-compliance costs below initial estimates.
- Assessment of noncancer health effects and ecological risks.
- Integration of QRA with CBA so that the health and ecological effects that risk assessors predict are those which the public cares about and which, therefore, are the ones needed to conduct a useful CBA.
- Activities like negotiated conflict resolution, which require not only that physical, biological, and health scientists work with economists and cognitive psychologists, for instance, but also that the lay public is involved in such a way that the valuation exercise that is part of cost-benefit analysis reflects the values of those who will bear the fruits of regulatory programs and both QRA and CBA have credibility among the public in whose interests regulations are being considered.

Research needs to be sharply accelerated and demonstration projects undertaken to expand the applications of incentive-based approaches to environmental

Ecological Effects

One important class of nonhealth outcomes is harm to nonhuman organisms and ecosystems. The EPA has taken the lead in developing a conceptual framework for conducting ecological risk assessment . . . and is preparing guidelines for this activity. Analysis is difficult because the effects may fall on individual animals or plants, on local populations of a certain species, on ecosystems (thus affecting many species), or on the survival of endangered species. At larger scales, effects on the distribution of ecological communities across the landscape are central to regional-scale ecosystem management . . . . There may be important ecological outcomes to consider and characterize at each of these hierarchical levels of ecological systems . . . .

Ecological impact analysis also demands an understanding of how the affected ecosystem functions. There are numerous interrelationships among taxa, across responses, and across organizational levels. In addition, some of the most important effects may be indirect, operating through several interrelationships. Many of these effects are inadequately understood, difficult to measure, or laden with uncertainty . . . . Some ecologists even dispute whether the concept of ecological risk (or its inverse, ecological health) is useful for policy analysis . . . . None of these scientific difficulties of estimation, however, negate the importance for policy decisions of considering ecological outcomes. Interested and affected parties may want to take account of ecological effects, even if the level of scientific understanding of them is poor. Qualitative assessments of relative ecological risks can provide useful insights for environmental decision making . . . . A critical need is to develop appropriate tools for assessing the value of ecological systems, including both economic and noneconomic (e.g., intrinsic) value.

SOURCE: National Research Council (NRC 1996b)

protection. Attractive opportunities for such applications include the control of nonpoint-source water pollution from agricultural fields and feedlots and from municipal storm-water runoff. Research should be focused on the institutional impediments that can often reduce the savings associated with incentive-based approaches. Promising avenues for such research include experimental economics (because "economics laboratories" have proved to be fertile grounds for testing the efficiency properties of markets for airport landing slots), telecommunications spectrum bandwidth, natural-gas pipeline rights-of-way, and other heretofore government-allocated goods and services, in addition to environmental problems.

Science and technology have important roles to play. If, for example, EPA will in the future give much more latitude to regulated parties in deciding how to meet pollution-reduction goals, it might be appropriate for it to step up its funding of research aimed at uncovering new and potentially attractive means of reducing emission. That would enable it to offer firms a menu of emission reduction, waste minimization, or other strategies from which they could choose in deciding how to comply with more flexible regulatory approaches, as opposed to "freezing technology" so that the best approach cannot be used for a given situation or to take continual advantage of technical progress.

Another virtue of an incentive-based approach is that it is the firms that must now take responsibility for reducing their emission from their products and processes. They probably know best how to develop the technologies and approaches that will provide them with the best opportunities at the lowest cost.

Conclusion

In recent years, many segments of society have become concerned about the economic impact of environmental regulation. Business, government, and individuals together spend an estimated $150-180 billion of their revenues each year to meet environmental regulations.

Present regulatory strategies do not sufficiently differentiate between minor and major risks. Furthermore, the costs incurred to reduce risks often do not bear a consistent relation to the magnitude of the risks involved and the number of people potentially affected. The nation's existing environmental goals could be met less expensively or faster by substituting incentive-based approaches to environmental regulation for command-and-control approaches. Incentive approaches provide a more flexible and cost-effective regulatory environment for industry, business, and government while maintaining or perhaps even improving environmental quality much less expensively.

The tools needed to implement incentive-based approaches—specifically, QRA and CBA (including some attention to distributional effects)—are not as well developed as they need to be if they are to be reliable aids in decision-making. Nor are these tools widely understood or accepted by decision-makers or other interested parties and the general public.

Recommendations

1. Research to improve the analytical tools available to decision-makers should be expanded. Several specific questions need attention by researchers

- With respect to cost-benefit analysis, what values do people attach to the services provided by ecosystems? To the protection of endangered species? To the extension of human life? To aesthetics and the quality of life? How do those values differ for different states of ecosystem and human health?

- Also, with respect to benefit-cost analysis, how can models better estimate the costs of regulatory proposals? How can models predict human behavior, as opposed to relying only on the more reliable predictive value of current technologies and practices? How should models account for lost opportunities, even when no out-of-pocket expenditures are made, and for continuous technological changes that bring down the costs of regulatory compliance? The distribution of risks and benefits involves issues that are inherently political, and cost-benefit analysis generally is silent on these distributional matters. In light of that, how can models more realistically take into account the distribution of risks, costs, and benefits among all affected stakeholders?
- What are the best ways to assess noncancer health risks, such as neurological and reproductive disorders? Disproportionate attention is paid today to collecting information on cancer risks compared with, for example, risks of neurological, reproductive, and other health disorders. Future research on quantitative risk assessment should be directed toward correcting this imbalance of emphasis.
- Can quantitative risk assessment and cost-benefit analysis be integrated so that the health or ecological end points that risk assessors predict are the ones that the public understands and cares about? To what extent could and should probabilistic techniques be used with worst-case assumptions? In the face of uncertainty, what degree of conservatism should be used in connection with probabilistic techniques?

2. Research should be increased and demonstration projects launched to expand the application of incentive-based approaches to environmental protection (these include pollution taxes, systems of marketable discharge permits, and deposit-refund schemes). The research should include the evaluation of institutional impediments that often reduce the savings associated with incentive-based systems. Experimental economics—in which a group of participants act out the decisions that they would make in given regulatory or other scenarios—is a particularly promising avenue for research. As more data on incentives become available, decision-makers can have a better intellectual rationale for choice between incentive-based approaches to environmental protection and command-control regulations.

3. Social-science research on comparative risk assessment (or risk ranking) can be helpful to all levels of government to establish regulatory and legislative priorities. Comparative risk-assessment activities should involve elected and appointed officials; members of business, environmental, and civic organizations; and lay persons. In addition, natural and social scientists who can provide information on the magnitude of various risks to health and the environment, the likely costs of mitigating these risks, and the uncertainties associated with both should be part of the process.

For more information and guidance, the reader should refer to the following:

EPA (Environmental Protection Agency), *Environmental Investments: The Cost of a Clean Environment*, EPA Document #EPA-230-11-90-083 (Washington, D.C.: November 1990).

NRC (National Research Council), *Improving Risk Communication* (Washington, D.C.: National Academy Press, 1989).

NRC (National Research Council), *Policy Implications of Greenhouse Warming: Mitigation, Adaptation, and the Science Base* (Washington, D.C.: National Academy Press, 1992).

NRC (National Research Council), *Keeping Pace with Science and Engineering: Case Studies in Environmental Regulation* (Washington, D.C.: National Academy Press, 1993).

NRC (National Research Council), *Science and Judgment in Risk Assessment* (Washington, D.C.: National Academy Press, 1993).

NRC (National Research Council), *Building Consensus Through Risk Assessment and Management of DOE's Environmental Remediation Program* (Washington, D.C.: National Academy Press, 1994).

NRC (National Research Council), *Issues in Risk Assessment* (Washington, D.C.: National Academy Press, 1994).

NRC (National Research Council), *Ranking Hazardous Waste Sites for Remedial Action* (Washington, D.C.: National Academy Press, 1994).

NRC (National Research Council), *Improving the Environment: An Evaluation of DOE's Environmental Management Program* (Washington, D.C.: National Academy Press, 1995).

NRC (National Research Council), *Understanding Risk: Informing Decisions in a Democratic Society* (Washington, D.C.: National Academy Press, 1996).

3

Focus on Monitoring to Build Better Understanding of Our Ecological Systems

Long-term monitoring serves as a basis for both the understanding of our ecological systems and the measurement of our progress toward ecological goals. The focus of this monitoring should not just be on the goal as the end point of these activities, but on the rate of change we are making toward achieving those goals (see Chapter 8).

People commonly make comparisons between today and the past on the basis of their memories—anecdotal comparisons like "winters aren't as cold as when I was a child." However, scientific comparisons of the physical or biological environment require solid evidence measured as quantitatively as possible. Data need to be recorded according to a carefully prescribed protocol so that measurements over extended periods or taken simultaneously in different locations can be plausibly compared with one another. When properly collected, these data can serve one or more of five useful purposes

- Detection of short-term changes and long-term trends from measurements of a particular scientific variable made at regular intervals.
- Establishment of existing, baseline conditions in advance of planned or unplanned changes in an environmental system.
- Determination of whether a particular planned activity has been properly implemented.
- Assessment of the effect of a planned activity.
- Provision of data for scientific research.

Some environmental variables have been measured for a long time. For example, extensive temperature records (maximum, minimum, average) exist for a large number of locations, including cities and ships at sea, for more than a cen-

> *The nation's future environmental goals should be to maintain the progress made thus far . . . . We would not want to see water or air pollutants increase, nor would we want to undo proper handling of hazardous wastes.*
>
> *Nonetheless, every area of current environmental activity should be reassessed and priority given to the areas most affecting public health—long-term, as well as short-term health. Ecological concerns that do not relate to public health deserve serious consideration but should not be presumed to have transcendent intrinsic worth. The loss of a subspecies must be weighed against the costs visited on ordinary citizens.*
>
> *All risks should be considered in the context of natural hazards and assumed risks. While there is no magic cutoff for acceptable risk, rankings of the cost of risk avoidance and of natural and man-made risks can help to defuse the hysteria that causes an allocation of resources to low-risk, high-cost concerns.*
>
> —Forum Participant Comment

tury. As particular environmental concerns became prominent, systems of measurement were put into place to provide the needed information. The interest in temperatures and rainfall led to the establishment of national weather bureaus, which then proceeded to add other dimensions as demands arose. Some of the data systems use relatively simple instruments and training, as with the thousands of amateur observers around the United States who report daily measurements. Others are highly sophisticated, such as the satellites that furnish the images that record the approach of tropical hurricanes. Some need to be global because the phenomena being investigated and recorded occur on a global scale—for example, satellite measurements of stratospheric ozone.

Although the desirability of directly comparable measurements has long been recognized, the great majority of environmental data collected over the last several decades fails to meet this criterion. Most scientific environmental research has the primary goal of understanding a particular system at one time and in one place, and only later and separately does the question of whether this understanding is substantially different in another place or in the same location at some future time arise.

> *Science and technology can contribute to achieving environmental goals by identifying and developing improved data-collection methods, systems and sensors that are faster, have wider collection area(s), and are less expensive, more accurate, and more remote.*
>
> —Forum Participant Comment

Environmental monitoring has been going on in a quantitative instrumental manner since before the 1870s. In general, the development of research efforts in long-term environmental monitoring originated first in the geophysical sciences and only much later spread to the biological sciences.

Knowledge Advances Needed for Development of Sound Environmental Management Goals

Setting Goals

1. *What is the distribution of biodiversity in the U.S.? How much complementarity exists between regions of high conservation value for genetic, species, and community diversity? How sensitive is the protection of biological diversity to the area protected? How much opportunity is there to protect biodiversity in disturbed landscapes?*
2. *What is the economic value of various services we obtain from ecosystems and what is the economic value of biological diversity? How should we weigh instrumental with intrinsic values of diversity?*
3. *How can planning tools be made to be more interactive and accessible to a broader cross-section of the public?*

Assessing Impacts and Managing Resources

4. *What species and communities occur in specific ecosystems and regions? What are their ecological requirements? How does the diversity of species influence various ecosystem services? What species play particularly important—keystone—roles in the system? How does fragmentation influence the key species? What methods can be used to successfully restore degraded habitats or ecosystem services?*
5. *What are the long term trends in the structure and function of given biological communities and what is the variability in various measures?*
6. *What are the chronic effects of chemicals being released into the environment on plant and animal populations?*
7. *How will species in various ecosystems respond to changes in temperature, precipitation, disturbance, and CO_2 levels as predicted under climate models?*
8. *What environmental indicators can be developed that bear on the achievement of environmental management goals?*

SOURCE: Walter Reid paper in Part II of this report.

The International Geophysical Year (IGY) in 1957-1958, and the polar year in the late 1800s, are examples of special measurement efforts by scientists from many countries to establish a baseline understanding of the physical characteristics of the terrestrial and marine environments. The introduction of Dobson spectrophotometers and balloon-borne temperature sensors into several locations in Antarctica for the IGY and their maintenance since then have provided baseline data that allow scientists to know that the stratospheric ozone concentrations over Antarctica each October in the 1990s are very much less than was characteristic of the 1950s and 1960s. More recent efforts include the World Ocean Circulation Experiment and the Tropical Ocean Global Atmosphere Program, and the measurement of CO_2 at the Moana Loa observatory in Hawaii. Comparable comprehensive global biological studies are still in the planning stages.

One of the consequences of these varying stages of development is that people perceive the monitoring of physical characteristics differently from that of biological characteristics. Many physical characteristics have been or could have been recorded for two or more decades—stratospheric ozone concentrations have been measured almost daily over Arosa, Switzerland, since 1931, and comparable weather measurements have been made globally with calibrated instruments for many decades. But, the situation in the physical sciences is often that measurements have been made for an extended period, but for one reason or another the results are not comparable. However, even when such comparisons over time or space are an important focus of a research effort, the results are not useful for answering comparative questions. The causes of the failures are many, such as changes in personnel without adequate opportunity for transfer of precise protocol procedures, slow drift in equipment response or breakdown and repair without appropriate recalibration, relocation of equipment to a nonequivalent site, and change in time of day of measurement. Whatever the cause, the result is a seriously flawed time series of data.

> *What would be necessary to conclude that an environmental goal was achieved, that measurable goals were to be consistently met? We need new metrics to define what a healthy environment is.*
>
> —Forum Participant Comment

The situation in the biological sciences is even more serious. First, biological systems are more complex and multidimensional than are physical systems. Although there are some standard descriptors of biological systems, such as species richness or population size and density, the fundamental decisions have yet to be made as to what characteristics are likely to be the most important for assessing whether a given change is an indicator of irremediable environmental degradation or loss of long-term ecosystem sustainability.

Second, unlike the physical sciences, there are very few long-term data sets on biological factors in even the most prominent ecosystems, and those which do exist are either species-specific (e.g., the monitoring of deer populations on the George Reserve in Michigan) or geographically restricted (e.g., the monitoring of the wolf populations on Isle Royale) and are not designed to integrate multiple ecosystem functions.

Finally, the data sets that do exist have many of the same flaws as the sets of physical data—such as changes in personnel, changes in protocols, failure to standardize collection methods among sites, and shifts in priorities at the collection site.

> *The barriers to achieving America's most important environmental goals are primarily political. Over-costly approaches or methods that ask a subgroup of citizens to bear the costs for the whole country are naturally opposed. A cooperative spirit would be fostered if approaches were flexible and fair. Landowners who set aside habitat for endangered species should be rewarded, not penalized. Manufacturers who can find superior technology to prevent pollution should be allowed to do so, rather than following a fixed technology approach. Cities should be trusted to protect their citizens' drinking water.*
>
> —Forum Participant Comment

ECOLOGICAL KNOWLEDGE

Of particular concern relative to gaining a better understanding of the state of the environment is ecology. Many of the environmental problems that challenge human society are fundamentally ecological. "Ecological knowledge and understanding are needed to detect and monitor changes, to evaluate consequences of a wide range of human activities, and to plan for the sustainable management of natural and human-dominated ecological systems" (Lubchenco et al. 1991). Scientific understanding of the structure and function of ecological systems and biodiversity has advanced tremendously over the last three decades. Still, the limits of scientific understanding of these phenomena are obvious: "Even where general trends, such as wildlife population declines or changing stream quality, are clear, scientists are often unable to determine the impact of a specific action on those trends with any precision (or even, whether the trends are a consequence of previous human actions or are natural). Problems of cumulative effects, lack of site-specific ecological knowledge, and the natural variability of ecological systems conspire to add substantial uncertainty to almost all uses of scientific knowledge in environmental decision-making. As a consequence, we must place as much emphasis today on techniques and policies for coping with uncertainty as we do with efforts to reduce that uncertainty" (see Reid paper in Part II).

A recent National Research Council (NRC 1994a) report found such research to be rather poorly organized: "Many national and local agencies have responsibilities for understanding and managing the nation's biological resources, but there is no effective cross-institutional framework for identifying and conducting research of the highest priority, coordinating activities, or making information available in a coherent and usable way to the many agencies and other organizations that need

> *The chief barrier to achieving environmental goals is ignorance—of how ecosystems interrelate, how to measure and observe systems, and how our society communicates and reaches consensus on specific actions to be accomplished.*
>
> —Forum Participant Comment

it." The absence of a national "organizational home" for ecological research hampers the integration of the results of research that is being conducted in various locations. Furthermore, such research is at a disadvantage in comparison with other research that has explicit "representation" in the federal budget process.

Another NRC (1992c) report observed that the lack of an integrated national environmental-research plan, among other things, weakens the ability of the United States to work creatively with governments of other nations to solve regional and global problems. It recommended strengthening environmental research by fundamentally advancing factual knowledge, maintaining disciplinary research at the same time that emphasis on multiscale and multidisciplinary studies is increased, and ensuring economical and high-quality research with stable funding bases.

Ecosystems exemplify an issue confronting society in many forms—complex, nonlinear dynamic systems. Better understanding of the functioning of natural systems might provide insights into how to interact with and control complex adaptive systems. Special attention should be given to developing new theoretical and practical approaches to characterizing natural systems in ways that improve our ability to manage them.

MONITORING THE STATE OF THE ENVIRONMENT

As concerns about the environment mount, an important role for science and technology is to provide quantitative data that characterize the state of the environment. In most human habitats, the ongoing state of the nearby supplies of air and water are of prime concern. Continuing efforts to monitor the purity of air and water are needed to determine qualitatively whether they are safe to breathe and to drink and quantitatively whether their purity is improving or getting worse. Such data are important not only as observations, but even more when decisions about the applications of economic resources toward such purity have been made or are under consideration. When steps have been authorized to produce a purer water supply, a proper role for science and technology is the quantitative determination of changes that have occurred in the water supply after implementation.

Retrospective examinations of the efficacy of an improved purification procedure often reveal that some critical component has not been measured or the

> *For the environment, science should be focused on providing clear and unambiguous data on the effects of pollutants on the health of people, other life forms, and the environment as a whole. It should also be attempting to clarify the areas where the data cannot yet be unambiguous, such as in global climate changes, and helping to project realistically the possible scenarios, with best-estimate probabilities.*
>
> —Forum Participant Comment

instrument not calibrated after repair, etc.; the result is that the available data are inconclusive. The long-known solutions to this problem involve devoting a minor fraction of the effort to calibration, to testing of blanks and unknowns—in short, to continuing quality control of the data as they arise.

Status of the U.S. Environmental-Monitoring System

Currently, a great deal of monitoring data is collected in the United States. However, the data are incomplete, especially in ecology (because of the late recognition of the importance of ecological sciences), of varied quality, and non-standardized in collection protocol (see Reid paper in Part II).

Because of those problems, combining data from different regions and different times is unreliable. Currently, "the scientific community does not unanimously agree on what the best indicators of environmental quality should be" (see Reid paper in Part II).

With standardization and sufficient quality control, some existing data sets could be more useful. "Federal, State, local and non-governmental organizations (NGOs) spend hundreds of millions of dollars each year on the collection, storage and use of environmental data. Much of these data are collected for specific purposes and are not designed for developing general measures of environmental quality" (see Ross et al. paper in Part II). This brings up a number of questions as to how such data could be collected and used

- Can we make better use of existing nongovernment monitoring, such as that required as part of permit requirements under existing or future law?
- Can we make better use of data collected by a variety of government entities around the country?
- Can we make better use of the data collected for the many environmental-impact statements carried out throughout the country?
- Can we make better use of trained and supervised volunteers without compromising the quality of the data collected or the rights of private landowners.

The U.S. Fundamental Ecological Research System

The United States has no national program addressing fundamental ecological research, to provide coordination and focus. One way of meeting that need could be to establish a program of focused fundamental research under the auspices of a federal lead agency. Properly structured, such a program could provide the direction and momentum that are missing in the nation's ecological research. The research program could address fundamental research questions on a variety of scales, from cell to organism to ecosystem to region, as well as, for example, determinants of diversity, the role of fragmentation and edges, and the maintenance or restoration of ecosystem function. Although the details of the research program could be determined by the competitive process that has served research

> *To conclude that our environmental goals were achieved, I would need scientific evidence that pollutants had diminished; that ecosystems were working; that there were no longer significant negative health effects from food, water, air, and soil; and that businesses and communities continued to function in a sustainable and successful manner consistent with ethical standards in a free and democratic society.*
>
> —Forum Participant Comment

in this country so well, the organization of research by problem would probably enhance effectiveness.

One project, for example, could address fundamental biological and ecological questions at the same time that it provides information relevant to important human activities, such as agriculture, biotechnology, and manufacturing. The selection of the focus could engage scientists from the ecological research community in a systematic assessment of potentially productive strategic-program directions. We offer the following examples as illustrations of where current research needs to be expanded:

- Chemical and nutrient cycles.
- Determinants of natural and human-induced variation in natural populations.
- Physiological responses of organisms to natural and human-induced stresses.
- Habitat fragmentation, scale and edge phenomena.
- Maintenance and restoration of ecological services.

Competitive-grant research programs could be organized to address such fundamental topics. At the same time, however, attention could be given to the practical aspects of the information being generated regarding such issues as deriving wealth from biological systems in sustainable ways, revealing the principal ways that human activities are stressing natural systems, and the ways such responses feed back to human society. The research could also improve the knowledge base for managing natural resources.

INDICATORS OF THE STATE OF THE ENVIRONMENT

Environmental indicators are a set of quantitative measures that provide a comprehensive picture of the condition of a nation's environment and that can be used to evaluate trends in environmental quality. Many agencies in the United States have begun attempts to develop such indicators. The agencies need to interact with the scientific community to review the indicators so that a consensus can be developed as to whether the indicators are sufficient or new ones are needed.

> *Environmental goals should include protection of natural resources, starting with clean air, water supply, and land. Environmental goals must be focused on maintaining or improving these vital resources.*
>
> —Forum Participant Comment

There is not unanimous agreement among scientists or among countries as to what these indicators should be, but rates and direction of changes in environmental characteristics that lead to changes in biodiversity in a given region are certainly the framework by which the indicators should be assessed. The Ross et al. paper in Part II reviews how indicators are developed. The Organization for Economic Cooperation and Development uses the pressure-state-response (PSR) model.[1]

The process of measurement of the state of the environment begins with the selection of a set of appropriate indicators that are major factors in any marked changes or are closely bound in behavior to some factor. The protocol specifies how the appropriate data should be collected, analyzed, checked for quality, and stored for future access. Adequate provision must be built into the process for continual assurance that the measurement procedures and equipment are in standard working order. Finally, examinations of the data for consistency must be carried out periodically to determine whether unanticipated problems have entered into the picture.

Because the preferred set of environmental indicators is likely to change over time, the nation's monitoring system needs to be of sufficient quality, content, and standardization to support a broad range of likely indicators. Data are needed not only for specific regulatory purposes, but also for developing general measures of environmental quality.

FINDINGS, CONCLUSION, AND RECOMMENDATIONS

Findings

If the nation is to have useful environmental goals, indicators of progress toward these goals are needed. The quantification of these indicators must be based on monitoring data of sufficient quality and duration that relevant trends can be differentiated from noise.

Reliable monitoring is also essential if the nation is to get early warning of emerging environmental problems (NRC 1991). The few long-term data sets that

[1]The PSR model asserts that human activity exerts *pressures* (such as pollution emissions or land-use changes) on the environment, which can produce changes in the *state* of the environment (such as changes in ambient pollutant concentrations, habitat diversity, and water flows). Society *responds* to changes in pressures or state with environmental and economic policies and programs intended to prevent, reduce, or mitigate pressures or environmental damage (see Ross et al. paper in Part II). Environmental quality is the reflection of the *state* component of the PSR model.

exist have already played key roles in alerting humanity to impending serious problems (NRC 1992c).

Current environmental data are being collected by disparate agencies and private groups. Much of this collecting is designed for other purposes (e.g., compliance) and not for developing general environmental indicators or providing early warning of environmental problems.

The existing environmental-monitoring data sets on which environmental indicators and early warnings must be based are incomplete in space and time, especially with respect to ecology (because of the late recognition of the importance of ecological sciences), and are not necessarily relevant to measuring progress or anticipating problems. These data are also of varied quality and need better standardization (see Ross et al. paper in Part II).

Managing the nation's environmental data is a complex task. In the NRC's (1994a) report on the National Biological Survey (NBS) titled *A Biological Survey for the Nation*, it was suggested that a distributed federation of databases be designed to make existing information more accessible and to establish mechanisms for efficient, coordinated collection and dissemination of new information (NRC 1993).

> *Far too much money is currently being wasted taking environmental data of dubious quality . . . in poorly chosen locations, and at temporal and spatial resolutions ill suited for comparison with the computer models used to . . . gauge progress.*
>
> —Forum Participant Comment

To establish indicators and identify emerging issues, only a small fraction of the data might be needed in some cases. The question is how we determine what fraction of data is necessary to support an indicators program. When evaluating such issues, it is important to take a hypothesis-driven approach and be responsive to clearly stated questions (NRC 1995d).

Monitoring is not given adequate priority in the nation's science efforts (NRC 1992c). Although a number of reports have addressed the nation's monitoring system and made recommendations for its improvement, little improvement has been made (NRC 1991).

The U.S. monitoring system is outdated and inadequate and would need to be modernized, expanded, and systematized to allow the calculation of useful indicators that match environmental goals. This monitoring system, which would provide the data for the indicators, should have sufficient quality control and standardization for the information yielded to be useful and the results from it to be believable. It could make possible the support of a compilation of indicators for a broad range of national and regional goals. This might allow for changes in goals and shifts in the choice of indicators.

There are other reasons for wanting monitoring data beyond their use in measuring progress toward environmental goals: long-term monitoring data are of great use in basic science. Trends in such data can serve as an early warning of problems not yet recognized. For example, the long-term monitoring of CO_2 data by Keeling has provided a key part of the information that scientists need to evaluate the potential for global warming. In addition, examinations of long-term data collections at natural-history museums has allowed scientists to show how plant life has responded in structure and form to changing atmospheric conditions. Currently, the scientific reward system tends to provide too little incentive for routine baseline data collection in relation to its potential value in giving early warning of possible environmental problems before they cause any appreciable harm.

One example of a national environmental monitoring and reporting system for air quality is embodied in EPA's *National Air Quality and Emissions Trends Report* (EPA 1993). This report presents data on monitored concentrations of what EPA calls criteria air pollutants—PM10 (particulate matter), SO_2, NO_2, CO_2, O_3, and Lead—in the 30 largest metropolitan areas in the United States and estimates of aggregate emissions of the same pollutants. Although the report has limitations—for instance, it reports virtually no information on emissions or concentrations of hazardous air pollutants, such as benzene, acrylonitrile, and asbestos and does not measure concentrations in all parts of the country or within a given city—it is exemplary in several respects. For example, the reported trends are based on a relatively consistent set of ambient-air monitors, the data have been collected for a long period, the data are clearly summarized and presented (including measures of variability, in addition to nationwide averages), and the statistical findings are described in a clear and graphically attractive manner.

The country would be fortunate to have comparable information on other environmental conditions and trends. For example, with respect to water quality, several monitoring systems would need to be taken into account.

Creating a mission to support monitoring of environmental indicators and to provide early warning can bring needed focus to the nation's diffuse data-collection system.

Conclusion

Science and technology provide quantitative data to characterize the state of the environment. Scientists analyze and interpret this information to provide society with a deeper understanding of its relation to the state of the environment. As part of this analysis, they compare past conditions with projections to determine the potential for achieving a desired state of environmental quality at a feasible rate of change.

One particular concern relative to achieving a desired state of environmental quality is our understanding of the ecological system in which plants, animals, and humans live. Current ecological data and understanding are inadequate to

- Detect, monitor, and characterize environmental changes.
- Evaluate the consequences of human activities.

- Provide an information base for sustainable management (i.e., "no loss") of both natural and ecological human-designed systems.

Therefore, it is difficult to conduct the comparative analysis of past, current, and future ecological states (as described earlier) to determine what actions are needed to achieve a desired end of environmental quality. Furthermore, current programs do not address these issues in a sufficiently coherent and comprehensive manner on a national basis. Indicators that are needed to measure the current status of ecological systems, to gauge the likelihood of meeting society's environmental goals, or to anticipate problems resulting from economic growth are not available. We are spending much money to collect data that are neither complete nor always relevant to the decisions that society needs to make about land use, transportation, industrial activity, agriculture, and other human activities. The system of monitoring the state of the environment needs to be improved to make it more relevant to decision-makers, and there is need for a more-sophisticated and better-informed discussion of what needs to be measured and why.

Recommendations

1. The White House Office of Science and Technology Policy should review and evaluate the quality of existing measurement and monitoring systems for relevance to and usefulness in meeting environmental goals. That would include establishing a system-design process to complete and maintain the monitoring system.
2. Congress should assign an existing or new federal research organization the mission of working with the scientific community to identify key subjects for ecological research and ensuring that this research is being pursued adequately somewhere in the overall environmental research system, which includes not only EPA and the Department of the Interior, but also the National Science Foundation, National Aeronautics and Space Administration, National Oceanic and Atmospheric Administration, Department of Defense, and DOE.
3. Research should aim at identifying and developing reliable indicators of the health and sustainability of the environment and ecosystems. Such indicators might include chemical nutrient systems, habitat fragmentation, and changes in biodiversity.
4. New systems of monitoring that meet society's decision-making needs should be identified and implemented. This includes provision of access to data sets via the Internet, integration of data from various scales (e.g., integrating satellite data with land-based measurements), and finding sources of historical data (e.g., glacial samples that have preserved the earth's history).
5. The nation's existing monitoring system needs to be reviewed and evaluated for its relevance to indicators of environmental progress and identification of emerging issues. This evaluation procedure is a first step toward improving the overall system.

For more information and guidance, the reader should refer to the following:

NRC (National Research Council), *Research to Protect, Restore, and Manage the Environment* (Washington, D.C.: National Academy Press, 1993).

NRC (National Research Council), *A Biological Survey for the Nation* (Washington, D.C.: National Academy Press, 1994).

NRC (National Research Council), *A Review of the Biomonitoring of Environmental Status and Trends Program: The Draft Detailed Plan* (Washington, D.C.: National Academy Press, 1995).

NRC (National Research Council), *Review of EPA's Environmental Monitoring and Assessment Program: Overall Evaluation* (Washington, D.C.: National Academy Press, 1995).

4

Reduce the Adverse Impacts of Chemicals in the Environment

Stresses on ecosystems and hence on humans have many important sources, including chemicals, mining, farming and forestry, and—perhaps even more serious—living conditions, levels of education and health care, and life-style choices. Clearly, this forum could not deal with all those sources and chose to concentrate on some that might be addressed with specific technology assessments and recommendations. Adverse impacts of chemicals in the environment constitute one source.

There is a desire in general on the part of societies to improve their standard of living. This inevitably results in greater consumption of water, food, mineral, and energy resources and of our manufacturing capability. Although wastes and products that follow such consumption result in a greater burden on the environment, it is also true that the higher the living standard, the greater the resources that can be, and generally are, devoted to protecting human health and the environment.

Thus, industrially advanced societies tend to have safer food, water, and air than do less advanced societies. Nevertheless, there remains a conflict between the desire of a growing population for more goods and services and the desire for a healthy environment. One of the many land-use and economic changes resulting from human expansion that generally cause impacts on ecosystems is the introduction of new products. New products that foster new desires or satisfy old needs are sometimes discovered to have environmental impacts that are unacceptable or become unacceptable relative to a continually raised environmental standard.

The last 50 years have seen the introduction of many new chemicals. Many have stood the test of time and shown their benefits to outweigh their environ-

> *We desperately need better tools to predict human risk from exposure to toxic chemicals. The information that will be useful will eventually arise from the development of a conceptual toxicity-evaluation scheme resulting from the recent advances in molecular genetics and biochemistry. This will enable scientists to target chemicals and substances of potential concern much more easily without needing a complex (and time-consuming) series of traditional toxicity tests.*
>
> —Forum Participant Comment

mental risks. For some, however, important adverse environmental effects emerged. The search to replace those without further environmental effects has become a strong driving force in industry, in the scientific community, and in the general public. The focus has been mostly on testing for acute human toxicity with surrogates and on estimating long-term chronic effects in humans, primarily emphasizing cancer, again with surrogates. Increasingly, researchers will strive to include effects on entire ecosystems, and long-term, multigenerational effects on fertility, reproductive quality, and hormonal functions. Of major interest will be chemicals with the potential to be persistent, toxic, and bioaccumulative (PTB). However, chemicals that are persistent but not toxic or bioaccumulative, such as CFCs, have also led to environmental problems, as have chemicals that are persistent and toxic but not bioaccumulative. Evaluations of such chemicals are also needed.

Some of the surprise effects of chemicals have been due to a failure to predict the scale on which technologies might be used once they were shown to be beneficial when used on a limited scale. For example, DDT has side effects that have increased nonlinearly with the scale of application; as a result, the incremental benefits of a seemingly benign technology reversed when it was applied on a larger scale. New technologies have to be constantly reevaluated in anticipation of scale effects.

Following are some examples of products or processes that created unforeseen environmental problems after their introduction.

1. Products
   - Pesticides, such as DDT, endrin, dieldrin, and benzene hexachloride (BHC).
   - Alkylbenzene sulfonate (ABS) synthetic detergents.
   - Polychlorinated biphenyls (PCBs).
   - Chlorofluorocarbons (CFCs).
   - Lead used in gasoline and paint.
   - Some chlorinated solvents.
   - Wood preservatives.
2. Processes
   - Chlorination for disinfection (in some situations).
   - Mercury release from chlor-alkali cells.

- Older coal gasification (now replaced with modern, but expensive, technologies).
- Dioxin release from incineration and some chemical reactions.

The problem of nondegradable (persistent) pesticides has been known for many years. Because of the excessive accumulation rate in higher species, a group of major pesticides used in the 1960s (DDT, endrin, dieldrin, and lindane) have been banned or greatly limited in use in the United States. Substitute pesticides that are readily degraded in the environment have been developed. New adverse environmental impacts continue to be found, even with some of the substitutes. In some cases, biodegradable substitutes proved to be much more toxic to humans and to require much more sophisticated handling than more persistent substances. At times, a parent pesticide is degraded in the environment but results in daughter products that persist. The search for alternative pesticides or other products that have less adverse environmental impact has been rewarding to society and industry. Such efforts need to be, and will be, continued and will be driven by the marketplace's responding to public and government interests.

Some persistent chemicals with a variety of uses do not bioaccumulate in birds or other higher species but instead partition readily into water and do not bind well to soils, so they migrate through the ground and cause groundwater contamination. Examples are the triazine group of herbicides, dibromo chloropropane (DBCP, a nematocide), and some industrial solvents.

PCBs were once widely used as coolants in electrical systems. Emissions from that use led to great concern over the potential for PCBs to accumulate in the environment, and cause problems similar to those caused by DDT and to accumulate in people, in whom some of the PCBs were known to be toxic.

Another problem persistent chemical was the synthetic detergent, ABS, which was widely used in the 1950s. It persisted in rivers, streams, and groundwaters, causing excessive foaming of water. It did not cause health effects, but it was aesthetically unacceptable in drinking water. Legislation outlawing its use or threats thereof led to substitution with a biodegradable alternative detergent in the early 1960s.

Major environmental problems resulted from lead in gasoline and paint, and its use in these products has been eliminated. That required the development of new formulations for gasoline, new designs for engines, and substitutes for paint pigments.

> *Creating safe indoor environments is emerging as an endeavor in need of much basic research, in that there appears to be very little known, at least to a layperson like me about the cumulative effects of the many materials, products, and other environmental impacts of working in office buildings and living in homes constructed of relatively new materials.*
>
> —Forum Participant Comment

The presence of excessive mercury in receiving waters due to use of mercury electrodes in plants producing chlorine and alkali was a major incentive in replacing that technology with membrane cells. Chlorinated solvents have been excellent for cleaning of clothing, machine parts, engines, and electronic components, but their disposal in landfills and their leakage from storage tanks have caused extremely expensive groundwater-contamination problems that have not yet been solved. Substitutes for chlorinated solvents are now widely used, and care in their disposal is required.

Substitutes for the CFCs that cause depletion of stratospheric ozone are being developed. Those which will be used in the near future (hydrochlorofluorocarbons [HCFCs]) are of concern because a decomposition product, trifluoroacetic acid, might be very persistent and, under extreme conditions, have the potential to cause an undesirable environmental impact.

Dioxin can be formed as a byproduct in some chemical processes, including one of the old routes to production of 2,4,5-T, a widely used herbicide.[1] This example illustrates that not only pure products must be evaluated, but also the contaminants that might be present in them, even if at low concentration. Methods of producing 2,4,5-T without producing dioxin are now in use. Dioxin is now known to be produced during combustion under poorly controlled conditions and when even very small amounts of chlorine-containing compounds are present.

Another example of byproducts of concern is the trihalomethanes that are formed from humic materials in drinking water when chlorine is added as a disinfectant. The trihalomethanes are potential human carcinogens, and their concentration in drinking water is now regulated. Other chlorinated byproducts are also formed by chlorination, but their health impacts are less well known. Modifications of water-treatment operations that reduce trihalomethane formation and alternatives to chlorination are therefore being sought.

How do we determine what adverse environmental impacts the byproducts of new technologies—whatever they might be—can cause? And how do these environmental problems compare with those of the alternatives? Global transport of persistent chemicals is increasingly recognized as an issue that must be dealt with by all, or at least a combination of, nations.

The above are only a few examples of the numerous major environmental problems that were created by the introduction of new products and processes. They should serve as reminders as societies develop new products and processes to satisfy their needs and desires. The potential of any chemical for environmental damage must be assessed before its commercialization, and our capability for doing so should be expanded, although we recognize the possibility that the new

[1]Dioxin exists in many congeners and isomers. "Congener," in this case, is a varying number of chlorine atoms on the dioxin molecules. One isomer of the congener containing four chlorine molecules is widely held to be extremely toxic. The other congeners and isomers are believed to be far less toxic, but the entire subject is far from completely researched.

chemical might replace another substance, natural or man made, already in use that could be even more damaging. Those cases demonstrate the need for continuous review of costs and benefits, which might not be the same for all countries and communities.

ANTHROPOGENIC CHEMICAL PRODUCTS

New chemical products intended as pharmaceuticals (and their important metabolites) are extensively tested under rules of the Food and Drug Administration (FDA). The FDA tests cover a wide variety of toxicity but concentrate on human effects. New discoveries intended for pest control or for use as agricultural insecticides, fungicides, or herbicides are subject to somewhat less rigorous testing required by EPA under the Federal Insecticide, Fungicide and Rodenticide Act (FIFRA). The FIFRA tests concentrate on predictors of human carcinogenic effects, with increasing attention to effects on "off-target beneficial species," such as birds. Costs of testing a chemical under FIFRA can be tens of millions of dollars; costs of FDA testing can be even more.

> *Science and technology can contribute by helping define risks to human health and the environment and defining cost-effective solutions to prevent risks or reduce risks to acceptable levels.*
>
> —Forum Participant Comment

New chemical products not under the FDA or FIFRA are covered by the Toxic Substances Control Act (TSCA). If these chemicals are intended to become articles of commerce, they are subject only to submission to EPA of a request for a "PreMarketing Notice" (PMN). EPA has 90 days to respond to such a request and often, in the absence of extensive data, relies on structure-activity relationship (SAR) predictions.

These chemicals can be subject to much more testing under the Occupational Safety and Health Act (OSHA) if they are known to be present in the workplace and a risk has been identified. In Europe, new chemicals are subject to more testing but still far less than that required for new drug or agricultural applications. International standardization of testing and international sharing of testing responsibilities and data would reduce costs and speed the availability of reliable and reproducible assessments.

ANTHROPOGENIC CHEMICAL BYPRODUCTS

Many anthropogenic chemicals end up in incinerators or wastewater treatment facilities. We need to be concerned about the reaction byproducts formed in such treatment facilities.

Chemicals that are disposed of in landfills (where they might leak from containment), are deliberately emitted (as in the case of hair spray or paint solvent), or are merely discarded will end up in the air, in surface runoff or in groundwater, or simply reside on land.

Today, these chemicals, their degradation products, and the byproducts of their production are, for the most part, investigated only when someone suggests an environmental hazard on the basis of anecdotal environmental monitoring, local tests, or calculations. Under TSCA and OSHA, much more testing of these chemical byproducts can be required, once they are identified. This identification is unlikely to occur without better knowledge of what byproducts might be formed from anthropogenic chemical production and the use and effects such byproducts may have on the environment.

FINDINGS, CONCLUSION, AND RECOMMENDATIONS

Findings

Acute air and water pollution caused by the release of chemicals or other wastes will require continued vigilance, but 25 years of progress have already been made in reducing such pollution. Solid waste pollution by chemicals is covered under the "Industrial Ecology" section.

Although considerable advances have been made, there is still a great need to improve the ability to predict the environmental consequences of a new chemical on a variety of scales before the great expenditure of getting it into the marketplace is undertaken. Even then, unforeseen environmental questions may arise after a product is introduced. Greater proficiency in addressing such environmental questions is needed and should greatly improve our ability to develop regulations that are appropriate to the problem. Industry needs to have greater assurance that its often expensive product development and commercialization will be successful and not quickly overturned by unforeseen human health or environmental problems. Thus, both to encourage the development of desirable products and to provide adequate safeguards against potential environmental liabilities after products are introduced into the marketplace, sound procedures for evaluating potential impacts of products on human health and the environment are essential, and this need is expected to grow.

We have reasonably good testing methods for acute toxicity. The tests use surrogate animals, and the correlation to humans is the weakest element. The quality of predictive modeling for acute effects, based on SAR, is only modest. For chronic effects, testing with surrogates for humans is modestly good, particularly for cancer. Tests for chronic toxicity in animals are only fair and for cumulative effects on ecosystems are very weak. Predictive modeling for chronic effects in general is poor. The quality of toxicity testing and modeling is assessed more fully in previous National Research Council reports (NRC 1994a). Modeling to predict persistence is fair and to predict bioaccumulation potential is moderately good, at least for many common classes of chemicals (e.g., chlorinated organics). This issue exists, not only for traditional organic chemicals, but also for organo-metallic compounds, ions, and complexes of the "heavy metals" mercury, cadmium, lead, copper, selenium, silver, beryllium, thallium, chromium, arsenic, nickel, and zinc.

There has been consideration of greater testing of chemicals covered only by premanufacturing notices, and it seems logical that a staged approach to testing similar to that required by FIFRA and based, incrementally, on exposure potential will be increasingly worthy of consideration.

For both the byproducts of various waste-treatment processes and the degradation products of intended products or processes that find their way into the environment, there will be the same need to assess potential ecological damage as there is to assess damage caused by specific intentional chemicals. "Daughter" (byproduct and degradation) chemicals pose additional complications in that their chemical formula and structure might not be known; they will be in lower concentrations than the industrial chemical products, making them and their impact harder to detect; and it will be far more difficult to obtain useful samples. They might also be in the presence of other materials that could confound tests. Small or laboratory-scale "miniecosystems" have been found highly useful for examining a wide range of conditions to sort out such variables so that the major biological processes affected by new chemicals on ecosystems can be better understood.

Ultimately, because of the sheer diversity and complexity of these potential chemical commercial products and their "daughter" products, modeling might prove to be particularly valuable. When sufficient specific industrial chemicals have been assessed for their potential for ecological damage, it should become possible to estimate this property for postulated chemicals. That would be similar to the prediction of thermodynamic properties from a combination of group contribution modeling and compound class-specific rules that have been developed and shown to be successful over the last 25 years. Similar modeling of treatment facilities and degradation processes should make it possible to estimate the potential effects of byproduct and degradation chemicals. Cross referencing those two projections ("daughter" products and ecosystem impacts) would help identify the facilities or chemicals that require more careful sampling and analysis and the need for containment or substitution.

Unexpected synergy between something already in the environment, whether anthropogenic or not and whether global or local, and a new chemical that was tested and shown safe by existing methods will remain a problem. Broader testing and testing in more "complete" simulated environments will help. However, in the long run, a better understanding of the basic biochemical process in the environment will be the strongest ally in deciding where to look and what to look for. The issue of unexpected synergy and its solution, will be similar to the problem of complex mixtures of chemicals that can be synergistic (i.e., more than additive or at least additive in their effects).

At least for the present, it seems impractical to hope that we will be able to identify, empirically, all the environmental impacts that should be tested for or to hope that our tests could detect the most sensitive potentially affected organisms at the low end in responses to "dose-level and exposure-time" testing. Increasingly, researchers are working on collections of cells, microorganisms and entire miniecosystems, in which a broad range of subjects and surrogates are combined with mathematical models in an attempt to develop a more-accurate and holistic prediction or measurement of the effect of a chemical on an ecosystem. Again, in the long run, a better understanding of the basic biochemical processes in the environment will be the strongest ally in deciding what to measure and how to measure it.

Conclusion

Concerns about chemicals in the environment have focused major attention on the possible consequence for humans, animals, and whole ecosystems. Substantial progress has been made, and some contaminated bodies of water have been restored to use. However, we still lack basic knowledge and procedures for evaluating the potential impacts of chemicals, compound mixtures, or artificial concentrations of natural substances that have an adverse effect on human health and the environment. Such knowledge will be essential for developing products with adequate safeguards against unwanted side effects.

Reasonably good methods are available for testing the potential carcinogenic effects of chemicals on surrogate species for humans, particularly rodents. The correlation between these surrogates and humans is far from proved and is hotly debated. However, on the basis of experience buttressed by significant testing, use of surrogate species seems to have been most helpful in reducing exposure to many suspect carcinogens.

However, there is a need for better tests to assess ecological damage potentially caused by single compound chemicals, the byproducts of various waste-treatment processes, and the degradation products of intentional products or unintentional process emissions that find their way into the environment. Better understanding of the basic biochemical processes occurring in the environment is necessary to decide where to look, what to look for, what to measure and how to measure it.

Recommendations

1. Better test methods should be developed to evaluate, model, and monitor the potential long-term environmental impacts of single compounds emitted as a result of new products or processes. Emphasis should be placed on compounds that degrade only very slowly.

2. Better test methods should be developed to define and ultimately to model and predict the byproducts and degradation products associated with production and use of materials.

3. Basic studies of biochemical effects and of the impact of various chemicals and other adverse effects on the biochemistry of sensitive plant and animal species should be strongly supported. It is from such studies and the monitoring program that the most-effective hypotheses about items of greatest concern and about the continual development of testing will arise.

4. Strong support should be given to innovative ideas for modeling and tests on lower-order surrogate species that help to reduce the cost of tests for potential adverse environmental health effects on humans or shorten the response time needed to obtain that information.

5. International standardization of testing and international sharing of testing responsibilities should be promoted to reduce costs and speed the availability of reliable and reproducible assessments.

6. The emerging concept of developing experimental "miniecosystems"—focused on controlled-exposure environments for testing and for developing mathematical simulations of ecosystem impact based on limited, specific tests—should be supported.

For more information and guidance, the reader should refer to the following:

NRC (National Research Council), *Opportunities in Applied Environmental Research and Development* (Washington, D.C.: National Academy Press, 1991).

NRC (National Research Council), *Issues in Risk Assessment* (Washington, D.C.: National Academy Press, 1993).

NRC (National Research Council), *Pesticides in the Diets of Infants and Children* (Washington, D.C.; National Academy Press, 1993).

NRC (National Research Council), *Ranking Hazardous-Waste Sites for Remedial Action* (Washington, D.C.: National Academy Press, 1994).

NRC (National Research Council), *Science and Judgment in Risk Assessment* (Washington, D.C.: National Academy Press, 1994).

NRC (National Research Council), *Carcinogens and Anticarcinogens in the Human Diet: A Comparison of Naturally Occurring and Synthetic Substances* (Washington, D.C.: National Academy Press, 1996).

NRC (National Research Council), *Understanding Risk: Informing Decisions in a Democratic Society* (Washington, D.C.: National Academy Press, 1996).

5

Develop Environmental Options for the Energy System

All aspects of the national and international system for the exploration, transportation, production, distribution, and use of energy have effects on environmental conditions. Thus both the supply of and the demand for energy must be discussed when exploring more environmentally sensitive options for the energy system. The future development of the energy system and the technologies and sources of energy that will be used, will determine the extent to which the system will be compatible with the environment.

Energy is crucial to the transformation of materials and essential to improving standards of living throughout the world. Energy is also a major driver in shaping the quality of the environment. On the one hand, energy creates environmental problems in its extraction, transportation, combustion, storage, and final consumption. Environmental problems are caused by activities as diverse as strip mining, oil spills, and nuclear-waste disposal. On the other hand, the wealth generated by economic returns allows a society greater economic opportunity to address environmental problems.

Energy's value to society is in the services that it provides: heating and cooling buildings, transporting people and goods, driving industrial processes, and powering the electronic information explosion. These services are created by capital and operating investments made both in energy and in the end use. For example, the costs of heating, cooling, and lighting a commercial building encompass investments made by both the electrical utility and the building owner. The latter's investments include insulation, computer-driven climate control, and high-efficiency lighting. The total efficiency of the energy system is a function of energy prices, overall economic performance, and technology development.

The one constant in energy prediction is surprise. At one time, energy ex-

> *Science and technology can contribute to the goal of clean air by developing technologies that reduce the cost of pollution control, so that we get cleaner air for our investment in pollution control than we do with existing technologies. These technologies might be*
>
> - *Better air-cleaning technologies (better, cheaper catalytic converters)*
> - *Better production methods (nonpolluting painting technologies)*
> - *Better products (lighter cars that go farther on a gallon of gas)*
> - *Different technologies (electric cars with reasonable range and acceleration).*
>
> —Forum Participant Comment

perts believed that energy and gross national product (GNP) were inextricably intertwined; today, we know that they can be uncoupled. At one time, most experts expected that oil prices would be two to three times higher than they actually are today. In recent years, energy systems have increased in efficiency and the supply of natural gas has been greater than projected.

THE ENERGY SYSTEM AND ITS ENVIRONMENTAL EFFECTS

The environmental effects of energy production and use occur on local, regional, and global scales. At local levels, energy production and use in autos, power plants, and industry are among the principal causes of urban air pollution, from particles to CO_2 to tropospheric O_3. Scientific understanding of the causes and developments of technologies have resulted in much-cleaner urban air.[1] Scientific understanding of the behavior of automobile emissions in the presence of sunlight illustrate the key roles of research and development in addressing environmental problems.

A particularly important part of the energy system at the local level is energy efficiency. A number of utility companies have initiated programs to improve the efficiency with which energy is used by residences, business, and industry. A number of technological advances in lighting, heating, cooling, and passive energy activities have also made it possible to keep energy demand constant without sacrificing lifestyle advances. In the case of transportation, automobile fuel economy has increased from 14.2 miles per gallon (mpg) in 1973 to 28.2 mpg in 1992—all through technological advances.

On the regional scale, energy production and use are the major factors in acid deposition. SO_x and NO_x emissions from power plants and NO_x emissions from automobiles are major contributors to acid deposition. Acid effects on aquatic

[1]For example, the smog in the Los Angeles Basin was originally attributed to the burning of trash. Scientific research led to the demonstration that the smog was actually largely associated with the products of combustion of automobile fuel. Science has been at least as important in identifying environmental problems as in solving them.

and forest ecosystems are of concern. Large-scale government and private investments in understanding this problem have helped to secure the passage of legislation that will have a substantial beneficial effect on the reduction of emissions that cause acid rain.

On the global scale, although great uncertainties remain with regard to the timing, geographic distribution, and magnitude of climate change, the Intergovernmental Panel on Climate Change (IPCC) has concluded that climate warming would result largely from the emission of CO_2 in the combustion of fossil fuels. The climate-change problem is inherently international because of CO_2, and the effects of climate change are globally distributed. For example, an international treaty on climate change has been agreed to by many nations of the world, and declarations of national intent to reduce emissions of greenhouse gasses to 1990 magnitudes are in place. It remains an open question as to whether national actions to bring about such reductions are feasible and will occur. In any case, such reductions would only postpone the projected climate change by a few years. Studies by the IPCC have shown that much deeper reductions—around 60-80%—in CO_2 emissions will be necessary, according to computer models, to stabilize CO_2 at current concentrations in the atmosphere.

Even the most basic uses of energy can have severe environmental effects, especially on ecosystems in developing countries. A good example is the use of wood as an energy source in developing countries. The harvesting of wood has destroyed ecosystems and led to serious soil erosion. With developing countries dramatically increasing their energy use, energy supplies could be increasingly dependent on fossil fuels in the future and that could increase the net contribution to CO_2 production.

One effect of energy production that has been most difficult to solve stems from the use of nuclear fission as a source of energy. Nuclear power has negative and positive environmental impacts as a replacement for large coal- and oil-burn-

> *Environmental problems touch all aspects of human existence. It is inappropriate to develop a simple laundry list of "one-size-fits-all" goals. We need, instead, to think in terms of broad themes. These include (1) mitigation of harm associated with residuals (air and water emissions and soil contaminants), (2) encouragement of increasingly efficient resource use, and (3) protection of ecosystems and wildlife. Which specific emissions should be targeted and which ecosystems deserve focus should be determined locally or regionally in most instances because most environmental impacts are local, not regional or national. Establishing a national laundry list of top priorities in specific is both a static approach and one that fails to take into account the degree to which many environmental problems are fundamentally local; priorities will vary by locality. This requires local flexibility in determining goals and priorities.*
>
> —Forum Participant Comment

ing plants. Use of nuclear power substantially reduces air pollution because nuclear power plants do not emit NO_x, SO_x, or particles. However, there is concern about the potential for accidents releasing the radioactive products of nuclear fission (from plant operations, transportation, or storage) as well as the disposal of nuclear wastes. The latter remains the largest environmental problem associated with nuclear power. The environmental aspects of the disposal of radioactive wastes from nuclear power plants remain a divisive issue in society. Legislation has been proposed for the disposal of such wastes. A site has been identified and other possible locations for repositories have been suggested. Billions of dollars have been spent, billions more will be spent, but the problem remains unsolved because of both scientific and political factors. The safe disposal of radioactive waste and the ability to demonstrate its safety in a way that commands credibility with the general public will determine the possibilities of further expansion and use of nuclear power systems in the United States.

Those are but a few examples of the ubiquitous adverse effects of energy production, distribution, and use. The environmental effects, however, are broad. The use of every energy source is part of a general fuel cycle. Environmental impacts are associated with resource extraction, fuel refining, storage, transportation, conversion, and end use.

ENERGY AND THE FUTURE

Energy is the key to physical transformations of the material world. A supply of plentiful energy is the key to an economic system that we can control to be environmentally benign, provided that the energy system itself is not the overwhelming source of the environmental problems. There is a key link between environmental impacts and the energy demands of people. Our ability to devise systems of transportation, agriculture, manufacture, and housing that use less energy will have pronounced effects on our environmental future.

Our ability to choose and control production processes, to purify wastes and remediate waste sites, to clean and move water, to extract minerals, and to perform all the other activities of human systems depends directly on the availability and cost of energy. At the same time, environmental responsibility increases, comparative energy costs for a new technology become lower, new possibilities for all human activities become available; the menu of alternatives from which we can draw expands. Expanding the availability of energy that is renewable, and thus indefinitely sustainable, that is not itself automatically a source of potential global warming, and that is not itself a source of uncontrollable polluting waste is one key to providing technological alternatives for sustainable human activity.

> *The nation's environmental goals for the future should be sustainable development across full spectrum of human activity.*
>
> —Forum Participant Comment

> *The key to achieving environmental goals is technology. It is important to emphasize that in the environmental arena, the application of technology and the introduction of products and processes into the commercial arena are essential. Development of new knowledge is important, but doing something with it is essential. It is critical to include both science and application, public and private sectors, and especially the governments of all industrialized nations in developing technology to address environmental concerns.*
>
> —Forum Participant Comment

Some have indicated that there is an important caution in moving toward a renewable-energy future. Both energy consumption and pollution could be associated with the deployment of the capital infrastructure needed to capture and distribute renewable energy. Any energy source has a capital cost per unit of energy service provided (even if the fuel is "free"). That implies potential environmental impacts associated with putting the capital structure in place and needs to be included in any analysis of this topic.

ENERGY SCIENCE AND TECHNOLOGY DIRECTIONS

The present federal and private investment in energy research and development (R&D) is large. Many aspects of the energy system are under investigation and development. Environmentally more benign modes of coal, oil, and gas extraction and exploration are under development. Extensive research and development regarding renewable sources of energy are being conducted by both public and private institutions. More-efficient electric-power production is a major R&D thrust of the electric utility industry.

The efficiency of the energy system on both the supply and the demand sides is under continuous improvement through investments in R&D and implementation of the results through timely investment in physical capital. Historically, the energy efficiency of the economy has systematically improved since the beginning of this century in terms of energy use per unit of gross domestic product (GDP), but not sufficiently to offset the growth of GNP even on a per capita basis.

The most environmentally troublesome aspect of the operation of the present energy system is its use of fossil fuels, which are today available at attractive market prices. Fossil-fuel use is at the core of many environmental problems, including urban air pollution, acid rain, ecological impacts of resource extraction, and global warming. The environmental externalities resulting from the use of fossil fuels, some of which were described above, could be greatly reduced if these externalities were properly incorporated into energy-pricing mechanisms. The important directions for long-term energy R&D will be those which will "defossilize" the energy system. During the course of the next several decades, less carbon-intensive fuels are likely to become essential components of the fuel supply. Natural gas can become the transition fuel to a less fossil-fuel-based economy. Worldwide defossilization will require much greater emphasis on three major R&D directions: (1) renewable energy sources, (2) energy efficiency and conservation, and (3) safe, publicly acceptable nuclear power.

Alternative Resources

Progress is being made in increasing the efficiency and reliability[2] of some forms of renewable energy resources. For example, the efficiency of photovoltaic cells has been continuously improving. Biotechnology could play a continuing role in increasing our ability to use biomass as a source of energy. In addition, improvements have been made in the cost and reliability of wind-generated power in areas suitable for this source.

Coal

Coal is the most abundant fossil fuel in the United States and many other countries. Despite its liabilities with respect to its impact on the environment, it is still and will probably continue to be the fuel of choice for electricity generation because of its low cost and abundance.

Efforts continue to be made to improve the efficiency of coal combustion and energy conversion. Although the average efficiency of all coal-fired power plants is around 33%, there are hopes of increasing it to 40% or more. Already, some plants are achieving 43% efficiency, and new generations of "clean-coal technologies" are expected to deliver efficiencies of 40-45% at costs comparable with those of conventional coal plants (NRC 1995a). In addition to improved plant maintenance and operation, some promising technologies are the integrated gasification combined cycle (IGCC) and the pressurized fluidized-bed combustion system (PFBC).

Nuclear Energy

The use of nuclear fission is now almost 50 years old. The acceptability of nuclear power has varied from nation to nation and has varied with time. Nuclear power in the United States faces formidable obstacles that are primarily social and economic. Fear of nuclear accidents and difficulties with waste disposal have resulted in social pressures that prevent siting of nuclear power plants. Regula-

[2]Reliability is important for a renewable source to become a base load power source. For example, in the case of wind energy, the sources would need to be operated in systems with gas turbines, cover sufficiently large areas or be interconnected to other sources of energy to average out local fluctuations for it to become a primary source of energy. An alternative would be to develop a sufficiently reliable and inexpensive system of energy storage.

tions and other requirements have contributed to increased costs of construction. To keep nuclear power as an open option, research focused on removing these obstacles should be maintained. A continuing R&D goal should be the development of innovative, safe, publicly acceptable nuclear-power systems. A previous NRC (1992b) report entitled *Nuclear Power: Technical and Institutional Options for the Future* has addressed the problems facing U.S. commercial nuclear power and what would be necessary to keep open this option.

There is now more than 35 years of R&D experience on nuclear-fusion systems. Nuclear-fusion power installations hold the promise of easier waste handling and less radioactivity because no spent fuel is discharged from a fusion plant. The structural materials used in containing the plant become radioactive and might ultimately have to be safely disposed of, although the materials would, in general, be considered as low-level radioactive waste. Expectations have always exceeded achievements in this field, but nuclear-fusion power holds the potential for cleaner energy production. Were such a power source successfully achieved, at a cost comparable with that of current energy sources, many of the environmental consequences of the present energy system could be dramatically reduced.

The difficulties facing the development of commercial power production from fusion systems will require many decades to solve and will require international cooperation. There is no certainty that commercial fusion power can be achieved, but the value of such an achievement is likely to be so great that it is worth continuing to pursue it.

Two recent studies addressed the need to maintain a strong program in fusion research while recognizing that several large facilities could not be supported. The study by the President's Council on Science and Technology (PCAST 1995) recommended a fusion budget of $320 million for FY 1996 and canceling the planned Tokamak Physics Experiment (TPX) project. Congress appropriated $244 million. The congressionally requested study by the Fusion Energy Advisory Committee (FEAC 1996) recommended closing the Tokamak Fusion Test Reactor (TFTR) by 1998 and endorsed the recommendations of the NRC (NRC 1995c) report *Plasma Science: From Fundamental Research to Technological Application*, which recommended a strong experimental program to build the necessary scientific base for a long-term fusion program.

Efficiency

Improved efficiency in both the production and use of energy must be a consistent long-range R&D goal. We can reduce environmental consequences to the extent that we are able to reduce energy use through increased efficiency. For example, research on batteries and fuel cells for automotive propulsion and on magnetic levitation systems for trains is worthy of emphasis. However, it should be kept in mind that the theoretical maximum of efficiency remains much larger than what has proved socially and economically practicable, given the highly dispersed and decentralized nature of the necessary investments and consumer decisions. The NRC (NRC 1990a) report *Confronting Climate Change: Strategies for Energy Research and Development* outlines a number of steps that can be taken to enhance our R&D effort in energy efficiency.

The federal government has taken a number of steps in this direction. Examples include the Partnership for a New Generation of Vehicles (PNGV), the Green Lights initiative to improve the efficiency of lighting, and the Clean Cities initiative.

Supporting Research and Development

The evolution of the energy system depends on R&D not only in the energy sciences and technologies previously cited, but also in many other fields such as materials for storage, energy carriers for distribution (e.g., hydrogen), and waste reduction. Some examples include R&D on materials that can make the storage, distribution, and use of energy more efficient. For example, high-temperature superconducting materials could improve energy storage. Especially important are the incentives and disincentives related to the use of environmentally advantageous energy forms. The effects of energy taxes and research on the economics of energy use warrant attention.

A very important R&D direction linked closely to the development of new energy sources is toward the use of hydrogen as an energy carrier. Prototype systems depending on a solar-energy/hydrogen-conversion cycle are in an experimental stage. The use of hydrogen as an energy carrier requires extensive R&D. If hydrogen turns out to be a feasible and economically sound mode of embodying and transporting energy, it can have a remarkable impact on the environmental consequences of energy use.

FINDINGS, CONCLUSION, AND RECOMMENDATIONS

Findings

Energy production and use underlie the growth of modern industrial society, but production and use are often replete with environmental problems. Of particular concern is the use of fossil fuels, which leads to environmental problems, such as urban air pollution, acid rain, resource extraction, and global warming.

Responses to those concerns can be actions on either the supply side (how energy is generated) or the demand side (how energy is used). In both cases, knowledge, technical, and social barriers need to be overcome before actions can be taken. These barriers can be steadily reduced by additional research and development.

Energy research and development should create options for an uncertain future of energy availability and the environmental impact of that energy. Development of cleaner and economically viable efficiency and production alternatives will be key to preserving options against a number of contingencies. One contingency is scarcity caused by rising energy demand and limited supplies. Another contingency would arise from new knowledge indicating severe environmental effects from CO_2, radionuclides, or other emissions from conventional energy sources. If either of these contingencies arises, alternative energy sources and end-use technologies will be critical.

> *The nation's environmental goal should be a good balance between economic development and a clean, healthy environment.*
>
> —Forum Participant Comment

Conclusion

Energy research and development should create options for an uncertain future of energy availability and the environmental impact of that energy. Development of cleaner and economically viable efficiency and production alternatives will be key to preserving options against a number of contingencies. One contingency is scarcity caused by rising energy demand and limited supplies. Another contingency would arise from new knowledge that indicated severe environmental effects of CO_2, radionuclides, or other emissions from conventional energy sources than is now expected. If either of these contingencies arises, alternative energy sources and end-use technologies will be critical.

Recommendations

1. The committee recommends sustained research and development that will lead to more options for energy generation and use, less emission of carbon into the atmosphere, and more-efficient use of natural resources. In particular, the following topics should be explored

- *Electricity.* Thus while the United States is still the largest national consumer of primary energy, its relative contribution to pollution from energy use in the world is declining. Per-capita electricity use remains strongly correlated with development as electricity continues to replace other forms of energy because of its environmental and other advantages. Further increases in electrification should be accompanied by research in non-fossil-fuel sources (described further below) load management, and in other conservation approaches.
- *Renewable energy sources.* Solar energy, especially as used in photovoltaic cells, and biomass are the leading options for renewable energy sources. Research efforts should focus on making these more economical. Transition to widespread use will not occur until the cost of electricity from these sources is so low that large public-sector subsidies are no longer required to make them cost-competitive.

• *Coal.* The United States, Russia, and China have vast reserves of coal. About 60% of U.S. electricity comes from coal plants. Coal is also a major energy source in other countries. Thus, even though efforts should be made to defossilize our energy sources, research and development to improve the efficiency and reduce the emissions of coal plants (clean-coal technology) can help the U.S. environment and be a major factor in the Asian market, where India and China will burn increasingly large amounts of coal.

• *Nuclear fission.* A major source of U.S. electricity (21%), nuclear plants do not emit carbon or other pollutants. However, the U.S. nuclear industry has been crippled by the high cost of plants, by the government's inability to solve the problem of safe and reliable disposal of nuclear waste, and by the resulting disenchantment of the public and investors. Nuclear research should focus on these problems and include designs with improved safety. Until such problems are solved, further expansion of installed nuclear power capacity is unlikely, at least in the United States.

• *Nuclear fusion.* For more than 35 years, researchers in the United States, Russia (previously the Soviet Union), Japan, and the European Community have sought to use the energy potential of fusion to generate electricity. The potential fuel source is vast; yet the scientific and technological problems remain daunting. Although it might offer substantial advantages over fission plants in waste, fusion is unlikely to be a major energy source within the next 30 years. Basic and applied research should be continued.

• *Hydrogen.* The committee recommends long-term R&D to investigate the feasibility of hydrogen energy cycles as a source of energy because of its potential as an efficient and clean carrier for the distribution of energy to users.

• *Transportation.* The largest present primary source of energy in the United States is petroleum products; more than 50% of energy from petroleum products is used in transportation (DOE/EIA 1995). The ubiquitous automobile influences our choice of jobs, where we live, and how we spend our leisure time. There has been enormous progress (including improvements in fuel efficiency) in the reduction of automobile emissions implicated in urban air pollution; but the automobile is still the major source of such pollution in part because growth in automobile ownership and in driving per vehicle has nearly offset this progress. Continued research to improve the fuel efficiency of automobiles will help, but further major improvements will almost certainly require switching from vehicles powered by the internal-combustion engine to electric (or possibly hybrid-electric) or hydrogen-fueled cars that need such technologies as fuel cells, flywheels, and greatly improved batteries. Research is needed if that transition is to occur.

2. The United States should continue to address ways of making energy use more efficient, including pollution reduction. It should also help to conduct the R&D and policy analysis required to take account of the severe needs of the developing world, where primary energy use for economic development is less

efficient, the growth in primary energy consumption is more rapid, and severe environmental damage per unit of energy is much greater than in developed countries. This will persist until better methods of energy use are developed and political and economic incentives for achieving their adoption sufficiently rapidly on a large enough scale can be devised.

For more information and guidance, the reader should refer to the following:

NRC (National Research Council), *Confronting Climate Change: Strategies for Energy Research and Development* (Washington, D.C.: National Academy Press, 1990).

NRC (National Research Council), *Automotive Fuel Economy: How Far Can We Go?* (Washington, D.C.: National Academy Press, 1992).

NRC (National Research Council), *Review of the Research Program of the Partnership for a New Generation of Vehicles* (Washington, D.C.: National Academy Press, 1994).

NRC (National Research Council), *Coal: Energy for the Future* (Washington, D.C.: National Academy Press, 1995).

NRC (National Research Council), *Plasma Science: From Fundamental Research to Technological Application* (Washington D.C.: National Academy Press, 1995).

NRC (National Research Council), *Nuclear Wastes: Technologies for Separations and Transmutation* (Washington, D.C.: National Academy Press, 1996).

6

Use a Systems Engineering and Ecological Approach to Reduce Resource Use

We seek an environmentally benign, indefinitely sustainable economic system. A key ingredient of the economy is industry, one of the key providers of goods and services. The current industrial system uses technology in ways that can lead to environmentally troublesome results: toxic materials, solid wastes, liquid effluents, and gaseous emissions that, when released into the environment, have a negative impact on our quality of life and can lead to excessive resource use. We have often dealt with these results after the fact by treatment and remediation, at great cost and effort.

NEW SYSTEM IDEAS

Industrial ecology uses a systems engineering and ecological approach to integrate the design, production, and consumption of products to reduce the use of resources. It is based on the concept that natural systems tend to reuse and recirculate materials. It suggests a different way of thinking that focuses on the flows of matter and energy in the industrial system and how a more benign and efficient system of creating products and services could be created. Industrial people have begun to look at their wastes, effluents, and products in a different way. Effluents and products formerly considered as leaving the industrial system are now considered to be a part of the system and a company's intrinsic responsibility. In Germany, for example, novel policies have been developed that require consideration of the entire product cycle from extraction to final disposal.

> *A. Develop a closed-loop system in which no resources are depleted, all materials are perpetually reused, and no waste is produced or discarded.*
>
> *B. Ensure that the environment left to each succeeding generation is healthier and more ecologically sound than before.*
>
> *C. Ensure that every human has equal access to the use and enjoyment of the natural world and that the practices of one group, nation, or race do not infringe on or diminish the access or enjoyment of any other.*
>
> —Forum Participant Comment

Industrial ecology includes thinking about ways to manage wastes and effluents; to prevent the use of materials that are toxic[1] or otherwise dangerous or difficult to use and handle, so that they are not released into the environment; and to use wastes and effluents products themselves. It envisions an economic system whose processes and products lead to outputs that are reusable and recyclable. Waste is waste. Having purchased material at the front end of a plant, only to throw part of it away at the back end suggests that efficiency of material use might be improved; having a product that embodies the energy and effort needed to produce it go out of the plant door forever suggests that reusing the material and embodied energy at the end of the useful life of the product might be more efficient than disposing of the product.

We do not "consume" materials; we use energy to transform materials, to create materials from other materials. Many of our materials, particularly metals, are neither changed nor consumed in use, but merely stored (e.g., copper wire in walls and iron in automobile engine blocks). Even the contents of landfills, particularly if they were so designed, could be considered to be materials-inventory sites.

Material products generally and many wastes and effluents specifically embody energy and effort that make them potentially less expensive to use to make new products than starting from dispersed virgin raw materials. But systems must be appropriately designed. We must include in the economics the energy and effort that would be required to dispose of otherwise-discarded wastes, effluents, and products in an environmentally benign and acceptable manner.

Products Seen as Services

A further dimension of new possibilities is a re-examination of what is really being provided to customers as the product. Is the customer interested in cleaning

---

[1]Two kinds of toxicants need to be considered: (1) wastes that contain toxic elements, such as heavy metals, which cannot be eliminated by means of chemical transformation alone; (2) toxic organic chemicals that can eventually be transformed into nontoxic components when sufficiently treated by incineration, biodigestion, chemical reaction, or some other means.

materials and implements or in a clean plant, in owning a washing machine or in washed clothes, in owning an automobile or in transportation? Changes in the answers to these questions might lead to changes in whether a material product is sold outright or leased and in whether a manufacturer has the right to replace, upgrade, or improve a product at times of its own choosing, as long as the service is continued or improved. Those possibilities build on old ideas but take on new dimensions when seen as ways for the industrial system to manage the environmental implications of material products and production. Some of the possibilities appear to be economically advantageous to both producer and consumer.

> *Science and technology can contribute by developing more "environmentally friendly" products.*
>
> —Forum Participant Comment

Leverage for Change: Service Products

The service industries (e.g., retailing, logistics, product distribution and transportation, hotels, and the financial industry) are large consumers of materials and energy. They also have great leverage on their suppliers of material products because they are very large buyers. Their interest in improving the efficiency of the material goods system and in choosing products that are more benign in the life-cycle effect of the product on the environment can be an important force for greater achievements in industrial ecology. It is important to bring them into thinking about these possibilities and, even more important, to learn how to develop market-based incentives to drive broad, creative, entrepreneurial efforts by the service sector.

> *We contend that investment in sustainability, environmental restoration, and environmental protection is the minimal contribution science and technology must provide if we are to achieve our environmental goals.*
>
> *As a tool for protecting the environment, the regulatory approach will become increasingly ineffective if the cost of compliance continues to rise and if global competitiveness and market forces take hold. In addition, it is unlikely that global environmental regulation will become a reality for dealing with environmental problems that cross national boundaries. The alternative for the United States is to deploy a science and technology strategy to decrease the cost of industrial-process modification (preventing waste and product stewardship) to protect our air, water, and food. Using scientific knowledge generated to provide the basis for risk-based decision-making offers a great opportunity to create a sustainable ecosystem.*
>
> —Forum Participant Comment

NEW DIRECTIONS

Industrial people have begun to work on a number of new directions, all requiring R&D. Efforts by some corporations are described in a book from the National Academy of Engineering (NAE), *Corporate Environmental Practices: Climbing the Learning Curve* (NAE 1994a). The vast majority of these efforts are based on industrial activity and R&D on its own—without or only supplemented by government involvement.

Design for Environment and Life-Cycle Analysis (LCA)

To carry out an industrial-ecology program, we require tools to help us to understand the environmental and system consequences of design choices. We need ways of deciding what the effects of product-design, and process-design, and materials choices will be on the material outputs of the manufacturing system, including the fates of the products when they are finally discarded by consumers. Where will they go, are they well adapted to reuse and recycling of parts and materials into new products, and how can they be reincorporated into the industrial system to the profit of the original manufacturer or of someone else in the industrial system? For the long-term, we must look for and encourage systems that

- Start from renewable sources.
- Maximize the useful life of the product.
- Reuse, recycle, and reclaim spent products at their highest possible value.
- Ultimately recover residual energy before returning the product to the environment in a form that permits another cycle to begin.

Such approaches must incorporate the ability to examine alternative possibilities and their effects and must become the means to make design choices so that products and processes can be both economically sound (in the context of including ultimate costs) and sound in their environmental consequences.

> *We believe that science and technology are essential to meeting current and future environmental goals. None of the goals related to environmental issues will be accomplished without extensive and creative application of innovative technologies. The combined dimensions of the technological basis of modern civilization and the remaining unmet human needs for current and future generations dictate the use of new technologies to replace or modify existing practices across a wide spectrum of activity. These new technologies must be sustainable, more efficient, and less resource-intensive and must work in concert with natural systems. Most of these technologies will be applied by and through the design and construction industry.*
>
> —Forum Participant Comment

A report from the National Academy of Engineering, *The Greening of Industrial Ecosystems* (NAE 1993) describes how corporations use some of those tools in their practice and possible tools that can be used.

> *Science and technology can contribute by making sure a systems approach is used that looks at all aspects of the life cycle.*
>
> —Forum Participant Comment

Alternative Manufacturing Processes

An industrial-ecology program should include an examination of alternative manufacturing processes and alternative paths to the same or similar processes. For example, what are all the alternative paths for organic synthesis of particular chemical products, and what are the implications of the various paths for the nature of intermediate process chemicals, energy consumption, and the required facilities and facility investments? This is a classic and fundamental problem of chemical engineering, but it requires new tools to explore a wider universe of possibilities and the consequence of alternative paths. Similar questions arise in metals processing and other industrial process streams.

Efficient Separation Technology

Many manufacturing and process technologies lead to increases in entropy via mixing of materials or their dissipation into the environment. When the materials in question are toxic, the result is particularly troublesome. In many cases, more-effective technology for using energy to unmix materials that have been mixed would be useful.

It is possible to separate materials with the brute-force expenditure of energy, it is sometimes possible to do it by paths that are more thermodynamically efficient than the obvious ones. Sometimes, it is a matter not of efficiency alone, but of convenience with respect to time or capital equipment. The search for effective and efficient separation technologies—particularly chemical-specific adsorption, membranes, and similar low-energy devices—is fundamentally worth while.

FINDINGS, CONCLUSION, AND RECOMMENDATIONS

Findings

A key component of industrial ecology is analyzing the environmental effects of all materials in manufacture, use, and disposal. Companies find that the use of the industrial-ecology approach in the design process provides them with more options for reducing the human health and ecological effects of their products and processes. However, its use is in its infancy. The information, planning,

> *The nation's environmental goals for the future should focus on (1) general stewardship of the environment, including industrial ecology, (2) much-improved public awareness and understanding of the risk-benefit and cost-benefit aspects of environmental stewardship, and (3) use of risk-benefit and cost-benefit methods for decisions pertaining to the environment.*
>
> —Forum Participant Comment

standards, and societal changes needed to implement this concept on a sufficient scale to make a difference are still lacking. The problem of implementation is made more difficult by the fragmentation of industry and, to some extent, by the present trend toward decentralization and devolution of hitherto vertically integrated industries.

Some form of "societal vertical integration" among many institutions and economic entities "from cradle to grave" will be necessary for a solution. How it can be achieved is an important topic for research and public-policy debate.

However, the actions implied by the industrial-ecology concept are not practiced widely. There are a number of barriers to its adaptation by industry, including such economic issues as the high cost of changing the existing infrastructure, concerns about the compatibility of such activities with the existing regulatory framework, fears of future legal liability, and inadequacy of information on and understanding of the industrial-ecology concept.

Conclusion

In our current efforts to reduce the pollution generated by and the ecological impact of society's industrial activities, we most often use "end-of-the-pipe" controls. However, end-of-pipe treatment is increasingly less likely to be the most cost-effective or the most-desirable means of pollution control.

In recent years, a new way of thinking about how to reduce environmental impacts has been developed. It is called industrial ecology, and it is influencing the thinking of many major corporations in how they handle environmental issues. Industrial ecology takes a systems engineering and ecological approach to integrate the producing and consuming segments of the design, production, and use of services and products to reduce environmental impacts.

Furthermore, industrial ecology requires substantial recycling, including the use of one plant's waste stream as feed for another plant, and therefore requires coordination, planning, and perhaps proximity—all of which could make it more difficult for it to achieve widespread use. One key challenge is to formulate effective economic incentives for developing a market-driven industrial ecology. Another is to alleviate the liability and regulatory barriers that inhibit the full application of industrial ecology.

Recommendations

1. Design of products and processes for environmental compatibility should make use of such mechanisms as life-cycle analysis, alternative manufacturing processes, and efficient separation technologies that use energy to unmix materials that have been mixed.

2. Products and processes should be designed to accommodate recycling and reuse more readily.

3. Regulations introduced for other purposes often create barriers to the use of economic incentives for promoting the adoption of the principles of industrial ecology. Research should be aimed at identifying and eventually removing such barriers.

4. The use of industrial-ecology approaches should be expanded to many industries through dialogue among an ever-widening circle of corporations, governments, academic institutions and environmental and citizen organizations.

5. Research should be conducted to develop methods for chemical species-specific separations that yield streams that are economically recoverable or dischargeable to the environment.

6. Research on new or improved catalytic systems that offer improved yields and improved specificity from more-benign chemicals should be promoted.

For more information and guidance, the reader should refer to the following:

NAE (National Academy of Engineering), *The Greening of Industrial Ecosystems* (Washington, D.C.: National Academy Press, 1993).

NAE (National Academy of Engineering), *Corporate Environmental Practices: Climbing the Learning Curve* (Washington, D.C.: National Academy Press, 1994).

NAE (National Academy of Engineering), *Industrial Ecology:* U.S.-*Japan Perspectives* (Washington, D.C.: National Academy Press, 1994).

NRC (National Research Council), *Tracking Toxic Substances at Industrial Facilities: Engineering Mass Balance Versus Materials Accounting* (Washington, D.C.: National Academy Press, 1990).

NRC (National Research Council), *Opportunities in Applied Environmental Research and Development* (Washington, D.C.: National Academy Press, 1991).

NRC (National Research Council), *Industrial Waste Production and Utilization* (Washington, D.C.: National Academy Press, 1995).

7

Improve Understanding of the Relationship Between Population and Consumption as a Means to Reducing the Environmental Impacts of Population Growth

The current and potential threats to environmental quality, of which there are many, are the results of the character and magnitude of today's economic activity and in combination with human population growth. As the experience of the United States, other industrialized nations, and developing countries indicates, birth rates and economic development are closely linked.

The world's population was 2.5 billion in 1950, it is 5.6 billion today, and it is estimated that it will be 8 billion by 2020. Current population growth worldwide is 1.7% per year.

Over the next 50 years, within the lifetimes of many of us, economic activity worldwide is projected to quadruple, and global population is expected to double to about 11 billion before leveling off. If population growth of this magnitude occurs with current industrial processes, agricultural methods, and consumer practices, the results could be both environmentally and economically disastrous.

According to the assessments of many environmental experts, the most critical need facing the world is the slowing of human population growth (National Commission on the Environment 1992; NRC 1994c). Continued global population growth of the current magnitude—1 billion more people every decade—has the potential not only to negate efforts to protect the environment, but also ultimately to overwhelm economic and social progress.

The reasons for giving the highest priority to reducing population growth are ethical, practical, and scientific. Many of the countries that are experiencing rapid economic growth and increasing consumption of goods aspire to a pattern of consumption like that of the United States and have government policies that encourage economic growth. But as economic growth and thus consumption increase, so do the environmental impacts of a population as it works and lives.

> *The nation's environmental goals should be,*
>
> *first, stabilization of the nation's, if not the world's, human population. Second, no net loss of water or air quality. Third, stabilization of our food supply, to include quantity and quality. Fourth, stabilization of plant and animal biodiversity. Fifth, no net loss of wetlands; watersheds; national forests, parks, wildlife preserves, and wilderness areas; or lake, stream, or ocean commercial and recreational fisheries stocks. Sixth, continued monitoring of global climate change, such as global warming, acid rain, and loss of the ozone layer.*
>
> —Forum Participant Comment

Thus stabilization of population growth is necessary if a country is to reach both its economic and its environmental goals. However, large-scale worldwide demographic and health surveys have demonstrated a large unmet need for family planning in nearly every country of the world. These needs are for low-cost, accessible, and safe means of contraception.

In developing countries, we cannot and should not deter economic development; but by recognizing the unmet need for contraception, we can influence the rate of population growth by helping people to manage their own fertility. People in developing countries want both economic development and smaller families. Many countries are already experiencing marked fertility declines; given additional resources and assistance, they can increase these declines.

Combining rapid economic development and increased resource consumption with rapid population growth will result in a compounding of the effects of economic development, which might overwhelm the capacity of a country to address them. As countries become more affluent, they are better able to address some aspects of industrial activity, such as air and water pollution, more effectively. If the resources of a country are needed to address the consequences of a rapidly growing population—such as depletion of potable water supplies, epidemics of infectious diseases related to increased population density, and increased need for fuel and nutrients—the environmental consequences of this growth cannot be adequately addressed.

The United States and all other countries have much to gain from efforts to stabilize global population and to improve living standards in developing countries, where 90% of the projected population growth will occur. The stress placed on the environment is a function of population and consumption. Therefore, the burden should not be placed entirely on developing countries. Priorities for developed countries, such as the United States,

> *What would be necessary to conclude that an environmental goal was achieved?*
>
> - *Slow rate of population growth.*
> - *Decrease the resource-intensiveness of developed countries.*
>
> —Forum Participant Comment

should be to switch to sustainable technologies to reduce waste in consumption and to assist the developing countries in their efforts to develop their economic sustainability and stabilize their population growth.

Population growth as an environmental concern, even in the United States, was recognized as an important issue before 1970. According to a 1970 public-opinion poll, 40% of respondents said that U.S. population growth was a major problem, and 46% considered population growth in the United States as "not a problem [now] but likely to be a problem by the year 2000" (see paper by Bowman in Part II).

SUSTAINABLE DEVELOPMENT

The combination of global population growth with better communications, improved interregional transportation, and new political and economic freedom in Asia, the former Soviet Union, Africa, the Middle East, and Latin America has resulted in the development of a world consumer society that uses an increasing fraction of the world's natural resources.

The National Commission on the Environment (1992) advocated a new direction for U.S. environmental policy and asserted that U.S. leadership should be based on the concept of sustainable development. Sustainable development is predicated on the recognition that economic and environmental goals are inextricably intertwined and mutually reinforcing. Therefore, these goals must be pursued simultaneously if sustainability is to be achieved (National Commission on the Environment 1992). The commission argued that by the close of the 20th century economic development and environmental protection must come together in a new synthesis: broad-based economic progress accomplished in a manner that protects and restores the quality of the natural environment, improves the quality of life of people, and broadens the prospects for future generations.

Economic growth cannot be sustained if it continues to undermine environmental quality and exhaust the earth's natural resources. Similarly, only healthy economies can generate the wealth and capacity to invest in environmental protection, to improve population health, and to enhance the overall quality of life. As observed globally in many impoverished nations, environmental protection is

> *Cheap, ubiquitous, low-technology birth control is needed, but there are, of course, enormous social, cultural, and political problems associated with widespread implementation of birth control.*
>
> *"Efficient resource production and use" can benefit from many science and technology contributions to agriculture, manufacturing, transportation, etc. Of course, massive technology transfer to developing countries is the key.*
>
> —Forum Participant Comment

> *The nation's environmental goal should be a sustainable biosphere, meaning one that maintains the products and services on which we depend. This goal involves stabilization of human population and our domesticated plants and animals at a level compatible with maintenance of ecosystem services (water, air, and natural resources) and minimization of the loss of unique biodiversity resources. As David Orr has said, our goal should be to learn to walk more lightly on the planet.*
>
> —Forum Participant Comment

not possible where poverty is pervasive and the quality of life degraded. In some African countries, population increases have outstripped not only local food supplies, but also the capacity of the environment to sustain that growth.

Economies of most developing countries are based on natural resources, which constitute their primary economic capital. Their long-term economic development depends on maintaining, if not increasing, such activities as agriculture, forestry, fishing, and mining—for both domestic use and export—as well as ecotourism. In doing this, primary attention must be given to protecting the environment.

There are varied relationships between population growth and density and environmental quality. Stabilizing the growth of human populations is necessary for future sustainable development, but not sufficient. Here one needs to consider the "ecological footprint" both of megacities like New York, Los Angeles, Hong Kong, or Mexico City and of resource-intensive small countries like the Netherlands and Japan. These cities and countries use resources from the land and sea that are multiples of their land mass (i.e., their "ecological footprint"). For example, the ecological footprint of the Netherlands is 14 times its land mass (Rees and Wackernagel 1994), and the number and size of such footprints will grow with continued population growth. Scientific and technological advances in energy and food technology have had both positive and negative effects on the environment in that, on the one hand, such advances have made energy production and distribution and food production and distribution more efficient, and on the other hand, advanced technologies have accelerated the depletion of some resources, such as the world's fisheries.

FINDINGS, CONCLUSION, AND RECOMMENDATIONS

Findings

As indicated at the International Conference on Populations and Development, the United States and other countries made a major commitment to cooperate with the world community to stabilize global population, recognizing the linkages between birth rates, child survival, economic development, education, and the economic and social status of women. To achieve a stable global population, universal access to effective family-planning information, contraceptives, and health care is essential. U.S. policies, both domestic and foreign, need to reflect a greater awareness of this interdependence and provide stronger support for international population programs and programs of economic assistance to developing countries.

Conclusion

The current and potential future threats to environmental quality, of which there are many, are the results of the character and magnitude of today's economic activity and human population growth. As the experience of the United States, other industrialized nations, and developing countries indicates, birth rates and economic development are closely linked. Although the extent to which threats to environmental quality and ecological resources will be intensified by future population growth is debated, there is agreement that continued population growth has the effect of narrowing the options available for meeting these threats. The predicted addition of billions of people to the global population in the next few decades could overwhelm programs aimed at enhancing energy efficiency, global monitoring, and industrial ecology. Many elements of social science, such as demography and sociology, and of medical research address issues that affect population growth. Environmental engineering develops strategies and devices that can be used to decrease the impact of population growth on the needs of developing countries.

The political milieu makes it difficult for U.S. federal agencies to recognize publicly the interdependence of environmental quality and global population growth. The federal budget often omits support for studies related to population growth—studies of demography, ecology of population growth, and contraception—even in the U.S. Agency for International Development and the Department of Health and Human Services. This omission reflects a serious constriction in the horizons of current U.S. environmental policy. As noted in *The Sustainable Biosphere Initiative* (Lubchenco et al. 1991),

> The issues associated with population growth are broad, involving such factors as changes in per capita income and resource distribution; increasing pollution and environmental degradation; problems of health and poverty; the effects of urban, industrial, and agricultural expansion; and especially the integration of ecologic and socioeconomic considerations. Even those factors that are primarily economic will have substantial environmental effects.

Recommendations

1. The United States, in its efforts to cooperate with the world community, should recognize the linkages between birth rates, child survival, economic de-

velopment, education, and the economic and social status of women in its environmental research efforts.

2. To cope with global population pressures, researchers should focus on ways to improve the potential for universal access to effective family-planning information, contraceptives, and health care.

3. U.S. policies, both domestic and foreign, need to provide support, through partnerships with developing countries, for the scientific and technological research needed by international population programs.

4. Interdisciplinary research should be conducted on the future environmental consequences of population growth, especially in vulnerable environments. This research should incorporate human biology, human behavior, epidemiology, and ecology to yield a better understanding of all aspects of the population-environment interface. Research should be conducted on the possible negative consequences of technology when introduced into a population, e.g., the possible adverse impacts of the use of artificial baby formula in developing countries.

For more information and guidance, the reader should refer to the following:

NRC (National Research Council), *Population and Land Use in Developing Countries: Report of a Workshop* (Washington, D.C.: National Academy Press, 1993).

NRC (National Research Council), *Population Summit of the World's Scientific Academies* (Washington, D.C.: National Academy Press, 1994).

8

Set Environmental Goals Via Rates and Directions of Change

When speaking about environmental goals, people often focus on achieving a specific level of environmental quality by a specific time. However, such formulations might not be the best way to evaluate our rate of progress. Measuring progress requires a metric and a path to determine the direction in which progress is going.[1] Earlier chapters have discussed the importance of monitoring—especially in relation to ecological systems.

Sustainability is an evolutionary concept, and there are inherent limitations in the rate of adaptation of human systems to technological and social change and in the speed of cumulative social learning. Population, resource use, and gross domestic product might be much more important in limiting development than is the carrying capacity of the biosphere. That is because carrying capacity is not an absolute, but a value that changes as technology evolves.

Populations and settlement patterns change much more quickly than they used to. Science and technology are major factors in environmental change and major instruments of human adaptation to it. It is important to ask what fraction of the agenda of science should be governed by "needs" formulated in societal terms and what fraction should be driven more by scientific opportunity, that is, the opportunity to build and generalize the conceptual structure of knowledge itself. It is not, of course, an either-or choice. The optimal system blends societal

[1]As was indicated by Daniel Bell, moderator of the panel on "The Management of Information of Knowledge" at the 1970 annual meeting of the House of Representatives Committee on Science and Astronautics Panel on Science and Technology: "Without the organization of information, we can no longer know where we are going to be going, and as an old Talmudic aphorism puts it, 'If you don't know where you are going, any road will take you there.'"

> *To believe that environmental goals are achieved, there would need to be indicators of progress, milestones would have to be established (which might more nearly be political than scientific), and an agreed on means of measuring progress would have to be put in place. In many cases—as with the question of sustainable rates of consumption—the absolute end point of success might never be achieved in fact, in which case measurable progress toward that end point becomes important.*
>
> —Forum Participant Comment

problems in search of scientific solutions with scientific solutions in search of societal problems in a highly complex mutually reinforcing system.

The advance of conceptual, curiosity-driven research is likely to turn up new options for solving problems in the future that we cannot even imagine in the light of today's knowledge. The annual and cumulative costs of future options are sensitive to the time scale on which the options are implemented. The cost-timing problem is complicated by the existence of benefits of learning by doing: slow implementation provides more opportunity for learning from experience and so for lowering costs over time, and it provides time for R&D, which can reveal more-efficient and less-expensive, technological and management-response options.

One needs to adopt the same approach to environmental improvement that has now been widely adopted in total quality management (TQM) and product and process improvement in general. One needs a dynamic system of continuous evolutionary improvement, not simply fixed end points. There is now considerable experience to demonstrate synergy between environmental improvement and product and process improvements that generate continuous gains in product value and continuous improvement in both labor and capital productivity in manufacturing or operations (in the case of services). The growing opportunities for exploitation of information technology made possible by its constantly declining cost and increasing capability present a theoretical opportunity for a much more dynamic and evolutionary approach to both product and process innovation. The potential for this idea has recently been put forward in considerable detail by Richard Florida, of Carnegie Mellon University, in a paper entitled *The Environ-*

> *A systems approach would establish definable, measurable, results-oriented indicators of progress toward goals. With such a system, attempted solutions would be evaluated against these indicators and rejected unless measurable progress were demonstrated. It is unlikely that any goals would be "achieved." However, a standard of continual improvement could be workable.*
>
> —Forum Participant Comment

ment and the New Industrial Revolution: Toward a New Production Paradigm of Zero Defects, Zero Inventory, and Zero Emissions (Florida 1995).

FINDINGS, CONCLUSION, AND RECOMMENDATION

Findings

Therefore, in setting environmental goals, rates and directions of change might be more important than endpoints, but it is endpoints that seem to be at the center of the public discussion of environmental policy.

Conclusion

Rates and directions of change should become the focused goals for current actions with end points as the tools for motivation (not just one-time goals). Both a desired endpoint and a rate-of-change are needed in order for society to be successful in achieving its goals. Without an endpoint, it can be difficult to mobilize the political will to take action; without a predetermined rate-of-change, it can be difficult to establish the specific actions needed to achieve the desired endpoint.

For example, at the Department of Energy's Hanford site, the environmental cleanup has as its societal goal returning most of the site to unrestricted use within 75 years. Without intermediate milestones, the rate of progress can be difficult to judge, and tens of billions of dollars can be spent over several decades without knowing the extent to which any substantial result has been achieved.

Recommendation

Rather than stopping at the selected specific end points being discussed in the federal government and elsewhere, environmental goals should be formulated in terms of an adjustable strategy for continuous evolutionary improvement in environmental performance that includes intermediate milestones.

Bibliography

Carnegie Commission on Science, Technology, and Government, *Enabling the Future: Linking Science and Technology to Societal Goals* (New York: Carnegie Corporation, 1992).

Clark, William C., and R.E. Munn, editors, *Sustainable Development of the Biosphere* (Cambridge, England: Cambridge University Press, 1986).

DOE/EIA (Department of Energy), *Monthly Energy Review* (Washington, D.C.: August 1995).

EPA (Environmental Protection Agency), *Environmental Investments: The Cost of a Clean Environment*, EPA Document # EPA-230-11-90-083 (Washington, D.C.: November 1990).

EPA (Environmental Protection Agency), *National Air Quality and Emissions Trends Report* (Washington, D.C.: 1993).

FEAC (Fusion Energy Advisory Committee), *A Restructured Fusion Energy Sciences Program* (Washington, D.C.: January 1996).

Florida, Richard, *The Environment and the New Industrial Revolution: Toward a New Production Paradigm of Zero Defects, Zero Inventory, and Zero Emissions*, Carnegie Mellon University Working Paper Series 95-31 (Pittsburgh, Pa.: Carnegie Mellon 1995).

Hulme, Mike, Sarah C.B. Raper, and Tom M.L. Wigley, "An Integrated Framework to Address Climate Change (ESCAPE) and Further Developments of the Global and Regional Climate Modules (MAGICC)," *Energy Policy* 23(4-5):347 (1995).

Lubchenco, Jane, Annette M. Olson, Linda B. Brubaker, Stephen R. Carpenter, Marjorie M. Holland, Stephen P. Hubbel, Simon A. Levin, James A. MacMahon, Pamela A. Matson, Jerry M. Melillo, Harold A. Mooney, Charles H. Preston, H. Ronald Pulliam, Leslie A. Real, Philip J. Regal, Paul G. Risser, "The Sustainable Biosphere Initiative: An Ecological Research Agenda," *Ecology* 72(2): 371-412 (1991).

National Commission on the Environment, *Choosing a Sustainable Future* (Washington, D.C.: World Wildlife Fund, 1992).

NAE (National Academy of Engineering), *The Greening of Industrial Ecosystems* (Washington, D.C.: National Academy Press, 1993).

NAE (National Academy of Engineering), *Corporate Environmental Practices: Climbing the Learning Curve* (Washington, D.C.: National Academy Press, 1994a).

NAE (National Academy of Engineering), *Industrial Ecology: U.S.-Japan Perspectives* (Washington, D.C.: National Academy Press, 1994b).

NRC (National Research Council), *Confronting Climate Change: Strategies for Energy Research and Development* (Washington, D.C.: National Academy Press, 1990a).

NRC (National Research Council), *Tracking Toxic Substances at Industrial Facilities: Engineering Mass Balance Versus Materials Accounting* (Washington, D.C.: National Academy Press, 1990b).

NRC (National Research Council), *Opportunities in Applied Environmental Research and Development* (Washington, D.C.: National Academy Press, 1991).

NRC (National Research Council), *Automotive Fuel Economy: How Far Can We Go?* (Washington, D.C.: National Academy Press, 1992a).

NRC (National Research Council), *Nuclear Power: Technical and Institutional Options for the Future* (Washington, D.C.: National Academy Press, 1992b).

NRC (National Research Council), *Research to Protect, Restore, and Manage the Environment* (Washington, D.C.: National Academy Press, 1992c).

NRC (National Research Council), *Population and Land Use in Developing Countries: Report of a Workshop* (Washington, D.C.: National Academy Press, 1993).

NRC (National Research Council), *A Biological Survey for the Nation* (Washington, D.C.: National Academy Press, 1994a).

NRC (National Research Council), *Review of the Research Program of the Partnership for a New Generation of Vehicles* (Washington, D.C.: National Academy Press, 1994b).

NRC (National Research Council), *Population Summit of the World's Scientific Academies* (Washington, D.C.: National Academy of Sciences, 1994c).

NRC (National Research Council), *Coal: Energy for the Future* (Washington, D.C.: National Academy Press, 1995a).

NRC (National Research Council), *Industrial Waste Production and Utilization* (Washington, D.C.: National Academy Press, 1995b).

NRC (National Research Council), *Plasma Science: From Fundamental Research to Technological Application* (Washington D.C.: National Academy Press, 1995c).

NRC (National Research Council), *Review of EPA's Environmental Monitoring and Assessment Program: Overall Evaluation* (Washington, D.C.: National Academy Press, 1995d).

NRC (National Research Council), *Nuclear Wastes: Technologies for Separations and Transmutation* (Washington, D.C.: National Academy Press, 1996a).

NRC (National Research Council), *Understanding Risk: Informing Decisions in a Democratic Society* (Washington, D.C.: National Academy Press, 1996b).

PCAST (President's Council on Science and Technology), *The U.S. Program of Fusion Energy Research and Development*, Report of the Fusion Review Panel (Washington, D.C.: July 1995).

Rees, William E., and Mathis Wackernagel, "Ecological Footprints and Appropriated Carrying Capacity: Measuring the Natural Capital Requirements of the Human Economy," in *Investing in Natural Capital: The Ecological Economies Approach to Sustainability*, AnnMari Jansson, Monica Hammer, Carl Folke, and Robert Costanza, editors, page 374 (Washington, D.C.: Island Press, 1994).

PART II

COMMISSIONED PAPERS

National Environmental Goals: Implementing the Laws, Visions of the Future, and Research Priorities[1]

RICHARD D. MORGENSTERN
Resources for the Future

CONTENTS

[1]*Note*: A previous version of this paper benefited from comments by Derry Allen, Terry Davis, Devra Davis, Roger Dower, Paul Portney, Peter Truitt, and Elizabeth Farber, who also provided able research assistance.

Conventional wisdom holds that one of the first steps in developing a government policy or program is to articulate the goals it is supposed to achieve (C.E. Lindblom, *The Policy Making Process*, 1978). Yet, at only slight risk of exaggeration, it can be argued that many of our major environmental statutes contain little more than hortatory phrases that offer scant guidance to the implementing agencies.[2] Thus, retrospective program evaluations often mask disputes over what initial program goals should have been.

While the goal of virtually all environmental legislation is to protect human health and the environment, such a broad statement is of little use because it begs various questions, including the following: How much protection is enough; who/what should be protected; and what kind of protection is appropriate? Most statutory provisions contain either acceptable risk goals, pollution reduction goals, or technology requirements. A few, such as the Clean Air Act, contain all three.

Over the past twenty years, considerable progress has been made in reducing risks, reducing pollution (in all media), and promoting diffusion of environmentally sound technologies. Yet the key issues of how much further to go, what to emphasize, and how to determine success remain largely unresolved. Almost without exception, environmental programs have failed to define targets that are both meaningful and measurable.

Within the past few years, a number of agencies have begun to fill that void. Most notably, in 1992, the EPA began a project to develop a set of goals to assist Agency management and the public at large in assessing the nation's future environmental progress. That project has gone through various iterations over the past three years and is now nearing completion. A draft version, referred to as the EPA goals project, is one of the three major projects reviewed in this paper.

Coincident with the question(s) of whether and/or to what extent we are making progress in meeting often vague, legislatively mandated goals, three other concerns have arisen in the policy community.

- The first involves the question of whether our legislative goals are really the right ones for the present and/or the future. Specifically, many have questioned whether our statutes are driving us to commit our nation's resources to certain high-cost problems/strategies, while at the same time ignoring other important environmental concerns, including some with potentially low-cost solutions.[3]
- The second policy issue concerns the absence of a strong linkage between our nation's environmental resource decisions and our economic and social de-

[2]Exceptions include provisions of the 1990 Clean Air Act Amendments for specified reductions in sulfur dioxide, nitrogen oxides, and stratospheric ozone depletors by various dates. For each of these pollutants, Congress reviewed analyses concerning the feasibility of attaining these goals prior to enactment of the statutes. In contrast, the Clean Water Act calls for the elimination of all discharges into navigable waters by 1985, which is, effectively, a hortatory goal.

[3]There is a growing literature on this subject. Early work includes *Unfinished Business* (EPA, 1987) and *Reducing Risk* (EPA, Science Advisory Board, 1990).

velopment decisions. This is related, in part, to the issue of inconsistent and/or duplicative government programs or policies. It has long been observed, for example, that one agency's pursuit of an environmental goal (e.g., reducing pesticide risks) may be undermined by another agency's pursuit of an agricultural goal (e.g., increasing agricultural output). Inconsistencies, of course, also occur within a single agency; we sometimes create a solid waste problem when we attempt to solve an air or water pollution problem.

- The third policy concern is focused on the coordination of research activities across the various agencies and subagencies of government. People question, for example, if federal research dollars are supporting an integrated strategy or, rather, if various parochial interests are the driving forces in individual agencies.

The first two concerns form the core of the agenda of the President's Council on Sustainable Development (PCSD). Comprised of leaders from the business, government, and nonprofit sectors, the PCSD was formed in mid-1993 to address a broad range of issues regarding the environment, the economy, and equity within our society. The third concern—pertaining to the integration of federal research efforts—is the province of the Committee on Environment and Natural Resources (CENR) of the National Science and Technology Council. Also established in 1993, CENR consists of representatives of the major environmental and natural resource agencies and key White House offices and is chaired by the Office of Science and Technology Policy.

This paper reviews and compares the three major efforts put forth by EPA, PCSD, and CENR to establish our country's environmental goals. EPA and PCSD goals are still under development; however, draft versions are available to the public. CENR goals are presented in its 1995 strategic planning document. Key conclusions of this paper are as follows:

1. The three goals reports address three fundamentally different sets of problems.
2. The three goals projects have different technical approaches including different scopes, time frames, completeness of metrics, clarity of policy tools, and use of interim milestones.
3. All three projects fail to address a number of important issues.
4. The state of the art in environmental goal-setting is still in its infancy; however, these three efforts clearly represent a major step forward.
5. An overwhelming strength common to all three projects is the implicit recognition that our environmental management system is in need of significant reform.
6. Follow-through is key to success.

Section I of this paper describes the three major projects in some detail. Section II addresses the issue of consistency among the projects. Section III develops a set of eight criteria upon which to compare and evaluate the projects' various goals and metrics. Section IV compares the goals projects on the basis of the comparison criteria. Section V highlights a number of issues not emphasized by the three goals projects. The final section attempts to draw some overall conclusions.

THE THREE MAJOR GOALS PROJECTS: A DESCRIPTION

The full texts of the projects' goals and metrics are presented in Appendixes A through C.

EPA's Environmental Goals for the Year 2005

As of this writing the EPA has not yet issued its final report, titled *Environmental Goals for America with Milestones for 2005*. EPA proposes 15 long-range environmental goals for the nation:

1. clean air;
2. climate change risk reduction;
3. stratospheric ozone layer restoration;
4. clean waters;
5. healthy terrestrial ecosystems;
6. healthy indoor environments;
7. safe drinking water;
8. safe food;
9. safe workplaces;
10. preventing spills and accidents;
11. toxic-free communities through pollution prevention;
12. safe waste management;
13. restoration of contaminated sites;
14. reducing global environmental risks; and
15. better information and education;

and clarifies each goal by providing

1. a one- or two-sentence description of the long-range goal (usually without a specified year for attainment);

2. a series of "ambitious but realistic" quantitative milestones, usually for the year 2005.

In total, there are 65 EPA milestones. Some of them are tied to specific outcomes, while others are necessary but not sufficient conditions for the realization of the outcomes. Some of the milestones are results-based performance measures, while others are emission or technology based. Some are national in scope, while others are regional. For expository purposes, it is useful to consider in detail several individual goals and their corresponding milestones:

EPA Goal: Clean Air

By 2010 and thereafter, the air will be safe to breathe in every city and community and it will be clearer in many areas. Life in damaged forests and polluted waters will rebound as acid rain is reduced.

The first milestone for "safe" air in 2005 is the reduction in the number of metropolitan non-attainment areas to six from the current level of 60. That is, "safe" air will have the effect of reducing the number of people living in areas that do not meet the ambient standards to 45 million from today's level of 120 million. Implicit in this first milestone is the notion that safety levels are defined as the national ambient air standards established under the Clean Air Act. "Safe," of course, does not really mean "no risk." Also implicit in this milestone and consistent with the Air Act is the notion that success—for the year 2005—does not necessarily mean that all Americans will live in areas meeting the ambient standards.

The second milestone for clean air uses an emissions goal from a large source category as a measure of success. It addresses one particular pollutant, volatile organic compounds, from the largest known source category, motor vehicles, and calls for a 65 percent reduction by the year 2005. The milestone description also calls for meeting the relatively prescriptive fuel and vehicle requirements of the Clean Air Act. However, no milestones are established for NO_x reduction or for stationary sources in general. Also, no specific emissions goals (other than a general call to meet existing ambient standards) are given for pollutants associated with other ambient air quality standards (e.g., particulates).

The third milestone for clean air on vehicle miles traveled is still under development. The fourth milestone addresses toxic emissions, as opposed to "conventional" pollution. It is a combined emissions-technology goal, calling for 174 categories of major industrial facilities, such as large chemical plants, oil refineries, and municipal waste incinerators, to meet toxic air emission standards. Unlike the second milestone which measures success by reductions in emissions of a particular pollutant, this milestone addresses "toxics" in general and does not specify how different toxics will be compared to one another (e.g., by volume or toxicity). In addition, unlike the second milestone, but consistent with the formulation in the Clean Air Act, success is measured in terms of compliance with a technology-based standard rather than on an output basis. Since not all the standards have been promulgated, the amount of toxics reduction has not been specified.

The fifth milestone focuses on the issue of acid rain and uses an emissions approach that includes virtually all source categories. Consistent with the CAA, it calls for a reduction in SO_2 by 32 percent from the 1994 level of 22 million tons.

The sixth milestone concerns the clearness of the air. Like milestone one, it is output based, specifying that annual average visibility in the eastern U.S. will improve by 10 to 30 percent. Further, it states that the greatest improvement will be found in a particular area, the central Appalachian region.

EPA Goal: Climate Change Risk Reduction

> *The United States and other nations will stabilize atmospheric greenhouse gas concentrations at a level that prevents dangerous interference with the climate system. The level should be achieved within a time frame that allows ecosystems to adapt naturally to climate change, that ensures food production is not threatened, and that enables economic development to proceed in a sustainable manner.*

The single milestone for this goal states that U.S. emission of greenhouse gases—carbon dioxide, methane, nitrous oxide, and halogenated fluorocarbons—will be reduced to 1990 levels by the year 2000. This is an emissions goal that President Clinton has endorsed, although it was originally established as a nonbinding target in the Framework Convention on Climate Change (1992). However, it is not clear how this goal relates to the larger goal of ". . . prevent(ing) dangerous interference with the Earth's climate system." Nor is there any linkage in the milestone between U.S. actions and those of other nations, as suggested in the goal statement.

EPA Goal: Safe Waste Management

> *The wastes produced by every person and business will be stored, treated, and disposed of in ways that prevent harm to people and other living things.*

The first milestone for safe waste management involves an emissions goal for dioxin emissions from hazardous, medical, and municipal solid waste incinerators. The second waste milestone concerns emissions of mercury and other harmful pollutants from the same source categories. Unlike the other milestones for safe waste management, which are primarily concerned with contamination of land, these milestones focus on releases into the air. These sources, which are subject to federal permit requirements managed by EPA's waste programs, represent an estimated 80 to 90 percent of known dioxin and mercury emissions.

Milestone three concerns confirmed releases from underground storage tanks. Milestone four involves toxic wastewater injected into deep Class I wells, while milestone five focuses on so-called high-risk wastewater injection in shallow Class V wells. All three utilize emissions goals and/or federally mandated practices/standards as a means to define "safe." In all cases, the emissions goals are quantitative. In one case, injection of toxic wastewater in high-risk shallow wells, the practice is to be eliminated. The notion of high risk, of course, is subject to further clarification.

PCSD's Eight National Goals to Put the U.S. on a Path Toward Sustainable Development

The following discussion is based on the PCSD interim report dated June 28, 1995.

The PCSD proposes eight "priority national goals" designed to "put the U.S. on a path toward sustainable development":

1. prosperity,
2. a healthy environment,
3. conservation of nature,
4. responsible stewardship,
5. sustainable communities,
6. cooperative democracy,
7. stable populations, and
8. international leadership.

For each goal, the PCSD provides

1. a one-sentence clarification of the goal; and
2. a number of possible indicators of progress, which are quantitative in nature and designed to measure the movement toward achievement of the goals.

In addition, the PCSD intends to propose policy recommendations for achieving each goal. (As of this writing, recommendations have not yet been released.)

In general, PCSD's priority national goals are broadly defined and long term in nature. Moreover, the goals are oriented toward the basic objectives of promoting efficiency, protecting the environment, and ensuring equity. This subsection explores three of these goals and their corresponding "indicators of progress" in an attempt to highlight some basic themes and characteristics of the PCSD's vision:

PCSD Goal: Economic Prosperity

> *Achieve long-term economic growth and prosperity that provides opportunity, meaningful jobs, and better living conditions for all Americans.*

Four of the six indicators of progress toward this goal (economic performance, savings rate, productivity, and environmental wealth) are designed to reflect the country's production and maintenance of wealth, and two indicators (income equity and poverty) are designed to reflect the country's distribution of income.

- The first indicator is the growth in GDP per capita and is designed to reflect economic performance.

- The second and third indicators, income equity and poverty, are measures of current income distribution.
- The fourth indicator, savings rate, is a key factor in assessing long-term economic growth.
- The fifth indicator, environmental wealth, is based on a new measure of wealth that reflects resource depletion and environmental costs.
- Finally, the sixth indicator, based on per capita production per hour worked, is designed to measure productivity.

As a group, these represent a balanced and measurable set of indicators. Baseline data for five of the six indicators are readily available. Because of both conceptual and practical problems, developing an indicator on environmental wealth, however, is more problematic. The Commerce Department has been attempting to construct a very similar indicator, often referred to as Green GDP. Currently, work has been halted by Appropriations Committee language, adopted in 1994, barring further development of this indicator.

PCSD Goal: A Healthy Environment

Ensure that every person can enjoy the benefits of clean air, clean water, safe food, and secure and pleasant surroundings.

Indicators of progress toward this goal are not all currently measurable:

- The first indicator, toxic materials per capita, is based on measures of long-lived and other toxic materials released into the environment as pollutants or waste; it does not consider distribution of toxic pollution across the U.S. population.
- The second indicator, life expectancy, is based on measures of expected life span covering various economic and demographic groups.
- The third indicator, infant mortality, is based on measures of infant mortality rates developed for various economic and demographic groups.
- The fourth indicator, safe drinking water, is based on measures of the percentage of the population whose safe drinking water does not meet safe drinking water standards. This indicator assumes the accuracy of the SDWA's definition of "safe."
- The fifth indicator, clean air, is based on a measure of the percent of U.S. population that lives in cities where air quality standards for one or more pollutants are not met. This indicator is consistent with the ambient standards of the Clean Air Act.

These five indicators cover a broad range of issues, yet they do so in a fairly general matter. In many respects, they correspond to the EPA milestones. However, EPA has a total of 65 milestones (the great majority of which are health related). The PCSD has clearly opted for breadth over specificity.

PCSD Goal: Sustainable Communities

> *Strengthen communities' capacity to engage their citizens in actions to enhance fairness, provide economic opportunity, and maintain a safe and healthy environment.*

Indicators of progress attempt to "allow for the cultural diversity among communities while recognizing key national trends":

- The first indicator, violent crime, is based on measures of the number of people who feel safe walking through their neighborhood in the evening. Establishing a baseline may be difficult for this indicator.
- The second indicator, community design, is based on measures of access to jobs, shopping, services, and recreation, nearby transportation choices, and housing through "alternative land designs." This indicator attempts to reflect a community's economic opportunity, one of the multiple components of the goal. In a world of advanced telecommunications, however, it is not clear that traditional measures of economic opportunity are accurate. Moreover, it is questionable whether additional access routes, shopping centers, recreation centers, etc., enhance or threaten the health of the environment (another component of the goal). Finally, it is not clear how this indicator is calculated.
- The third indicator, public parks, is based on the amount of urban green space or park space.
- The fourth indicator, public participation, is based on the percentage of registered voters who cast ballots in the past two national elections and the percentage of individuals within a community who participate in social, recreational, charitable, and other civic activities. This indicator attempts to measure the extent to which citizens are "engaged" in the maintenance/betterment of their communities; it does not address the effectiveness of such efforts.
- The fifth indicator, investment in future generations, is based on the amount of community resources dedicated to its children, including maternal care, childhood development, and K-12 education. How the amount of resources dedicated to children is determined is not clear (e.g., is time spent by parents at home somehow included in this calculation?).
- Finally, the sixth indicator, transportation patterns, is based on the average mass transit miles, vehicle miles traveled per person, and the number of trips made possible by alternatives to personal motor vehicles. Weighting of the four subindicators and interpretation problems may be issues for this indicator.

Unlike the literature on either economic or health-related issues, the literature on sustainable communities is not well developed. Thus, in many ways, the authors are breaking ground with these measures. While such measures represent a solid effort, issues related to baseline data availability and metric comprehensiveness and specificity are apparent.

The PCSD also proposes goals and indicators specific to the energy, transportation, and agriculture sectors.

CENR's Environmental and Natural Resource Goals for Research for Fiscal Year 1996

CENR's goals for environmental and natural resource research are presented in the context of five overall goals for Science and Technology:

- improved environmental quality;
- a healthier, safer America;
- a stronger economy;
- enhanced national security; and
- improved education and training.

CENR's goals for improved environmental quality cover seven areas:

1. air quality;
2. biodiversity and ecosystem dynamics;
3. global change;
4. natural disaster reduction;
5. resource use and management;
6. toxic substances/hazardous and solid waste; and
7. water resources and coastal and marine environments.

For each of the seven goal areas, CENR provides a description of the current state of understanding; a characterization of the themes of the current research; proposed areas of enhanced emphasis; selected milestones for 1995-1998; and a proposed budget for fiscal year 1996, reflecting the Administration's priorities. Five of the seven research areas show at least slight budget increases. Two of the areas (resource use and management and natural disaster reduction) show slight declines from the previous year.

In addition to the seven research areas, CENR presents five crosscutting topics for Integrated Environmental Research and Development:

1. ecosystem research;
2. observations and data management;
3. social and economic dimensions of environmental change;
4. environmental technology; and
5. science policy tools: integrated assessments and characterizations of risks.

These crosscutting topics span the seven environmental research areas. Each topic, in turn, has an environmental goal, key policy objectives, areas of enhanced emphasis, and selected milestones, 1995-1998. No separate budgets are presented for the crosscutting topics.

The expository material emphasizes the process by which strategic planning and coordination of environmental research and development occurs across a dozen cabinet level agencies, several of which subsume strong subagency activities, as well as a number of separate White House offices. In developing FY 1996 goals, CENR sought extensive consultation from outside the Executive Branch (e.g., from Congress, from interest groups, from the public at large). The whole report was subject to outside peer review. Such consultations emphasized the need for competitive awards, strengthened academic research, merit review, and international cooperation.

For our purposes it is useful to examine several of the individual research areas:

CENR Goal: Air Quality

> *The goal of the federal air quality research program is to help protect human health and the environment from air pollution by providing the scientific and technical information needed to evaluate options for improving air quality in timely and cost-effective ways.*

The discussion of air quality begins with reference to the legislated mandates to produce assessments and make important policy and regulatory decisions within the coming years. Emphasis is given to ground-level ozone, acidic deposition, airborne particles, toxic compounds, and visibility. In addition, concerns are raised about the quality of indoor air.

The current research program is characterized as including

1. long-term observations and analysis to evaluate the effectiveness of recent regulatory initiatives;
2. identification of emerging health or environmental problems;
3. characterization of the processes involved in air quality changes; and
4. assessment of the state of knowledge on air quality issues.

Two topics are listed as areas for enhanced emphasis:

1. understanding the formation of ground-level ozone in urban and rural areas and
2. characterizing the health impacts of airborne fine particles.

Seven milestones are listed for the period 1995-1998. These typically involve completing particular studies ranging from on-the-ground efforts to national assessments. Overall, CENR proposes that air quality research receive an increase in funding for FY 1996.

CENR Goal: Global Change

> *The goal of global change research is to observe and document global environmental changes and identify their causes, predict the responses of the*

earth system, determine the ecological and socioeconomic consequences of these changes, and identify strategies for adaptation and mitigation that will most benefit society and the environment.

CENR describes the policy context for global change as including the Framework Convention on Climate Change, the Montreal Protocol, the Clean Air Act Amendments, and various conventions related to global environmental issues (e.g., forestry, desertification, protection of oceans, and biodiversity). Research issues concern the ultimate impact of the current buildup of greenhouse gases on climate, as well as the effects of stratospheric ozone depletion on humans and ecosystems.

The current research program is characterized as including

- climate change and greenhouse effect research;
- stratospheric ozone and UV effects research;
- seasonal to interannual climate fluctuations research; and
- large-scale ecosystem productivity research.

Research areas designated for enhanced emphasis are

- evaluating the socioeconomic driving forces of global change;
- understanding the consequences of global environmental change;
- developing adaptation and mitigation options; and
- conducting integrated assessments.

Six milestones are listed for the period 1995-1998, the topics of which range from completing individual studies, to making certain scientific measurements, to fulfilling U.S. commitments to participate in various international cooperative research programs. Overall, CENR proposes that climate change research receive a slight increase in funding for FY 1996.

CENR Goal: Toxic Substances, Hazardous and Solid Waste

The goal of federal toxic substances and hazardous and solid waste research is to prevent or reduce human and ecological exposure to toxic materials, such as pesticide residues, polychlorinated biphenyls (PCBs), and lead, and their adverse consequences by providing the scientific and technical information needed for informed decision- and policy-making and effective problem solving.

CENR describes the policy context for toxic substances and hazardous and solid waste as including nine major statutes, including TSCA, RCRA, and CERCLA, among others. The overall goal of the federal research programs in this area is to provide the scientific and technical information needed for informed decision and policy-making to prevent or reduce human and ecological exposure to toxic materials. Current research can be segmented into the categories of risk assessment and risk management. Future research is designed to enhance capabilities in both areas. Five specific milestones are listed for 1995-1998, including the completion of specific studies, the conduct of cooperative research with industry partners, and the implementation of a national program for verifying performance of innovative environmental technologies. Overall, CENR proposes that research in this area increase in FY 1996.

CONSISTENCY AMONG GOALS PROJECTS

This paper cannot compare every aspect of each of the goals projects for technical consistency. In broad terms, the three projects are roughly consistent. However, as would be expected in undertakings of this magnitude, there are some apparent inconsistencies. Reducing airborne exposures to fine particles, for example, is not emphasized in either the EPA or the PCSD efforts, but is an area of priority in the CENR plan. In contrast, global warming is a key issue in both reports sponsored by the Administration (EPA and CENR), yet it is not accorded major importance in the more broadly based PCSD report. The same general pattern holds for toxic wastes, which are emphasized in both the EPA and CENR reports but treated somewhat less prominently in the PCSD effort.

There are good explanations for some, if not all, of these differences. In the area of fine particles, recent research has raised questions about the potential for serious health effects at relatively low concentration levels. It is thus appropriate that the research agenda (CENR) focus on this question, but it is not appropriate that EPA make it a priority for implementation beyond compliance with the current standard. While revising its ambient standard on fine particles could be an EPA objective, it perhaps is too specific an issue to appear in the goals project as a major milestone.

In the case of global warming, the Administration has clearly made this issue a priority both in terms of meeting its goal of holding greenhouse gas emissions in the year 2000 to 1990 levels and in terms of conducting more research on the issue, including mitigation research. In contrast, many of the members of the PCSD, especially those from the business sector, have not fully embraced global warming as a problem that merits major action at this time. Thus, it is not surprising that the PCSD has only addressed the global warming issue in very general terms.

Differences in the area of toxic wastes are analogous to those in the area of global warming. Consistent with current laws, both the EPA and the CENR give considerable emphasis to the problem. Yet, over the longer term it is not at all clear how much emphasis to accord the issue. While the PCSD does propose a specific indicator on toxic accumulation (the amount of long-lived and other toxic materials released into the environment), toxic wastes do not loom as a major part of its goals or indicators.

CRITERIA FOR COMPARISON AND EVALUATION

In order to compare these various goal schemes, it is helpful to develop and apply some uniform criteria. This section briefly outlines eight comparison criteria.

Scope of goals—Goals are relatively ***broad*** or ***focused***, depending on their scope.

Time frame—***Short-term*** goals are characterized as those with a 3-year-or-less time frame, ***medium-term*** goals as those with a 3-to-25-year time frame, and ***long-term*** goals as those with a 25-plus-year time frame.

Measures of success—An organization's goals typically target resources expenditures (inputs), qualitative performance (outcomes), and/or quantitative performance (outputs). Measures of success are thus characterized as either ***input based***, ***outcomes based***, or ***output based***.[4] Input-based measures of success are commonly criticized for their inability to reflect results. Outcomes-based measures are often difficult to track over time. Output-based measures are typically criticized as lacking clear measures of effort and accountability.

Completeness of metrics for assessing progress toward goals (***high, medium,*** or ***low***)—Arguably, progress assessment is as important as goal-setting itself. Metrics chosen for progress assessment, however, may be problematic for a number of reasons. First, target levels may be unrealistic. Second, metrics chosen may not be appropriate, reliable means for assessing progress toward a particular goal. Third, because multiple metrics may be designated to assess progress toward a single goal, how to interpret them collectively may not be clear.

Clear policy tools for achieving goals (***yes*** or ***no***)—Some schemes for achieving goals are clearly regulatory in nature, while others prescribe grant-giving, direct action, or other policy mechanisms. At the same time, however, some goal schemes do not prescribe any specific approaches to achieve desired results.

Interim milestones of success (***yes*** or ***no***)—Goal schemes with metrics for progress assessment may or may not specify interim milestones of success. Without such concrete milestones, it is generally difficult to evaluate strategies and make necessary midcourse corrections.

Explicit assessment of trade-offs in goal choices (***yes*** or ***no***)—An effective goal scheme must be grounded in reality and at the same time directed toward ideals. Most goal statements reflect ideals. However, many fail to address the real trade-offs that must be confronted in order to reach the ideals.

Environmental policy analysts as formulators of goals and metrics (***yes*** or ***no***)—Goals and metrics may be formulated by expert environmental policy analysts and then reviewed by interest groups or the general public. Alternatively, they may be formulated by interest groups and the public and then reviewed by the

---

[4]These performance measurement terms are defined in an OMB memo dated September 23, 1994.

experts, interest groups, and the public. Use of the latter approach may reflect the degree of "hands-on" involvement of those outside the community of environmental experts.

THE THREE MAJOR GOALS PROJECTS: A COMPARISON

Table 1 characterizes EPA, PCSD, and CENR goal schemes according to the eight comparison criteria introduced in the previous section. The following discussion highlights significant differences.

Goal Scope and Time Frame

The missions of EPA, PCSD, and CENR vary considerably: CENR seeks to conduct relevant and useful research for the achievement of environmental goals; the EPA aims to protect the environment (principally as mandated by governing statutes and treaties); and the PCSD attempts to enhance the public welfare (including the environment). As a result, the goal scopes and time frames differ. CENR goals are focused and very short term; EPA goals are focused and generally medium term; PCSD goals are broadly defined and long term.

Measures of Success

CENR, EPA, and PCSD utilize different measures of success. CENR defines success in terms of its participation in international cooperative research pro-

TABLE 1 A Comparison of EPA, PCSD, and CENR Goals Projects

	EPA	PCSD	CENR
1. Scope of goals	focused: environmental protection	broad: public welfare	focused: environmental research
2. Time frame	medium term	long term	short term
3. Measures of success	outcomes or output based	outcomes based	input or output based
4. Completeness of metrics for assessing progress toward goals	medium	low	high
5. Clear policy tools for achieving goals?	yes	no[a]	yes
6. Interim milestones of success?	yes	no	N/A
7. Explicit assessment of trade-offs in goal choices?	no	no	no
8. Environmental policy analysts as formulators of goals and metrics?	yes	no	yes

[a]Expected, not yet available.

grams (input) and in another case defines success in terms of its completion of specific studies (output). Similarly, EPA sets both technology (outcome) and ambient (output) milestones for clean air, and PCSD suggests public participation and violent crime (outcome) indicators for sustainable communities.

Completeness of Metrics for Progress Assessment

The completeness of metrics for progress assessment appears to vary inversely with goal time frame. PCSD goals are longest in time frame and the least complete in metrics; CENR goals are the shortest in time frame and the most complete in metrics; and EPA lies somewhere in between:

- PCSD provides multiple indicators for single goals, but does not explain how to combine them; numerous PCSD indicators are not clearly measurable; and some indicators only partially assess progress toward desired goals.
- Progress toward EPA's milestones is measurable, but when all milestones are met, it is not clear how much further effort is required to achieve the overall goals of "clean air," "safe drinking water," etc.
- CENR progress is easily measurable due to the explicit and short-term nature of its goals.

Clear Policy Tools For Achieving Goals

EPA sets out to achieve its goals through various methods, many of them regulatory; CENR attempts to meet its objectives through proposed budgets. The policy mechanisms for achieving PCSD goals currently are not specified. However, PCSD has indicated that policy recommendations are forthcoming.

Interim Milestones of Success

Probably because of the short-term nature of its goals, CENR does not develop interim milestones. EPA provides some measures of expected progress over time in most cases and has stated its intention of issuing updates over time. PCSD mentions neither interim nor final milestones of success.

Explicit Assessment of Trade-offs in Goal Choices

None of the goal schemes prioritize individual goals or acknowledge the basic trade-offs between desired outcomes. PCSD, for example, fails to describe an optimal mix of biological diversity and meaningful jobs (components of conservation and prosperity goals, respectively). EPA does not state whether it is more important to prevent oil spills or achieve stratospheric ozone layer restoration. CENR only implicitly acknowledges trade-offs through its budget decisions.

Environmental Policy Analysts as Formulators of Goals and Metrics

EPA and CENR goals and metrics were designed primarily by expert environmental policy analysts, whereas PCSD goals and metrics were designed by professionals with more diverse backgrounds from the government, business, and nonprofit sectors. The "hands-on" involvement of leaders from the government (not limited to environmental agencies), business, and nonprofit sectors is a key element of the strategy to establish broad "ownership" of the PCSD conclusions.

Other Observations

Viewed through one set of lenses, clear and explicable differences among the three goals projects are apparent. The CENR strategy—which addresses research as opposed to policy goals—is clearly the most focused effort. Goals and time frames are established with clear metrics. Budgets are proposed to accomplish those goals. The EPA goals—which roughly follow statutory mandates and current or proposed program activities—are somewhat less focused, but do contain a number of key metrics with specific time frames. Like the EPA plan, the PCSD proposal addresses policy goals. Unlike the EPA plan, the PCSD proposal is not constrained by statutes or ongoing programs and therefore is able to focus on broader societal questions. Not surprisingly, the PCSD is less precise, and contains fewer metrics, time frames, and implementing mechanisms.

Viewed through another set of lenses, some potentially significant shortcomings are evident. Some participants in the CENR effort, for example, have complained that the effort is little more than a "stapling together" of agency initiatives with little top-down direction or priority setting—more akin to a budget "cross-cut" than a meaningful exercise in prioritization. Similarly the public participation process has been criticized as "show and tell" with little opportunity for informed discussion. Defenders of the CENR project note that it is the first attempt to conduct such a government-wide environmental research exercise and therefore it should not be judged too harshly. Breaking down agency barriers, they claim, is a difficult process that should be evaluated over a several-year period.

EPA's effort has also been criticized. Commenting on an earlier draft of the report, the Chairman and Ranking Minority Member of the Senate Appropriations Subcommittee have expressed concern that "instead of setting priorities, the . . . plan appears to include almost everything of interest in the agency." Outside reviewers of the earlier draft have raised questions about the rationale for setting targets at specific levels, about the mechanisms for reaching the targets, about the costs and benefits of the goals, about who will pay to reach the goals, and about the need to address special populations (e.g., Native Americans). Also, various questions have been raised about specific metrics and, more importantly, about the overall vision implied by the effort. Several reviewers have questioned whether these goals are really consistent with notions of sustainable development. Agency defenders note that this project is a first effort and that many of the criticisms are also criticisms of the way the Agency has done business over the past twenty-five years. EPA has responded constructively to these criticisms by making major revisions to reflect the various concerns, and by seeking review by the Agency's Science Advisory Board.

Critics of the PCSD project complain that the process has been too long and complex, has involved too many stakeholders, and has taken on too grand an agenda to expect concrete results. Moreover, critics are concerned that goals are being defined at too high a level of generality to be operational. Others complain that economic assessment has not been performed and goal-achievement mechanisms have not been specified. Defenders argue that the scope of the problems facing the environmental agenda is so great that a major review is needed. Further, they point out that if the key stakeholders are not involved there is no hope of solving these long-run problems. One of the innovations of the PCSD is the bringing together of stakeholders to formulate the goals and metrics. Thus, the process is a key part of the project. Also, they note that the PCSD was first established two years ago when there appeared to be a broader set of constituencies prepared to consider more sweeping pro-environmental changes.

KEY ISSUES NOT EMPHASIZED BY THE THREE GOALS PROJECTS

While the scope of the three goals projects is certainly broad, there are, nonetheless, some issues not emphasized. Eight such issues are as follows:

1. *Systematic measurement and monitoring:* There is widespread agreement among experts that one of the greatest impediments to environmental progress is our lack of detailed measurement and monitoring information. The EPA report has highlighted this issue, but the others have not made it a priority.

2. *Program evaluation:* A number of experts have called for independent studies to evaluate past program successes and failures in order to better design future programs. In general, agencies have been reluctant to undertake such evaluations for a variety of reasons, including budgetary constraints. None of the projects makes review of past efforts a priority, although EPA does call for periodic updates of progress toward attaining goals.

3. *Interdisciplinary social science research:* Various expert groups, including EPA's Science Advisory Board in the 1990 *Reducing Risk* report, have called for greater emphasis on interdisciplinary social science research to increase understanding of the behavioral and institutional aspects of environmental protection. Heretofore, such research has received little priority in federal budgets. None of the projects reviewed makes this a priority.

4. *Regulatory reform:* While many recent public policy discussions have

focused on issues of regulatory reform, none of the three goals reports has given priority to this issue. Among the three, only the PCSD has emphasized economic growth and the use of market mechanisms as goals. In general, discussions of economic efficiency, the use of benefit-cost analysis, and review of past as opposed to future regulatory decisions are either absent from or not emphasized by the three reports.

5. *Devolution to the states:* While not usually seen as a goal in and of itself, devolution of responsibilities to state or local governments can have significant implications for environmental goals. The Administration, Congress, and various experts have called for greater responsibilities to flow to the states as a way to tailor both goals and program design to local needs. None of the three goals reports has formally incorporated state or local decisionmakers into its vision for the future, although EPA does acknowledge the need to do so.

6. *Preservation of nature for its own sake:* A number of environmental organizations advocate environmental protection for "biocentric" or "ecocentric" reasons as opposed to the generally "anthropocentric" basis implicit in the three goals projects. Various economic studies have demonstrated through both survey techniques and the presence of citizen-supported environmental groups, that there is some "willingness to pay" for such efforts. None of the goals projects emphasizes this issue, although the PCSD emphasizes ecosystem protection with a nonanthropocentric orientation.

7. *Public access to environmental information:* There is widespread belief among experts that responsible reporting of scientifically sound environmental information can be a powerful influence in bringing about desired changes in behavior by individuals and by firms. The EPA report has highlighted this issue but the others have not made it a priority.

8. *Acknowledgment of the trade-offs among goals:* Certainly there are trade-offs among environmental goals themselves and among the environmental, economic, and equity goals of the PCSD. Yet none of these reports has articulated what the key trade-offs are, and none has really spelled out how or on what basis some of the key decisions were made in the individual reports.

CONCLUSION

The task of this paper has been to compare and contrast the three most recent efforts to establish environmental goals and thereby to assess the present "state of the art" in environmental goal-setting. A number of conclusions can be drawn from this effort.

1. *The three goals projects address three fundamentally different sets of problems.* The EPA project is oriented to implementation of current environmental statutes and treaties and it begins to look longer term. A great deal of emphasis is placed on measurable results. The PCSD tries to fashion a vision for the next

century by incorporating the themes of environment, economy, and equity, and by encouraging better integration of the different segments of society in pursuit of environmental management. In bringing together the government, business, and nonprofit sectors for the actual formulation of goals and metrics, the process itself becomes a key part of the strategy for obtaining "ownership" of the conclusions. The CENR project is fundamentally different in that it is a research strategy rather than a game plan for environmental policy. It seeks to improve coordination among federal agency research budgets and to balance competing bureaucratic agendas.

2. *The three goals projects have different technical approaches including different scopes, time frames, completeness of metrics, clarity of policy tools, and use of interim milestones.* The CENR project is the most focused, shortest in time frame, and most complete in metrics. The EPA project is also focused, yet somewhat longer in time frame and less clear about priorities among goals. The PCSD project is broad in scope, long in time frame, and incomplete in metrics. As indicated, the participatory nature of the process is integral to the PCSD project. The PCSD is expected to design policy tools for achieving its goals; however, such recommendations are not yet available. PCSD provides no interim milestones for assessing progress.

3. *All three projects fail to address a number of important issues.* The PCSD and EPA fail to develop a clear rationale for achieving their stated goals; benefit-cost assessment and specific strategy articulation—two crucial steps for goal achievement—are not emphasized. With regard to the CENR project, questions have been raised about the extent of top-down integration of goals and activities, as opposed to a "stapling together" of individual agency programs. In general, the projects fail to highlight the need for program evaluation of past as well as future activities, interdisciplinary social science research, regulatory reform, and public reporting of information, as well as the need to assess trade-offs among goals. Only EPA emphasizes the need for systematic measurement and monitoring.

4. *The state of the art in environmental goal-setting is still in its infancy; however, these three efforts clearly represent a major step forward.* Despite serious concerns about the rigor, completeness, clarity, and vision of these projects, it is clear that all three goal-setting efforts are groundbreaking. The federal government has never been so explicit about what it is trying to accomplish in the environmental arena, either in terms of measurable environmental results or in terms of specific research outputs. Similarly, although multi-stakeholder efforts have been assembled previously, none has incorporated such a broad set of constituencies or taken on so bold an agenda as the PCSD.

5. *An overwhelming strength common to all three projects is the implicit recognition that our environmental management system is in need of significant reform.* A common denominator to all efforts is a less hortatory and more pragmatic framework than is evident in many of our basic environmental statutes. After 25 years of a growing federal role in environmental protection and with

environmental programs that account for more than 2 percent of GDP, we are currently grappling with tough management issues. Such difficulties are consistent with the observation that environmentalism is in transition from a social movement to an accepted component of our society and economy. The notion that "we're all environmentalists now" implies that the key issue is no longer to win over the hearts and minds of the American public, but to deliver on now-widely-held expectations, and to do so in an efficient and equitable manner. Collectively, these three diverse reports implicitly acknowledge the problems of the current system. They should be viewed as early steps down a long road of redesigning and reshaping environmental programs and policies over the coming years.

6. *Follow-through is the key to success.* Almost twenty years ago, President Carter's zero-based budget engendered much discussion about change in the federal government—both policy and process change. There were, in fact, many virtues in the zero-based budgeting idea, yet in the absence of strong follow-through, the effort faded quickly. The CENR needs to generate periodic report cards of agencies' progress in meeting their research goals, with both timing and content geared to the ongoing federal budget process. Similarly, EPA needs to finalize its goals and then move quickly to set up internal as well as public progress reports. In both cases there is a strong need to institutionalize the efforts so they can be better insulated from the political process. Perhaps the economic statistics agencies are a useful model for such efforts. The PCSD has an even greater challenge. The entire activity is largely a voluntary enterprise involving literally hundreds of participants from government, business, and the nonprofit sector. Clearly, strong, organized follow-through is critical to the realization of the PCSD goals.

APPENDIX A

PROPOSED EPA GOALS AND MILESTONES

Long-range Goal	Milestone(s)
Clean Air: By 2010 and thereafter, the air will be safe to breathe in every city and community, and it will be clearer in many areas. Life in some forests and polluted waters will rebound as acid rain is reduced.	• By 2005, the number of metropolitan areas not meeting air quality standards will be reduced from 60 in 1995 to 6, which corresponds to reducing the number of Americans living in nonattainment areas to 45 million. • By 2005, emissions in smog-causing volatile organic compounds (VOCs) will fall 65 percent per automobile from 1990 levels. • *Vehicle miles traveled milestone being developed.* • By 2005, all 174 categories of major industrial facilities, such as large chemical plants, oil refineries, and municipal waste incinerators, will meet toxic air emission standards. • By 2005, sulfur dioxide emissions, the primary cause of acid rain, will be reduced from the 1994 level of 22 million tons. • By 2005, annual average visibility in the eastern United States will improve from 10 to 30 percent from 1995 levels.
Climate Change Risk Reduction: The United States and other nations will stabilize atmospheric greenhouse gas concentrations at a level that prevents dangerous interference with the climate system. The level should be achieved within a time frame that allows ecosystems to adapt naturally to climate change, that ensures food production is not threatened, and that enables economic development to proceed in a sustainable manner.	• By 2000, total U.S. greenhouse gas emissions—carbon dioxide, methane, nitrous oxide, and halogenated fluorocarbons—will be reduced to the 1990 level.
Stratospheric Ozone Layer Restoration: By 2045, stratospheric ozone concentrations will return to the levels found prior to the discovery of the "ozone hole" over Antarctica.	• By 2005, ozone concentrations in the stratosphere will have stopped declining and slowly begun the process of recovery. • By 2005, adjusted chlorine concentrations in the stratosphere will be reduced from 1995 levels of 4.1 parts per billion (ppb) to [4.0?] ppb. • By 2005, atmospheric concentrations of the ozone-depleting substances CFC-11 and CFC-12 will peak at no more than x and y parts per trillion, respectively. • By 2005, U.S. production of all ozone-depleting substances, except HCFCs, will be eliminated.

Source: EPA Draft Goals Report dated July 19, 1995.

Clean Waters: Our waters will support human health and uses such as swimming, fishing, drinking water supply, agriculture, and industry. Our waters will also support ecosystem health by sustaining healthy communities of plants, fish, insects, and other animals that depend on the aquatic environment. We will conserve remaining wetlands and restore others to health.

- By 2005, there will be no overall net loss of wetlands.
- By 2005, half of the aquatic species currently designated as threatened or endangered will have stable or increasing populations.
- By 2005, 65 percent of rivers and streams and 80 percent of estuaries will support healthy biological communities.
- By 2005, 89 to 93 percent of the nation's surface waters will support aquatic life.
- By 2005, 88 to 98 percent of the nation's fish and shellfish harvest areas will provide food safe to eat.
- By 2005, 93 to 95 percent of the nation's surface waters will be safe for recreation.
- By 2005, 50 percent of the wells monitored for ground water quality will fully support each state's intended uses of the water, such as for drinking water, agricultural irrigation, or industrial processing.
- By 2005, the annual rate of sediment erosion from agricultural croplands will be reduced 20 percent from 1992 levels to a total of 950 million tons per year.
- By 2005, annual discharge of pollutants of concern to surface waters will be reduced by 1,668 million pounds from CSOs, by 19 million pounds from sewage treatment plants, and by 700 million pounds from industrial sources.

Healthy Terrestrial Ecosystems: The United States will maintain a mosaic of lands capable of sustaining existing or greater numbers and diversity of indigenous plants and wildlife, and providing ecological, economic, and recreational benefits.

- By 2005, the current acreage of American lands that are managed with good conservation practices that protect soil, native vegetation, and wildlife will double from [?] to [?].
- By 2005, ecological restoration will generate at least one significant improvement per major U.S. region in terrestrial ecosystem function, total area, or other measure of quality.
- By 2005, ecosystem protection efforts will eliminate area loss of ecosystem types considered critically endangered and will reduce the net area loss rates and increase the total protected acreage of the most highly beneficial terrestrial ecosystem types.
- By 2005, 45 percent of the terrestrial species currently designated as threatened or endangered under the Endangered Species Act will have stable or increasing populations.
- By 2005, the number of migratory bird species with increasing populations will grow by __ percent.

Healthy Indoor Environments: The environment inside every home, school, and office will be healthy, comfortable, and productive.

- By 2005, the total number of children in the United States between 6 months and 5 years old whose blood lead levels exceed 10 mg/dL will be no more than 730,000, compared to approximately 1.7 million children in the late 1980s. By 2005, the total number of children in the United States between 6 months and 5 years old whose blood lead levels exceed 15 mg/dL will be no more than 250,000, compared to approximately 500,000 in the late 1980s.
- By 2005, 27 million homes will have been voluntarily tested for radon with corrective actions taken in 1 million homes, and 1.5 million new homes will have been built with radon-resistant features.
- By 2005, children's exposure to environmental tobacco smoke will decrease through voluntary actions in the home. The proportion of households in which young children are regularly exposed to smoking will be reduced to 15 percent from over 39 percent in 1986.
- By 2005, EPA will have reached agreements with manufacturers to substantially reduce emissions from 10 or more products whose emissions create a relatively high adverse impact on indoor air quality and public health.
- By 2005, 3,000 or more of commercial and school buildings will have building air quality management plans promoted by EPA.

Safe Drinking Water: Every public water system will consistently provide water that is safe to drink.

- By 2005, 95 percent of people served by drinking water systems will be provided water that meets health requirements throughout the year.
- By 2005, 90 percent of the nation's rivers, streams, lakes, and reservoirs designated as drinking water supplies will provide water that is safe to use as a source for drinking water.
- By 2005, 50 percent of community water systems will have surface and ground water protection programs.

Safe Food: Chemical residues can be introduced into foods at any point in their production, procession, marketing, storage, transportation, and preparation for consumption. Protection of the food supply for consumers is a stated federal policy and goal in EPA programs.

Safe Workplaces: All people will work in places that are safe from exposure to hazardous chemicals.

Preventing Spills and Accidents: Accidental releases of substances that endanger our communities or wildlife will be reduced to as near zero as possible, and those that do occur will cause negligible harm to humans, animals, and plants.

- By 2005, there will be 25 percent fewer accidental releases of oil, chemicals, and radioactive substances than in 1993.
- By 2005, there will be a 50 percent increase over 1993 levels in the number of industrial facilities in high-risk areas that have reduced hazardous substance inventories to minimum levels or eliminated them altogether.

Toxic-Free Communities Through Pollution Prevention: Our communities will grow increasingly clean as people learn how to efficiently produce, use, and recycle materials in ways that do not damage the environment.

- By 2005, pollution prevention practices will contribute to a 25 percent reduction from 1992 levels of toxic wastes reported by industrial facilities to the Toxic Chemicals Release Inventory.
- By 2005, more than 99 percent of new chemicals approved during the previous ten years will be shown to have been safe.
- By 2005, 10 percent of public and private consumer purchases will be for environmentally preferable products and services.
- By 2005, the generation of municipal waste per capita will be reduced to 4.3 pounds per day, and 30 percent of the municipal solid waste generated will be recycled.
- By 2005, the presence of the most persistent, bioaccumulative, and toxic constituents in hazardous waste will be reduced by 50 percent from 1991 levels.
- By 2005, essential toxicity test data will be available for 30 percent of the major production chemicals in commerce. In addition, by 2005, the amounts and types of data available on chemicals in commerce will be more than double those available in the early 1980s.
- By 2005, capital investments in prevention technologies will grow from 20 percent of environmental investments in 1992 to 50 percent.
- By 2005, the number of commercial chemicals determined to be safe for use by industry and consumers will double from 13,000 in 1995 to 26,000.

Safe Waste Management: The wastes produced by every person and business will be stored, treated, and disposed of in ways that prevent harm to people and other living things.

- By 2005, dioxin emissions from hazardous, medical, and municipal solid waste incinerators will be reduced 97 percent from 1994 levels.
- By 2005, emissions of mercury and other harmful pollutants from hazardous, medical, and municipal solid waste incinerators, will be reduced by at least 60 percent from 1994 levels.
- By 2005, confirmed annual releases from underground storage tanks will be 80 percent lower than in 1994.
- By 2005, the amount of toxic wastewater injected into deep Class I wells will be reduced by 75 percent from 1988 levels.
- By 2005, new high-risk wastewater injection in shallow Class V wells will be eliminated.
- By 2005, 100 percent of municipal solid waste management facilities will have approved controls in place to prevent releases of harmful pollutants to soil and ground water.
- By 2005, 100 percent of hazardous waste facilities will have approved controls in place to prevent releases of harmful pollutants to soil and ground water.
- By 2005, 100 percent of hazardous wastes disposed of on land will be treated to make them less harmful prior to their disposal.
- By 2005, 95 percent of known underground storage tank systems to be replaced, upgraded, or closed.

Restoration of Contaminated Sites: Places currently contaminated by hazardous or radioactive materials will no longer endanger public health or the natural environment, and they will be restored to uses desired by surrounding communities.

- By 2005, 70 percent of the 1,300 contaminated sites on the EPA's National Priorities List will be cleaned up, have the contamination contained, or have cleanup or containment work under way.
- By 2005, 80 percent of the estimated 5,000 sites that warrant further EPA action will have contamination removed or cleanup completed.
- By 2005, at least 10 percent of contaminated federal lands will be cleaned up and restored to uses desired by surrounding communities.
- By 2005, actions to stabilize the further spread of contamination and/or protect people from further exposure to contamination will be under way at 1,275 industrial waste facilities (32 percent). These actions will be under way at 100 percent of facilities where actual human exposures have been identified.
- By 2005, cleanups will be completed at 200,000 leaking underground storage tank sites.
- By 2005, radioactivity will be cleaned up or contained at 6 percent of radioactively contaminated sites.
- By 2005, the 10 percent most severely contaminated sediment sites in 1995 are to have point sources of contamination controlled.
- By 2005, Responsible Parties will continue conducting 70 percent or more of the remedial work and 40 percent of the removal work at Superfund sites.
- By 2005, 99 percent of the currently known universe of potential Superfund sites (38,000) will have had an assessment decision made to determine whether the site will require federal action.

Reducing Global Environmental Risks: Global and transboundary environmental threats to U.S. interests will be eliminated.

Better Information and Education: All people will be informed and educated stakeholders in environmental quality and active participants in environmental decisions at the personal, local, national, and global levels.

- By 2005, information will be available on toxic chemical releases into the environment from all major industrial pollution sources.
- By 2005, the public will have access to comprehensive, integrated environmental information on individual facilities.
- By 2005, information on environmental programs will be available through electronic means that citizens and local organizations can access in homes, schools, and libraries.
- By 2005, EPA will make available comprehensive and integrated information and statistics on national, regional, and local environmental conditions and trends.
- By 2005, there will be substantial growth in the number of environmental education programs in schools, colleges, and communities. The communities will teach educators, students, and the general public how to make informed and responsible decisions about their actions that impact the environment.

APPENDIX B1

PROPOSED PCSD NATIONAL GOALS AND INDICATORS

National Goal	Possible Indicators of Progress
Prosperity: Achieve long-term economic growth and prosperity that provides opportunity, meaningful jobs, and better living conditions for all Americans.	• Economic Performance: Growth in GDP per capita. • Income Equity: Ratio of the income of the top 20 percent compared with the bottom 20 percent of the United States population. • Poverty: Number of children living below the poverty line. • Savings Rate: Per capita savings rate. • Environmental Wealth: New measures that reflect resource depletion and environmental costs. • Productivity: Th : level of per capita production per hour worked.
A Healthy Environment: Ensure that every person can enjoy the benefits of clean air, clean water, safe food, and secure and pleasant surroundings.	• Toxic Materials: Measures of long-lived and other toxic materials released into the environment as pollutants or waste. • Life Expectancy: Measures of expected life span covering various economic and demographic groups. • Infant Mortality: Measures of infant mortality rates, developed for various economic and demographic groups. • Safe Drinking Water: Measures of the percentage of the U.S. population whose drinking water does not meet safe drinking water standards. • Clean Air: Percentage of population that lives in cities where air quality standards for one or more pollutants are not met.
Conservation of Nature: Protect and seek to restore the health and biological diversity of ecosystems.	• Vulnerable Ecosystems: Measures of the vulnerability of natural ecosystems to degradation from present land use patterns, involving such resources as forests, grasslands, wetlands, and coastal lands. • Conservation Status: Measures of lost natural systems or species (incorporating measures for soil loss, wetland loss, threatened and endangered species, remaining old growth forests, threatened and endangered rivers). • Nutrients and Toxics: Measures of nutrients and toxic pollutants that endanger or harm waters. • Exotic Species: Measures of ecological risk due to the introduction and spread of exotic species.

Source: PCSD report dated June 28, 1995.

Responsible Stewardship: Create an ethic of stewardship and community that encourages Americans to reduce resource use and take responsibility for the environmental and social consequences of their actions.	• Material Consumption: Consumption per capita by type of material. • Toxics Accumulation: Amount of long-lived and other toxic materials released into the environment. • Virgin Material Use: Raw or virgin material input, per dollar of GDP output, by sector. • Renewable Material Use: Market share of renewable, recoverable, and recycled material inputs. • Water Use: Net amount of water used, compared with its recharge capacity.
Sustainable Communities: Strengthen communities' capacity to engage their citizens in actions to enhance fairness, provide economic opportunity, and maintain a safe and healthy environment.	• Violent Crime: Number of people who feel safe walking through their neighborhood in the evening. • Community Design: Measures of the access in rural and urban areas to jobs, shopping, services, and recreation, nearby choices for transportation, and housing through alternative land designs. • Public Parks: Amount of urban green space or park space. • Public Participation: Percentage of registered voters who cast ballots in the past two national elections and the percentage of individuals within a community who participate in social, recreational, charitable, and other civic activities. • Investment in Future Generations: Amount of community resources dedicated to its children, including maternal care, childhood development, and K-12 education. • Transportation Patterns: Average mass transit miles, vehicle miles traveled per person, and the number of trips made possible by alternatives to personal motor vehicles.
Cooperative Democracy: Change the process of government to involve more fully citizens, businesses, and communities in collaborative resolution of natural resource, environmental, and economic decisions that affect them.	• Social Capital: Measures of social capital, such as investment in education, and civic awareness. • Citizen Participation: Voter turnout and community participation in such civic activities as professional organizations, PTA, sporting leagues, and charity work. • Collaborations: Measures of characteristics that contribute to successful collaboration.

Stable Populations: Move toward stabilization of U.S. population.	• Population Growth: Rate of population growth in the United States and the world. • Status of Women: Measures of the national and global social/economic status of women. • Unintended Pregnancies: Number of unintended pregnancies in the United States. • Teen Pregnancies: Number of teenage pregnancies in the United States.
International Leadership: Practice globally the values of sustainability we espouse as a nation.	• Treaty Commitments: Adherence to U.S. commitments under international environmental treaties, such as those signed in *Agenda 21*. • International Assistance: Level of U.S. international assistance, including Official Direct Assistance as a percentage of GDP. • Environmental Assistance: U.S. contribution to the Global Environmental Facility and other environmentally targeted development aid.

APPENDIX B2

PROPOSED PCSD SECTOR GOALS AND INDICATORS

Sector Goal	Possible Indicators Of Progress
Energy: Improve the economic and environmental performance of energy use to enhance national competitiveness and social well-being.	• Energy Use: Amount of energy input per dollar of GDP output by sector. • Renewable Energy: Share of renewable and non-renewable energy use in U.S. energy supply. • Electric Efficiency: Average efficiency of electricity generation. • Greenhouse Gas Emissions: U.S. annual emissions of greenhouse gases.
Transportation: Provide a U.S. transportation system that optimizes the performance and use of each type of transportation and enables access to regional and public transit that is reliable, affordable, and convenient.	• Congestion: Congestion levels in urban areas. • Oil Imports: Oil dependency. • Transportation Emissions: Rates of annual greenhouse and other pollutant emissions (including carbon monoxide, lead, nitrogen oxides, small particulate matter, sulfur dioxide, and volatile organic compounds) from transportation. • Transportation Patterns: Average mass transit and personal vehicle miles traveled per capita per year.
Agriculture: Achieve long-term social and economic viability of farm communities and ensure a healthy and affordable supply of food and fiber.	• Average percentage of household income spent on food and fiber. • Percentage of GDP spent on food production and distribution. • Level of concentrations of nutrients and pesticides in ground and surface water. • Trade balance in agricultural products. • Comparison of the per capita income and wealth of rural populations with national and non-rural averages, adjusted for differences in cost of living. • Comparison of unemployment rates among rural population groups, with national and non-rural averages.

Source: PCSD report dated June 28, 1995.

APPENDIX C

CENR GOALS AND MILESTONES

Research Goal	Selected Milestones, 1995–1998
Air Quality: The goal of the federal air quality research program is to help protect human health and the environment from air pollution by providing the scientific and technical information needed to evaluate options for improving air quality in timely and cost-effective ways.	• Provide scientific input to air quality management planning for the highly stressed Great Smoky Mountains National Park by completing an extensive field study with diagnostic modeling to understand the extent, causes, and processes involved in local visibility problems. • Characterize the roles of production and movement of ground-level ozone formation in a region of high natural hydrocarbon emissions and a region of high complexity (Nashville Field Campaign-Southern Oxidants Study) to help formulate more effective emission abatement applications for specific regions of the country. • Quantitatively compare the effects of anthropogenic fine particles to those of coarse, windblown dust particles on human health. • Conduct a National Acid Precipitation Assessment Program (NAPAP) assessment of (1) the reduction in deposition rates necessary to prevent adverse ecological effects and (2) the costs, benefits, and effectiveness of the current acid deposition control strategies mandated under Title IV of the CAAA of 1990. • Create Great Waters and Urban Toxics Inventories to characterize the major risks faced by Native Americans from their basic fish stocks and by inner-city individuals from airborne toxics in the urban environment. • Conduct a comprehensive state-of-science assessment of surface ozone that is summarized in policy-useful terms and that is prepared by the broad scientific community as well as other communities, sponsored by the relevant agencies, reviewed by peers and stakeholder communities, and timed to aid decisions associated with midcourse corrections in the state implementation plans required by the CAAA. • Standardize indoor air tests, develop instrumentation, and evaluate standard procedures that will lead to commercialization of monitoring equipment to improve the "health" of the nation's residential and commercial buildings.

Source: CENR Strategic Planning Document dated March 10, 1995.

Biodiversity: The goal of federal research on biodiversity and ecosystem dynamics is to ensure the sustainability of the ecological systems and processes that support our social needs in areas such as agriculture, forestry, fisheries, recreation, medicine, and the preservation of natural areas.

- Publish common standards and protocols needed to classify and map ecological units and their biological and physical attributes.
- Publish an ecosystem map for the United States at a scale that allows land-use planners, resource managers, industry, the public, and policymakers to incorporate spatially explicit social, economic, and environmental factors into urban planning and resource management decisions.
- Complete establishment of a network of representative long-term sites to determine how various ecosystem management approaches can be achieved.
- Calculate the social and economic impacts (local, regional, and national) of alternative management scenarios; track cumulative social and economic effects of various ecosystem management regimes, such as impacts on fishery management and agricultural programs.
- Determine the functional characteristics to be used to group species so that data essential to successfully maintaining or restoring the population of a species can be extrapolated from studies of a few representative species to entire groups of species.

Global Change: The goal of global change research is to observe and document global environmental changes and identify their causes, predict the responses of the earth system, determine the ecological and socioeconomic consequences of these changes, and identify strategies for adaptation and mitigation that will most benefit society and the environment.

- Contribute to the international effort to develop a long-term comprehensive global earth observation system by launching the first in a series of earth observing system satellites, and establish a global change data and information system to make high-quality global change data accessible to researchers worldwide.
- Incorporate new understanding of atmospheric radiation processes (including the role of clouds and aerosols), ecosystem processes, and social and economic driving forces into improved coupled ocean-atmosphere-land surface models to predict future long-term changes in climate.
- Observe and document changes in the earthl s stratospheric ozone layer through both space- and surface-based observation systems, and observe and document corresponding changes in UV radiation at the earth's surface through development of an intercalibrated network for monitoring radiation. Observe changes in human and ecosystem health related to changes in surface UV radiation and evaluate processes leading to health and environmental changes from UV radiation. Evaluate the effects on the ozone layer and the health and environmental risks of alternatives to CFCs and other halons.
- Provide regular forecasts of the timing and distribution of extreme climatic events (flooding, droughts, etc.) related to seasonal to interannual climate variability (from such phenomena as the El Niño-Southern Oscillation) to communities to assist them in developing plans for preventing damage from climate-related disasters.
- Develop regional assessments of vulnerability to climate change and evaluate the potential social, economic, and human health effects on communities, and the effects on local natural ecosystems, agricultural, forest, fishery, and water supply resources, that would occur if climate changes consistent within the range predicted by the IPCC were to occur.
- Fulfill the U.S. commitment to establish strong international cooperation on global change research by supporting credible, internationally developed research programs (e.g., the International Geosphere-Biosphere Program, World Climate Research Program, and Human Dimensions of Global Environmental Change Program), the development of regional research institutes (such as the Inter-American Institute and the SysTem for Analysis, Research, and Training), and international assessments (such as that of the IPCC).

Natural Disaster Reduction: The goal of federal research in natural disaster reduction is to provide the scientific information necessary to make our society resilient to natural disasters by reducing the loss of life, property damage, and economic disruption caused by earthquakes, floods, hurricanes, tornadoes, fires, and volcanoes.

- Make publicly available through Internet an information system to support the National Mitigation Strategy.
- Identify weather-sensitive industries, and work collaboratively with them to assess the economic impacts of severe weather; implement a framework under the U.S. Weather Research Program for increasing the benefit of severe weather forecasts to these industries.
- Complete a plan for national risk assessment that will guide U.S. planning for natural disaster avoidance and response.
- Develop and distribute improved hazard warnings, and increase the effectiveness of hazard warnings in ensuring human safety through mechanisms for stakeholder feedback (including policymakers, community planners, emergency response personnel, the general populace, and special populations).
- Develop new technologies and engineering techniques for the seismic safety of new and existing buildings and lifelines, and implement new guidelines to enhance public safety and building resilience.
- Develop an interagency system to provide real-time, accurate, and reliable observations of geomagnetic storms and solar wind, which can seriously compromise satellite operations and electric power delivery, and develop the capability to make 10-year geomagnetic storm forecasts.

Resource Use and Management: The goal of federal R&D on resource use and management is to promote the management, conservation, and use of natural resources in ways that sustain and enhance terrestrial and marine ecosystems and the quality of life. This broad goal has three subcomponents: (1) link research to resource management at various temporal and spatial scales; (2) develop the science base and the technologies for determining the mix of resources that will promote sustainability; and (3) determine how best to sustain and use a given resource across landscapes and the seascape.

- Define the protocols and techniques needed for integrated resource assessments.
- Establish an integrated data base that links socioeconomic factors to measures of natural resource conditions and trends.
- Define and implement site-specific and regional natural resource condition indicators that reflect the local and regional impacts of management activities.
- Provide new or modified methods and management systems, for both renewable and nonrenewable resources, that are cost effective and that minimize environmental damage associated with consumptive and non-consumptive uses.
- Determine ways to extend the service life of materials, or improve recycling technologies to reduce consumptive use of renewable and nonrenewable resources.

Toxic Substances and Hazardous and Solid Wastes: The goal of federal toxic substances and hazardous and solid waste research is to prevent or reduce human and ecological exposure to toxic materials, such as pesticide residues, PCBs, and lead, and their adverse consequences by providing the scientific and technical information needed for informed decision- and policymaking and effective problem solving.

- Produce a national research strategy on endocrine-disrupter chemicals.
- Finalize the reassessment of the health and ecological effects from exposure to dioxin and related compounds.
- Conduct cooperative research with industry partners to develop technological improvements to reduce inefficiency, substitute cleaner and less toxic chemicals, reduce costs, and improve environmental performance.
- Provide improved exposure models for hazardous air pollutants.
- Improve ecological risk characterization by better defining the responses of communities and ecosystems to toxic chemical stresses.
- Implement a national program to verify performance of innovative environmental technologies.

Water Resources and Coastal and Marine Environments: The goal of research on water resources and coastal and marine environments is to provide the scientific basis for managing water resources and aquatic environments to ensure adequate, quality water resources for domestic, industrial, agricultural, fishery, transportation, recreation, and other uses to meet equitably and efficiently the needs of present and future generations and to ensure the integrity, productivity, diversity, and vitality of lake, stream, estuary, and ocean coastal ecosystems.

- Complete mapping and change detection of coastal land cover for all major coastal areas of the United States, including the coordinated management and dissemination of the change-detection data sets and management applications derived from them.
- Complete data collection, interpretation, and report preparation for the first 20 National Water-Quality Assessment Program sites, and initiate detailed planning for the final 20 sites.
- Provide new, regional algorithms for remotely monitoring water mass movement, algal pigments, and productivity in coastal and estuarine water from satellites and aircraft.
- Complete a peer-reviewed, comprehensive national assessment of the U.S. coastal environment that integrates evaluations of the state of the natural environment with assessments of the effectiveness of current governance mechanisms and structures and the social and economic effects of environmental change.
- Provide improved assessment and field tools for predicting the cumulative effects of multiple stressors and carrying capacities in U.S. coastal and estuarine systems.

Measurement of Environmental Quality in the United States

N. PHILLIP ROSS, CARROLL CURTIS,
WILLIAM GARETZ, AND ELEANOR LEONARD
Office of Policy and Planning
Environmental Information and Statistical Division
U.S. Environmental Protection Agency

CONTENTS

The purpose of this paper is to provide an overview of environmental quality assessment activities in the United States. The topic is extremely broad and complex; this paper only touches on a limited number of issues and initiatives.

BACKGROUND

The rapid growth of the American society has brought with it the concomitant pollution of the environment. By the early 1960s it was apparent that additional government regulation was needed to deal with the growing levels of detectable anthropogenic pollution in our ambient environment. The United States Congress passed the National Environmental Protection Act (NEPA) in 1969 (42 U.S.C. 4341). This landmark legislation was the precursor to the creation of the U.S. Environmental Protection Agency. The NEPA required that the executive branch create the President's Council on Environmental Quality (CEQ) to formulate and recommend national policies to promote the improvement of the quality of the environment. Additional responsibilities were provided by the Environmental Quality Improvement Act of 1970 (42 U.S.C. 4371 et seq.).

The CEQ has statutory responsibility for overseeing the implementation of NEPA. The Council also develops and recommends to the President national policies that further environmental quality; performs continuing analysis of changes or trends in the national environment; reviews and appraises programs of the federal government to determine their contributions to sound environmental policy; conducts studies, research, and analyses relating to ecological systems and environmental quality; and assists the President in the preparation of the annual environmental quality report to Congress.

In its annual report, CEQ uses data obtained from a number of federal agencies to report on:

1. the status and condition of the major natural, man-made, or altered environmental classes of the nation, including, but not limited to, the air, the aquatic (including marine, estuarine, and fresh water) and the terrestrial environment (including, but not limited to, the forest, dry land, wetland, range, urban, suburban, and rural environment);
2. current and foreseeable trends in the quality, management, and utilization of such environments and the effects of those trends on the social, economic, and other requirements of the nation;
3. the adequacy of available natural resources for fulfilling human and economic requirements of the nation in light of expected population pressures;
4. a review of the programs and activities (including regulatory activities) of the federal government, the state and local governments, and nongovernmental entities or individuals with particular reference to their effect on the resources; and

5. a program for remedying the deficiencies of existing programs and activities, together with recommendations for legislation.

The Annual Report requires that CEQ obtain considerable support from other federal agencies with environmental responsibilities. CEQ coordinates this input through its Inter-agency Committee on Environmental Trends (ICET) which is co-chaired by CEQ and EPA. Even with the support of other agencies, the Annual Report does not provide a comprehensive annual state-of-the-environment picture, but it does provide a compilation of a selected set of analyses of environmental concerns focused towards policy interests in that year.

The United States is one of a very few countries in the world that does not produce a comprehensive publication on the state of the environment. The U.S. environmental community is presently moving in the direction of state-of-the-environment reporting through the development of a set of "environmental" indicators that will give a comprehensive picture (i.e., spatial) of the condition of the nation's environment and can also be used to evaluate temporal trends in environmental quality. These "environmental indicators" would be used much in the same way that we use economic indicators to assess the state of the economy and forecast economic trends.

The scientific community does not unanimously agree on what the best indicators of environmental quality should be. Unlike economic statistics in which the universe of concern is usually well defined (i.e., defined operationally by economists) and directly accessible via surveys and questionnaires, the environment does not provide such parameters. What is an ecosystem? What is the border of the wetlands? How do we assess air quality on a national scale, on a local scale? We cannot question the trees. We must design instruments that indirectly measure the parameters of interest. The collection of environmental data for inferential purposes is difficult and extremely expensive, and as such, not many data have been collected for purposes of describing the universe; most information is collected for purposes of compliance and enforcement. Only recently have we started to examine the impacts that our regulatory efforts have had on the quality of the ambient environment.

FRAMEWORKS FOR ORGANIZING ENVIRONMENTAL INDICATORS

Before undertaking the task of identifying environmental indicators for state-of-the-environment assessments, it is absolutely necessary to develop a framework with which to approach the process of selection and development. There are potentially thousands of environmental indicators. In order to develop relevant sets, some conceptual framework for a unified system of environmental information is necessary. Such a framework would provide the basis for identifying a set of environmental indicators (i.e., core set) that can be used to assess the quality of the nation's environment. The USEPA has developed a conceptual framework for environmental information.[1] The framework is based on a Pressure-State-Response model (PSR), which is presently used by the Organization for Economic Cooperation and Development (OECD), Canada, the Netherlands, and a number of other countries and NGOs. The following discussion is a extracted from the USEPA document:

Federal, state, local, and non-governmental organizations (NGOs) spend hundreds of millions of dollars each year on the collection, storage, and use of environmental data. Many of these data are collected for specific purposes and are not designed to be used for developing general measures of environmental quality. A framework will provide a structure for organizing this vast quantity of primary data into an integrated system of compatible spatial and temporal statistics, indices, etc., which can facilitate secondary uses of environmental information for indicators and decision making. The basic PSR framework was originally developed by the Organization for Economic Cooperation and Development's Group on the State of the Environment.[2] The PSR model asserts that human activity exerts *Pressure* (such as pollution emissions or land use changes) on the environment, which can produce changes in the *State* of the environment (for example, changes in ambient pollutant levels, habitat diversity, water flows). Society then *Responds* to changes in pressures or state with environmental and economic policies and programs intended to prevent, reduce, or mitigate pressures and/or environmental damage.[3]

Environmental quality is the reflection of the *State* component of the PSR/E model. Both pressure and responses impact environmental quality. State is the most difficult measure to obtain, and in many instances, measures of response or pressure are used as indicators of quality based on the underlying causal relationship of the PSR/E model. The *State* of the environment is concerned with ambient physical, chemical, biological, and ecological conditions; changes in ecosystem composition, structure, and function at various spatial and temporal scales (including the "built" environment); human health; and environment-related welfare.

The USEPA model builds on the base OECD PSR model in the following ways:

1. A derivative category called "Effects" is added, for attributed relationships between two or more Pressure, State, and/or Response variables, resulting in a "PSR/E" frrmework (Fi rure 1).

2. Human driving forces of environmental change, and pressures of non-human origin are also included in the framework. Distinctions are made in terms of specific subcategories in which the State of the environment can be measured, and the types of entities making Responses.

3. Each subcategory is elaborated with a generic menu designed to facilitate linking environmental information collection efforts to common sets of environmental values, goals, and priorities.

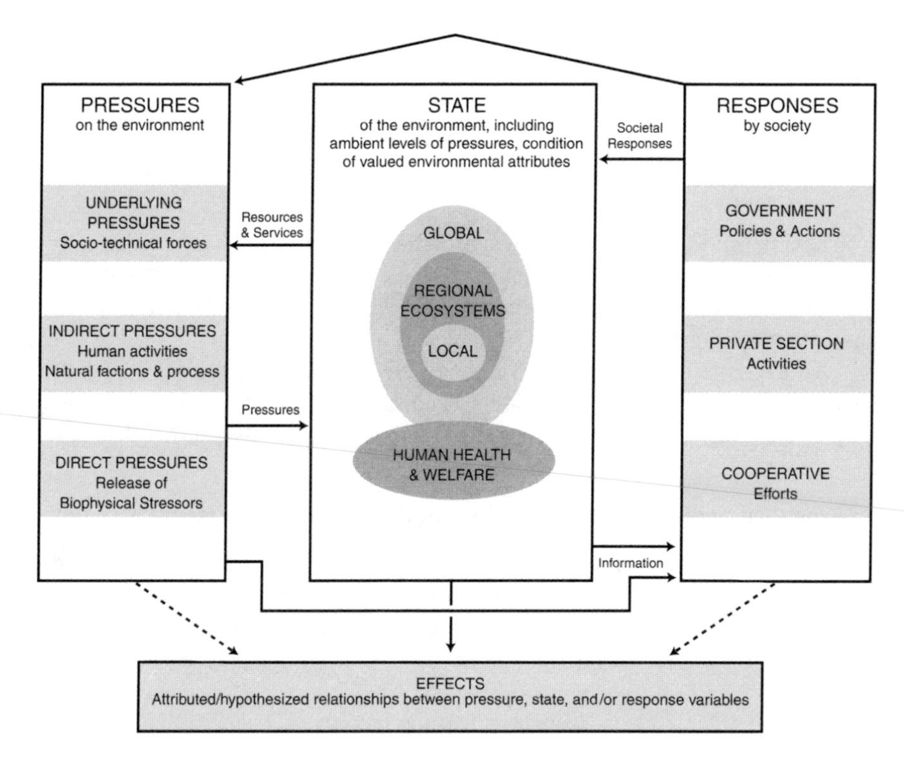

FIGURE 1 Pressure-State-Response/Effects (PSR/E) Framework.

4. The framework is consistent with a hierarchical view of ecosystems, allowing for the spatial nesting of environmental information, compatible with community- or ecosystem-(place-)based approaches to environmental management.
5. It is compatible with assessment-driven approaches to indicator selection.[4]

Indicators of Environmental Quality

In the last several years a number of organizations have been focusing on the development of environmental indicators that can be used to measure environmental quality, conditions, and trends. Like economic indicators (e.g., unemployment rates, cost-of-living index), environmental indicators hope to provide the public and decision makers with directional measures of change that will allow for a more informed public and improved environmental planning and decision making. There are many definitions for environmental indicators that appear in the literature, however the operational definition that we use in this paper is:

> An environmental indicator is an environmental or environmentally related variable or estimate, or an aggregation of such variables into an index, that is used in some decision making context:

- to show patterns or trends in the state of the environment;
- to show patterns or trends in the human activities that affect, or are affected by, the state of the environment;
- to show relationships among environmental variables; or
- to show relationships between human activities and the state of the environment.

This definition of an environmental indicator is purposely very broad to reflect the diversity of assessment and reporting contexts in the term as used. Thus the definition includes both measured or observed variables and composite indicators that aggregate a number of variables into a single quantity.

An example of a possible indicator of environmental quality is suspended sediment concentrations in the nation's rivers and streams. Figure 2 is a graphical display showing this measure as obtained from the USGS's NASQAN data base.

Suspended Sediment Concentrations in the Nation's Rivers and Streams

About 10 percent of NASQAN stations showed decreased suspended sediment concentrations over the sampling period 1980-1989. The quantity of suspended sediment transported to coastal waters decreased or remained the same in all but the North Atlantic region.

Indicator/Data Selection

Assessments of environmental quality can utilize either primary data collected specifically for the purpose for which they are used or secondary data that

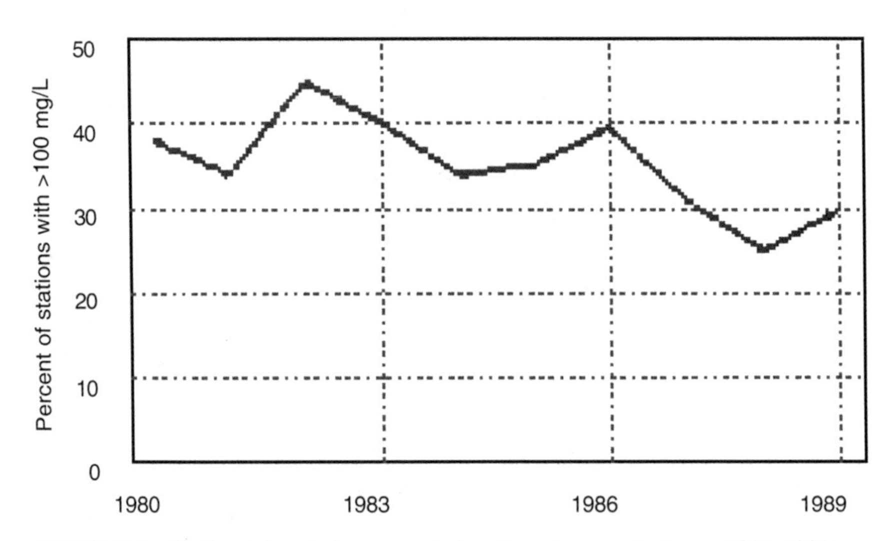

FIGURE 2 National trends in suspended sediment concentrations, 1980-1989.

were originally collected for other purposes. When basing assessments on existing data, analysts do not have the kind of control over factors such as data quality that they would if a new data collection effort were undertaken. In spite of these shortcomings, time and resource constraints frequently dictate the use of existing data. Various assumptions, models, and extrapolations are then applied in an effort to "adjust" the data so they can be used in a new assessment context (see, for example, Kineman[5]).[6] The USEPA has developed a draft proposal for a process to be used in selecting environmental indicators for a variety of purposes (policy development, program assessment, state of the environment, etc.). See Appendix A.

As discussed earlier, it is the state of the environment (SOE) that provides the most direct measure of environmental quality. Unfortunately, the U.S. does not collect a lot of data that will allow for easy development of SOE indicators. Tables 1–4 attempt to provide a comprehensive listing of *State* measures organized via the PSR/E model. Although the listings are not inclusive of all possible measures, they provide the focus necessary to define a "core" set of quality measures that can provide some assessment of the state of the environment in the context of the model.

As you can see from the tables, there is potentially a large number of indicators that one would want to have to make an overall assessment of environmental quality. It is critical that a core set be defined: a set of indicators that will provide decision makers and the public with the baseline information they need to manage the environment, and at the same time not bankrupt the system. Environmen-

TABLE 1 State of the Environment—Global Ecosystem

Valued Environmental Attributes (VEAs)

Stability of global climate: atmospheric composition, temperature, precipitation patterns, storms, droughts, ocean currents
Integrity of the stratospheric ozone layer
Global scale genetic and species diversity
Global environmental diversity
Biogeochemical cycling (and storage) of carbon, nitrogen, phosphorus, and other elements
Energy fixation/primary productivity
Topsoil quantity and quality
Management of species migration

Environmental Conditions and Changes of Human and Natural Origin

Atmospheric levels of greenhouse gases; ozone depleting substances
Global temperature
Global habitat alteration and destruction, including deforestation
Global levels of soil erosion/degradation
Globally transported pollutants in air or water (e.g., to polar regions)
Global changes in species occurrence and distribution
Proliferation of introduced (non-native) species

TABLE 2 State of the Environment—Regional Ecosystems

Valued Environmental Attributes (VEAs)

Regional genetic diversity, species diversity
Regional environmental diversity (i.e., types of habitat)
Biological integrity/health (e.g., Karrs Index of Biotic Integrity)
Primary productivity/energy fixation
Productive capacity of land for agriculture, forestry; soil quantity and quality (e.g., diversity of soil biota)
Air quality
Water quality
Productivity of valued plant or animal species
Stocks of nonrenewable resources: minerals, metals, fossil fuels, etc.
Hydrologic functions of landscapes: flood regulation; groundwater recharge; water supply; water filtration; river flows to support aquatic species, irrigation, recreation, transport, power
Geomorphological functions of landscapes: wind and wave buffering; erosion control; sediment retention
Stability of regional climate: precipitation, temperature, humidity, storms, etc.
Contaminant/pollutant detoxification, dilution, storage by media (including air, water, soil and sediments) and biota
Biogeochemical cycling, including eutrophication
Discrete landscape features valued for aesthetic, cultural, spiritual reasons: particular mountains, waterfalls, etc.
Habitat for wildlife, including migratory corridors
Natural pest control
Wilderness, open space

Conditions and Changes of Human and Natural Origin

Winds, ocean currents
Precipitation, flooding, droughts
Regional temperatures, humidity
Hurricanes, tornadoes, dust storms, other extreme weather events
Solar radiation; cloud cover
Glaciation, sea ice
Sea level
Landform geology; erosion, sedimentation, landslides and land subsidence, earthquakes, volcanic eruptions; soil types
Drainage basins; changes in river flows; groundwater depletion
Soil erosion, compaction, salinization, and other degradation
Import/export of soil, nutrients, etc., to/from ecosystems; various non-point source pollution
Regional ambient levels of pollutants in media; long-range transport of pollutants in air, water
Forest and grass fires
Ecosystem types, land cover/land use types (extent and spatial pattern)
Distribution of native species and communities; species loss, changes in species range
Feeding areas, habitats, migration routes of wildlife
Regional habitat destruction, fragmentation; succession/retrogression
Distribution, proliferation of exotic species, less desired native species, pests, disease vectors

TABLE 3 State of the Environment—Local Ecosystems

Valued Environmental Attributes (VEAs)

Safe drinking water (quantity and quantity)
Maintenance of hydrological and geomorphological functions (see regional menu)
Food safety (freedom from contaminants, undesired organisms)
Air quality (visibility, outdoor, indoor, workplace)
Pleasant climate (e.g., temperature, precipitation)
Tree cover
Natural control of pest and exotic (non-native) species
Pollination
Nutrient flows/cycles
Productivity of commercially, recreationally valued species
Local biodiversity and biotic integrity; healthy populations of local keystone and other desired species
Local environmental diversity
Proximity of homes to jobs, shopping, schools, parks, civic facilities
Access to local and regional transport (roads, public transport); safe routes for non-motorized traffic (sidewalks, bike paths)
Land availability for various uses: residential and commercial construction, agriculture, transportation corridors, parks, etc.
Utilities (electricity, communications networks, etc.)
Sanitation (disposal, treatment, recycling options)
Recreationally, aesthetically valued locations/sites/vista
Other aesthetically and culturally valued attributes
—Quiet
—Absence of noxious odors
—Cultural and historical sites and districts

Conditions and Changes of Human and Natural Origin

Quantity and distribution of land and water suitable for various human uses
Local climate
Pollutant levels, proliferation of disease vectors in air, water, soil, food
Proliferation of unwanted exotic species, less desired native species
Local habitat alteration/fragmentation, destruction
Trophic structure and functioning of ecosystems, including energy transfer, nutrient flows, etc.
Biological community structure: species diversity, niche structure, etc.
Condition of key species (individuals and populations), body burdens of chemicals; population size and dynamics
Extent and distribution of paved surfaces, etc.

TABLE 4 State of Human Health and Welfare VEAs

Human Health and Health-Related Economic Welfare

Longevity (i.e., avoidance of premature death)
Appropriate physiological function of body systems (i.e., avoidance of morbidity for each of the following systems):
—circulatory
—respiratory
—nervous
—digestive
—musculoskeletal
—endocrine
—immune
—reproductive systems, etc.
Psychological health (i.e., avoidance of unnecessary environmental stress)

Health-Related Economic/Welfare Values

—Adequate income
—Time for family, work, and leisure
Value of marketed environmental goods: crops, livestock, timber, fish, shellfish, fur-bearing animals, other species valued for use as food, pets, etc.
Non-animal commercial inputs (chemicals, fertilizer, peat, metals, minerals)
Fossil fuels
Livestock forage
Water supply for domestic consumption, agriculture, energy production, industrial/commercial uses, waste disposal.
Land and water use for human settlements, transport (e.g., navigation channels), etc.

Other Use Values

Recreation and tourism: camping, hiking, boating, swimming, sightseeing, photography, fishing, hunting, meditation, etc.
Other aesthetic values (e.g., scenic views in residential areas)
Scientific and research value

Non-Use Values

Existence value
Historical, cultural, heritage, and spiritual value
Bequest value
Intrinsic value
Scarcity/uniqueness value

Value of Ecosystem Services (marketed or not, see Tables 1-3)

tal data collection is extremely expensive, difficult, and time consuming. It is estimated that the USEPA spends in excess of $500 million a year on data collection, most of which is for enforcement and compliance data. Other federal agencies, as well as states, local governments, the regulated community, and environmentally focused NGOs, also spend significant dollars on environmental information collection.

ONGOING ENVIRONMENTAL QUALITY ASSESSMENT

As discussed earlier in the introduction, the CEQ has the responsibility for reporting to the President on the quality of the nation's environment each year. The CEQ publication *Environmental Quality* relies heavily on information to be input from several federal agencies covering a large variety of environmental areas and issues. Aside from this publication, there is no official U.S. publication (Note: except the U.S. national report for UNCED) that provides comprehensive information on the state of the U.S. environment. National environmental statistics are not collected in a centralized manner and there is no single source that one can use to assess the state and quality of the environment. The U.S. is one of a few countries in the world that does not have a centralized statistical system for collecting and analyzing environmental information and statistics. No single agency has the responsibility to provide information on the overall quality of the environment. Different federal and state organizations have focused responsibilities and produce a number or statistical summaries that can provide a limited picture of environmental quality, trends, and conditions.

The federal agencies that collect, publish, and disseminate environmental statistics are discussed below. For each agency a brief overview of the kind of information available is provided along with some examples of statistical summaries and graphics.

ENVIRONMENTAL RESPONSIBILITIES OF THE U.S. EXECUTIVE BRANCH

The executive branch of the U.S. government is responsible for developing environmental policy and implementing and enforcing federal environmental statutes. This responsibility is vested in the various executive offices, departments, independent agencies, and associated organizations (see Appendix B for more detail):

- *Council on Environmental Quality*—Formulates and recommends to the President national policies to promote the improvement of the quality of the environment and carries out other responsibilities as provided by NEPA.
- *Department of Agriculture*—Lead agency for natural resources and the environment, which includes the Forest Service and Soil Conservation Service.

These two services work to enhance the environment and maintain high production capacity by helping landowners protect the soil, water, forests, and other natural resources.

- *Department of Commerce*—The National Oceanic and Atmospheric Administration in Commerce collects data and works to improve the understanding and benefits of the earth's physical environment and oceanic resources.
- *Department of Energy*—As part of its mission to provide a framework for a comprehensive and balanced energy plan for the nation, the Department directs programs designed to increase the production and utilization of renewable energy and improve the energy efficiency of transportation, buildings, industrial systems, and related processes.
- *Department of Health and Human Services*—Plays a large role in defining and pursuing environmental health goals through research, administration, and service programs and provides assistance and support to state environmental departments and health agencies. The following institutions have responsibilities for environmental health activities: Centers for Disease Control and Prevention; National Center for Environmental Health; Agency for Toxic Substances and Disease Registry; Food and Drug Administration; National Institutes of Health, and the National Institute of Environmental Health Sciences.
- *Department of the Interior*—As the nation's principal conservation agency, the Department is responsible for most of the nationally owned public lands and natural resources. Its responsibilities include fostering sound use of land and water resources; assessing and protecting fish, wildlife, and biological diversity; preserving the environmental and cultural values of national parks and historic places; and providing for outdoor recreation. The Department assesses mineral resources and works to ensure that their development is in the best interest of the nation by encouraging stewardship and citizen participation in their care. The Department also has a major responsibility for Native American communities and for people who live in island territories under U.S. administration. Lead agencies for the environment are the United States Fish and Wildlife Service, National Park Service, National Biological Survey, United States Geological Survey, and Bureau of Land Management.
- *Department of Justice*—Through its Environment and Natural Resources Division, the Department enforces civil and criminal environmental laws in order to protect citizens' health and the environment. It defends environmental challenges to government programs and activities; and represents the United States in all matters concerning the protection, use, and development of the nation's natural resources and public lands, wildlife protection, Native American rights and claims, and the acquisition of federal property.
- *Department of Labor*—Through its Occupational Safety and Health Administration, the Department administers a variety of federal laws guaranteeing workers' rights to safe and healthful working conditions.
- *Department of State*—Through its Bureau of Oceans and International

Environmental and Scientific Affairs, the Department formulates proposals and implements U.S. policy on international issues and significant global problems related to environment, oceans, fisheries, population, and space and other fields of advanced technology.

- *Department of Transportation*—Environmental responsibilities include environmental impact assessments, and analyses of current and emerging transportation issues related to energy and the environment. It also enforces various laws related to transportation and discharge of oil and hazardous materials.
- *Environmental Protection Agency*—The mission of this independent agency is to control and abate pollution in the areas of air, water, solid waste, pesticides, radiation, and toxic substances. Its mandate is to mount an integrated, coordinated attack on environmental pollution in cooperation with state and local governments.

Several independent commissions and government corporations also share some of the responsibility for our environment. Chief among these are the Federal Emergency Management Agency (emergency planning, preparedness, mitigation, response, and recovery from natural disasters and human-caused emergencies); National Aeronautics and Space Administration (studies global climate change and integrated functioning of the earth as a system); National Science Foundation (supports research for improved understanding of the fundamental laws of nature); Nuclear Regulatory Commission (licenses and regulates civilian use of nuclear energy to protect public health and safety and the environment); Federal Energy Regulatory Commission (transmission and pricing of electricity and hydroelectric licensing); U.S. Information Agency (encourages international discussion and cooperation on fundamental concerns, including the global environment); Marine Mammal Commission (protection and conservation of marine mammals); Tennessee Valley Authority (in support of its mission to advance economic growth in the Tennessee Valley, conducts research and development programs in forestry, fish and game, and watershed protection).

WHAT INITIATIVES ARE UNDER WAY FOR THE FUTURE?

Environmental Strategies for the 1990s and Beyond

The United States is entering a new phase in the evolution of environmental protection, one that recognizes that effective environmental protection and control involve more subtle and complex variables than have been considered under earlier, centralized, command-and-control approaches. It also one that emphasizes the positive relationship between a healthy environment and a prosperous economy. The following strategies embrace these concepts:

Ecosystem Approach to Resource Management. Public concern about the environment, together with new thinking by scientists and resource managers, has led to a new philosophy about how to manage resources in the United States. This philosophy says that we can manage resources to sustain their full array of values and uses through a broader understanding of their associated ecosystems. This approach calls for a shift of focus from more traditional single-resource, single-species management to a collaborative, developed, holistic approach that integrates ecological, economic, and social factors affecting a management unit defined by ecological, not political, boundaries. This approach requires knowledge of the composition, structure, and function of ecosystems, their relationships and influences on each other, and their capacity to support multiple uses and to produce goods and services for society without sacrificing health, sustainability, or biodiversity.

The elements of an ecosystem approach are not new, but their endorsement by the federal government represents an advance in coordinated resource management. Much of the genesis for the evolving federal approach to ecosystem management came from the Administration's 1993 Report of the National Performance Review, which strongly supported the concept of cross-agency ecosystem planning and management and led to the establishment of an Interagency Ecosystem Management Task Force to develop principles and guidelines for ecosystem sustainability.

In addition, statutes such as the Forest and Rangeland Renewable Resources Planning Act, the National Forest Management Act, and the Federal Land Policy and Management Act, which outline various procedures to follow in federal public land planning, also authorize the employment of principles intrinsic to ecosystem management. For example, they call for planning to be interdisciplinary, coordinated among agencies, and based on available science. The National Forest Management Act explicitly directs that the diversity of plant and animal species be considered in planning. Moreover, the Endangered Species Act directs the Secretary of the Interior, and the Secretary of Agriculture with respect to National Forest System lands, to establish and implement a program to conserve fish, wildlife, and plants, including those listed as threatened or endangered.

There are many additional examples of ecosystem-based management being practiced or planned at a variety of geographic scales and by a broad range of agencies and cooperators. Many of these are described in a recently released Congressional Research Service (CRS) Report for Congress entitled *Ecosystem Activities: Federal Agency Activities*.[7]

In recognizing that techniques for applying an ecosystem approach may vary according to the natural resource issues and ecological systems involved, several ongoing ecosystem activities were selected by the Interagency Ecosystem Management Task Force in 1994 as case studies. These "survey and assist" ecosystems (Anacostia River Watershed; Great Lakes; Pacific Northwest forests; Prince William Sound; South Florida; Southern Appalachian Highlands; and Southern Louisiana wetlands) were selected to learn from the experience of those implementing an ecosystem approach and to help determine what the Task Force or

Congress can do to support efforts in the field and facilitate more effective performance by the federal agencies. In addition, the Great Plains, Mojave Desert, and Monterey Bay ecosystems were selected as "new initiative laboratories" to employ a collaborative process in documenting historical ecosystems, developing a vision of the range of desired future conditions, and considering how current stakeholders will address key concerns. This work is in progress.

Sustainable Development. "Sustainable development" is broadly defined as economic growth that will benefit present and future generations without detrimentally affecting the resources or biological systems of the planet. In many respects, sustainable development is a corollary of ecosystem management—management approach that integrates ecological, economic, and social factors in restoring and/or maintaining the health, sustainability, and native biological diversity of ecosystems to support human communities and their economic base. Sustainable development operates on the tenet that the economy and the environment are inextricably linked; that an economy will not remain healthy if renewable resources are consumed faster than they can be replenished, or non-renewable resources faster than substitutes can be developed.

In 1993, the President's Council on Sustainable Development was established by Executive Order 12852 to develop a national sustainable development strategy. The 25-member Council brings together leaders from industry; government; and environmental, labor, and civil rights organizations. It is charged with developing bold new approaches to integrate economic and environmental policies.

In 1994, the Council issued a draft Vision Statement in which it states that

> . . . a sustainable United States will have an economy that equitably provides opportunities for satisfying livelihoods and a safe, healthy, high quality of life for current and future generations. Our nation will protect its environment, its natural resource base, and the functions and viability of natural systems on which all life depends.

The Council also issued draft Principles for Sustainable Development related to the vision described above. A final report of the Council was issued in the fall of 1995 after the Forum took place.

To support the activities of the Council, an Interagency Working Group on Sustainable Development Indicators was initiated in late 1993 to encourage cooperation among federal agencies in the creation of indicators for sustainable development. This effort also supports the recommendations of Agenda 21 from UNCED that sustainable development indicators be developed at the national level in a holistic fashion.

Sustainable development indicators are numerical measures that indicate the extent to which the current path of economic development is affecting the following three basic types of resources on which the well-being of both present and future generations depends:

1. renewable and non-renewable resources that directly support economic activity;

2. environmental resources—the reservoirs of air, water, and land into which wastes from human activity are discharged and through which humans and other living things are exposed to waste products; and

3. ecological resources—the plants and animals and their habitats that make up the biosphere and the processes by which they interact and evolve, providing in the process both essential support for human life and many valuable goods and experiences.

Patterned somewhat after widely used economic indicators, sustainable development indicators are being designed to measure performance in relation to sustainability goals and policies, and to provide the feedback needed to promote improvement of management techniques in a timely manner. Work in this area is progressing.

Environmental Goal Setting. EPA has launched a project to produce a set of realistic and measurable U.S. environmental goals to be achieved early in the next century. To generate broad national input into the process, EPA sponsored a series of public meetings around the country in 1994. Participation by other federal, state, and local government agencies is also being sought, and the project is being coordinated with the President's Council on Sustainable Development. The effort also complements the efforts of other countries, such as Canada, Norway, New Zealand, and the Netherlands, in developing goals and plans for sustainable development.

Ultimately, the EPA goals will contain three tiers of measurable targets. Tier 1 goals will specify a condition of the environment that the nation is seeking to achieve by a certain year. Tier 2 objectives will specify reductions in pollutant loadings or other source-related causes that must be achieved to reach a Tier 1 goal. Tier 3 "action targets" will identify the specific work that EPA and others must complete to accomplish the overall goals.

The three tiers of goals will provide direction for the design of more effective, efficient programs to fulfill national priorities. Together with congressional mandates, the goals will drive EPA's planning, management, and budget.

Pollution Prevention. There is a growing realization that in many instances it is far cheaper to prevent pollution than to clean it up. In the Pollution Prevention Act of 1990, Congress stated a national policy that "pollution should be prevented or reduced at the source whenever feasible" and that "disposal and other releases into the environment should be employed only as a last resort and should be conducted in an environmentally safe manner." This statute also requires development of a national source reduction strategy, and it calls on the states to promote the use of source reduction techniques by businesses, which some states are already doing.

Since Earth Day 1993, a number of new initiatives and executive orders have been implemented to establish the federal government as a leader in advancing pollution prevention. Most notably, federal facilities are now required to develop written pollution reduction strategies incorporating source reduction in facility management and acquisition programs. Each agency must begin immediately to minimize the acquisition of the most potent (Class I) ozone-depleting substances and to maximize the use of safe alternatives. Similarly, agencies must also establish a plan and goals for eliminating or reducing the unnecessary acquisition of products containing extremely hazardous substances or toxic chemicals. Federal facilities that manufacture, process, or use toxic chemicals are now required to reduce toxic emissions and report publicly on toxic wastes and releases under the Emergency Planning and Community Right-to-Know Act. Energy efficiency in the workplace will be enhanced by government purchase of Energy Star computer equipment (which saves energy by automatically entering a low-power, standby state when inactive) and other energy-efficient products. Also toward this end, agencies are now required to set goals of reducing energy consumption, increasing energy efficiency, auditing their facilities for energy and water use, increasing the use of solar and other renewable energy sources, designating "showcase" facilities, and minimizing the use of petroleum-based fuels. In addition, federal agencies must implement affirmative acquisition programs for products less harmful to the environment when possible, including alternative-fueled vehicles and products containing pre- and post-consumer recycled materials. These and other federal initiatives will help augment the importance of adopting pollution prevention principles at every level of government and throughout the private sector. They are also aimed at encouraging new technologies and building markets for environmentally preferable and recycled products.

Enhanced Regional and International Environmental Cooperation

The United States has taken several steps toward greater cooperation on many of the most pressing environmental challenges facing the world. In addition to its work on a U.S. sustainable development plan, the United States has fulfilled several other commitments that grew out of the 1992 United Nations Conference on Environment and Development (UNCED), also known as the Earth Summit. At the Earth Summit, the United States joined other countries in signing the Framework Convention on Climate Change, an international agreement whose ultimate objective is to achieve stabilization of greenhouse gas concentrations in the atmosphere at a level that would prevent dangerous anthropogenic interference with the climate system. Since then, the United States has released the "Climate Change Action Plan" (1993), which details the initial U.S. response to climate change, and the "Climate Action Report" (1994), which describes the current U.S. program and represents the first formal U.S. communication under the Framework Convention on Climate Change. In June 1993, the United States signed the Convention on Biological Diversity, indicating its commitment to help stem the loss of the earth's species, their habitats, and ecosystems. The United States has committed to provide $430 million over the next four fiscal years toward the replenishment of the recently restructured Global Environmental Facility (GEF), which will act as the interim institutional mechanism for both the biodiversity and the climate change conventions.

The United States has also succeeded in getting regional and international trading systems to begin to address environmental issues. The North American Free Trade Agreement (NAFTA), which went into effect in 1994, expressly endorses the principle of sustainable development and includes environmentally positive provisions on dispute settlement and investment. In 1993, the United States joined the governments of Canada and Mexico in signing the North American Agreement on Environmental Cooperation, an integral part of NAFTA. The objectives of the Environmental Agreement are to promote improved environmental conditions throughout North America and to improve national enforcement of laws relating to environmental protection. It also provides for monitoring the environmental effects of the NAFTA. In addition, the United States reached agreement with Mexico for two new institutions devoted to environmental improvement in the border area: the Border Environment Cooperation Commission, which will work with local communities to develop and arrange financing for vitally needed environmental infrastructure projects, and the North American Development Bank, which will use 90 percent of its capital—to be contributed equally by the United States and Mexico—to leverage private funds in order to finance the construction of border environmental projects through bonds and other instruments.

The United States demonstrated its commitment to integrating environmental protection and international trade by endorsing the establishment of a Committee on Trade and Environment as part of the new World Trade Organization (WTO) under the General Agreement on Tariffs and Trade (GATT). The Committee, with powers and functions equivalent to those of other standing committees established in the WTO, will have broad terms of reference that include making recommendations on changes to existing trade rules. It can recommend changes in trade rules to govern the use of trade measures for environmental purposes, and to safeguard environmental agreements. It can also examine how the trading system should treat environmental packaging and labeling, and what rules should govern the use of taxes, such as energy taxes, for environmental purposes. With the creation of this committee, the trading system can now become fully involved in promoting sustainable development.

NOTES

1. USEPA, Environmental Statistics & Information Division, "A Conceptual Framework to Support Development and Use of Environmental Information in Decision-Making", EPA 239-R-95-012, April 1995.

2. Adriaanse, Albert, *Environmental Policy Performance Indicators*, p. 11, April, 1993.

3. Ibid 2, page 6.

4. Ibid 2, page v.

5. Kineman, J.J., 1993, "What is a Scientific Database? Design Considerations for Global Characterization in the NOAA-EPA Global Ecosystems Database Project", in *Environmental Modeling with GIS*, M.F. Goodchild, B.O. Parks, and L.T. Steyaert, eds. Oxford University Press, 1993, New York.

6. Ibid 2, page 20.

7. CRS (Congressional Research Service).

APPENDIX A

Process for Selecting Indicators and Data and Filling Information Gaps: Criteria for Selecting Indicators

In selecting environmental indicators, it is important to have clear selection criteria. Previously developed criteria are available from several sources including the Environmental Monitoring and Assessment Program (EMAP), the International Joint Commission for the Great Lakes (IJC), and the Intergovernmental Task Force on Water Quality Monitoring (ITFM). The choice of selection criteria depends in part on the intended use of the indicators. Therefore, the list of criteria suggested here, in Table A-1, has been adapted from other sources (primarily the ITFM criteria).

The selection criteria are grouped based on considerations of validity, interpretability, timeliness, understandability, and cost considerations. These considerations include the following:

- Indicators should be valid measures of the valued attribute. Validity is a qualitative association between the concept embodied in the value and the measurable quantity represented by the indicator. Validity is established if there is a close scientific or logical link between the indicator and the value. Three factors are listed in Table A-1 that contribute to a close logical link between the indicator and the valued attribute. First, indicators that have wide scope provide a balanced measure of the valued attribute. Second, indicators that respond to the cumulative effect of multiple stressors will be more generally applicable than those that are responsive to only a few stressors. Third, indicators that are highly correlated with other measures of the valued attribute will be generally applicable to the environmental system being measured. Indicators must be sensitive enough to measure changes over a reasonable time but not so sensitive that they fluctuate substantially between time periods. The signal-to-noise ratio for an indicator is in part determined by the data used to assess the indicator. Expert knowledge and peer review can be used to assess the sensitivity of different indicators.
- Indicators should be interpretable in terms of the end point in the assessment process. They should be able to distinguish unacceptable from acceptable environmental conditions.
- Timely indicators that anticipate future changes in the environment are preferred over those that are not anticipatory. To the extent that an indicator does not anticipate future conditions, the indicator with the least time lag would be preferred. The time lag depends on both the characteristics of the indicator and the time lag between the data collection and when the data are available to calculate the indicator.
- Indicators should be understandable by the public and perceived as relevant. Understandability is in part a characteristic of the indicator and in part a function of how the indicator is presented. EPA may need to educate the public

TABLE A-1 Criteria for Selecting Indicators

Criterion	Explanation
Validity	
Scientific Relevance to the Valued Attribute[a]	Scientific theory links the indicator to societal or environmental values.
Scope/Applicability	The indicator responds to changes on an appropriate geographic (e.g., national or regional) and temporal (e.g., yearly) scale.
Integrates Effects/ Exposures	The indicator integrates effects or exposure over time and space and responds to the cumulative effects of multiple stressors. It is broadly applicable to many stressors and sites.
Representative	Changes in the indicator are highly correlated with other measures of the valued attribute.
Signal-to-Noise Ratio	The indicator is able to distinguish meaningful differences in environmental conditions with an acceptable degree of resolution.
Interpretability	
Interpretable[a]	There is a reference condition or benchmark against which to measure changes and trends. The indicator can distinguish acceptable conditions in a scientifically defensible way.
Comparability	The indicator can be compared to existing and past measures of conditions to define trends and variation.
Timeliness	
Timely/Anticipatory	The indicator provides early warning of changes.
Understandability	
Understandable[a]	Indicator is in, or can be transformed into, a format that is understandable to the target audience.
Perceived Relevance to the User	The measured quantity is seen by the audience as being important or relevant to their lives.
Cost Effectiveness	
Cost Effectiveness	Information is available or can be obtained with reasonable cost and effort. Provides maximum information per unit effort.
Minimal Environmental Impact	Sampling produces minimal environmental impact.

[a]Indicates critical criteria.

on the importance of some indicators. If possible, indicators should be "attention grabbers" in that they communicate to the audience why the value is important, e.g., information on the number of fish is generally more interesting to the public than data on macroinvertebrates in the food chain. Keeping data presentations simple, graphic, and consistent will help. When there is uncertainty as to how an indicator will be understood, the use of focus groups may help EPA to understand how the public perceives the indicator and to provide guidance on improvements to the indicator.

- Finally, indicators should be cost effective relative to alternatives, and to the effort and expertise to collect the data, if required, and monitor the indicator over time.

CRITERIA FOR SELECTING EXISTING DATA SETS TO QUANTIFY INDICATORS

Table A-2 sets forth proposed criteria for evaluating the usefulness of a data set for quantifying an indicator. Critical criteria for selecting data sets would include the availability of data on the selected parameters, appropriate temporal and spatial coverage, documented quality, and accessibility. Because changes in the data collection procedures might affect the technical credibility, the magnitude of the estimation error (and the associated sample size), and the cost, another critical criterion for consideration of a data set is that minimal standards of technical credibility, estimation precision, and cost can be achieved by either the present data collection procedures or reasonable modifications.

It is likely that either sampling procedures or laboratory analysis procedures will change over the time that a data source is used to quantify an indicator and monitor progress. These changes will result from advancements in technology and changes in budgets and uses of the data sets over time. The effect of these changes can be minimized by using (1) measurements for which changes in technology are likely to improve the precision but not affect the measurement bias and (2) procedures for which the measurement bias is relatively insensitive to the magnitude of the collection effort. To the extent that this cannot be achieved, a comparability study can be used to compare the indicator before and after the change. The value of both the original and the revised indicator can be used for some time to provide information on how the two indicator compare. This same procedure can be used if a entirely new data set is used for the revised indicator.

PROCESS FOR SELECTING INDICATORS AND DATA

The proposed process for selecting indicators and data and filling information gaps consists of four basic steps. Characteristics of how these four steps interrelate is provided in Figure 1 on p. 139. These four steps are discussed in more detail below:

TABLE A-2 Criteria for Selecting Existing Data Sets to Quantify Indicators

Criterion	Explanation
Availability of Data[a]	Data set provides measurements of the parameter(s) or variable(s) specified in the indicator.
Appropriate Temporal Coverage[a]	At a minimum, information should be available for the present and for future years. In addition, temporal coverage within reporting cycles (usually annually) may have gaps but should not exclude data that will significantly affect the indicator.
Appropriate Spatial Coverage[a]	Information should be available on a national (regional) basis for a national (regional) program, or, if the information is compiled from local or regional data, the information will need to be aggregated using scientifically and statistically valid procedures.
Documented Quality	The information should be of known quality, i.e., there should be (1) documented QA/QC procedures for the collection, analysis and presentation of data; (2) documentation of any deviations from the procedures; and (3) quantitative information on both sampling and non-sampling errors.
Accessibility[a]	The information should be retrievable and analyzable using existing data retrieval and analysis procedures. EPA would not be prohibited from using the data due to confidentiality concerns, etc.
Technical Credibility	The procedures used to manage and analyze the data should follow accepted professional practices. In addition, the sample and data collection procedures should not be inconsistent with the use of the data as a measure of the indicator, as judged by technical experts in the field who are familiar with the data. The calculated bias in the indicator should be insensitive to the magnitude of the data collection effort and to political pressures. In general, this criterion will eliminate self-reported data from consideration.
Acceptable Estimation Error	The precision and bias of the indicator should be acceptable given the desired precision specified by the program.
Acceptable Cost	Cost of data collection, management, and analysis are within programmatic guidelines.

[a]Indicates critical criteria.

Step 1: Identify and Recommend Indicators for Reporting

Step 1a: Identify possible indicators. Compile two lists of candidate environmental indicators. The first list includes indicators developed to measure valued attributes, without any consideration for the availability of data. The second list includes currently used indicators. In many cases these indicators are meaningful and informative summaries constructed from available data. Many indicators will appear on both lists.

Step 1b: Score each indicator, using the criteria in Table A-1, to identify candidate indicators. Review each indicator in the two lists and select a range of possible indicators for further consideration. Indicators will be rejected if they do not satisfy all of the critical criteria shown in Table A-1. Of the remaining indicators, use a combination of peer review, literature review, and expert knowledge to select the canhidate indicators for further consideration. The availability of data would be used as a criterion only for selecting between otherwise similar indicators.

Make a preliminary determination (using the descriptions of each indicator and the short forms for each data set) of which data sets are useful for each indicator. (See step 3a below.) This will provide additional information for each indicator on what data may be available.

Step 1c: Review the values, the information about the indicators, the criteria for selecting indicators, and the data available for each indicator (as a secondary criterion) to select proposed indicators. The process will use a combination of peer review, literature review, and expert knowledge to select the proposed indicators.

The process of making a preliminary choice, gathering more information, and making a more refined choice of indicators is iterative. Additional iterations may be necessary to refine the selection of indicators and to incorporate new information as it is gathered.

Step 2: Inventory and Describe Existing Data Sets That May Be Suitable for Quantifying Indicators

This step consists of (1) an ongoing inventory of existing data held by EPA, other federal agencies, and other groups that may be suitable for use in reporting and (2) summarizing information about the data sets to facilitate the use of appropriate data sets to measure the selected indicators. This step is broken into three substeps shown in Figure A-1 and described below.

Step 2a: Inventory existing data sets for their potential suitability for quantifying the candidate indicators. Prepare an assessment of existing data sources potentially available for reporting. Continually revise the assessment as new information is gathered through discussions with other federal agencies, non-government organizations, etc.

Step 2b: Complete a preliminary characterization of data sets potentially suitable for the candidate indicators. After the candidate indicators have been selected, conduct a preliminary assessment of each data set that might potentially be used to assess one of the candidate indicators. To do this, prepare a "short form" to screen each data set. The "short form" will summarize the most important information needed to decide if a data set is potentially appropriate for a selected indicator. Two draft short forms appear on pages 172 through 178.

Step 2c: Complete a detailed characterization of candidate data sets. After an initial selection of data sets that might be appropriate for the preliminary selection of indicators (see step 3a), do a more extensive examination for each data set that might be useful for the candidate indicators. For example, a form provides additional details beyond that provided by the short form if it can be used to determine if a data set (1) provides adequate data for a selected indicator or (2) provides data that, if augmented or modified, can be made adequate.

Step 3: Identify and Recommend Information/ Data for Selected Indicators

Once the indicators to be used in reporting have been selected, the next step is to select the information/data to be used to quantify the indicators. This requires examining existing data collection and analysis programs to determine if appropriate information is or will be available. To the extent that characteristics of the data collection procedures affect the evaluation of the criteria for selecting indicators—validity, interpretability, sensitivity, timeliness, understandability, and cost effectiveness—the program should evaluate these criteria in light of the proposed data set. The process will be accomplished in the following two steps:

Step 3a: Identify candidate data sets for each indicator. Use the information from the short form, published EPA documentation, and other sources to identify data sets that might be appropriate for use with each indicator. Several data sets might be appropriate for use with an indicator. If several data sets could be used, all would be considered unless one or more were clearly inferior to the others (i.e., being similar on most criteria but clearly worse on some). Additional information would be collected on all data sets that were being considered for any of the candidate indicators (step 2c). In some cases it may not be possible to identify any data that might be appropriate.

Step 3b: Identify proposed data sets for each indicator, if possible. Use additional information to identify data sets that either are appropriate for use with each indicator or, if not, could be made acceptable with additional data collection or changes in procedures. If several data sets are appropriate for use with an indicator, the best one would generally be chosen. In some cases it might not be possible to identify any data that are appropriate. For those indicators where the available data are either inadequate and can be improved or are not available, a data gap exists.

Step 4: Fill Information Gaps

For indicators that lack adequate information:

1. Document the information gap.
2. Review existing indicators and data to see if some can be used as interim indicators to at least provide some information on possible progress.
3. Develop strategies for filling information gaps, including improvements to existing programs in data collection, data analyses, and information management. Developing strategies includes determining if data can be made available by modifying existing data management and analysis procedures. For example, this could include the reanalysis of existing data or the integration and harmonization of two or more separate data sets.

a. If the information can be made available by changes in existing data management or data analysis procedures, develop a strategy for making the needed changes.

b. If the information cannot be made available by changes to existing data management or data analysis procedures, determine if there are validated test methods, statistical methods, etc., at the levels of accuracy and levels of reliability required:
   —For each indicator with validated methods, identify the type of data required (including statistical design) and design a data collection and data analysis program. If feasible, implement the program.
   —For each indicator without validated methods, set up a process to develop these. If needed, set priorities for developing these methods. Once appropriate methods are developed, identify the type of data required (including statistical design and data analysis) and design a data collection and analysis program. If feasible, implement the program.

APPENDIX B

DEMONSTRATION GRAPHICS FOR STATE OF THE ENVIRONMENT REPORTING

I. Examples of Graphics Used in Air Quality Status and Trends Reports

A. Ambient Air Quality. Graphics from the EPA, OPPE, ESID, Compendium of Selected National Environmental Statistics in the U.S. Government (Draft: 8/24/94 with updates by agency).

Ambient concentrations of sulfur dioxide, 1984-1993

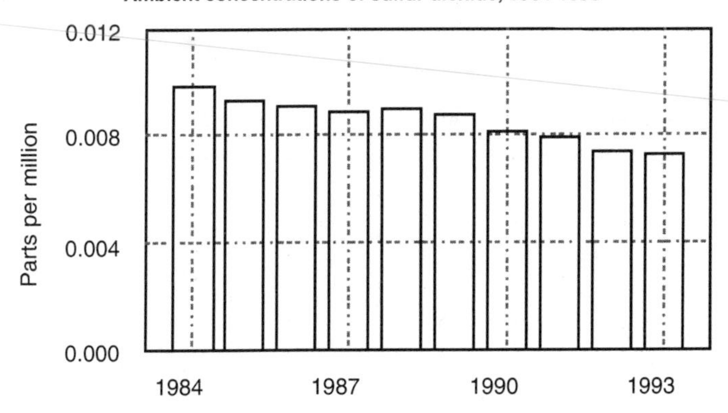

Source: EPA/OAQPS. National Air Quality and Emissions Trends Report, 1993.

Ambient concentrations of carbon monoxide, 1984-1993

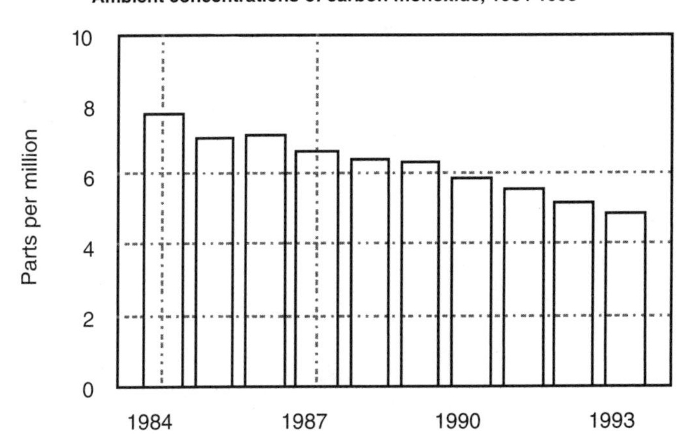

Source: EPA/OAQPS. National Air Quality and Emissions Trends Report, 1993.

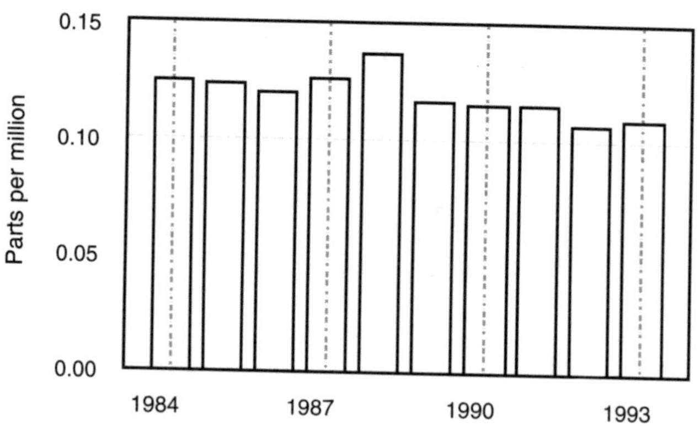

Source: EPA/OAQPS. National Air Quality and Emissions Trends Report, 1993.

Ambient concentrations of nitrogen dioxide, 1984-1993

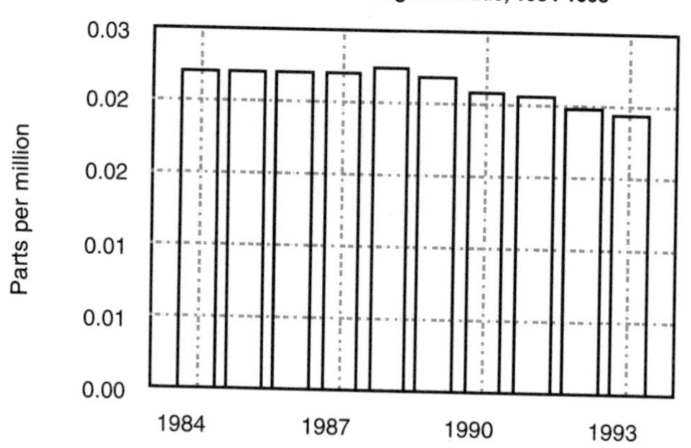

Source: EPA/OAQPS. National Air Quality and Emissions Trends Report, 1993.

Ambient concentrations of lead, 1984-1993

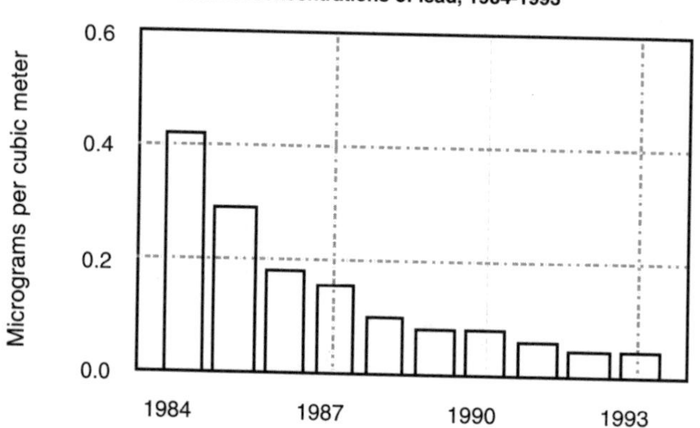

Source: EPA/OAQPS. National Air Quality and Emissions Trends Report, 1993.

B. Metropolitan Areas Not Meeting National Ambient Air Quality Standards (NAAQS). Graphic from EPA, OPPE, Proposed Environmental Goals For America With Benchmarks For The Year 2005: Summary (Draft For Government Agencies' Review, February 1995).

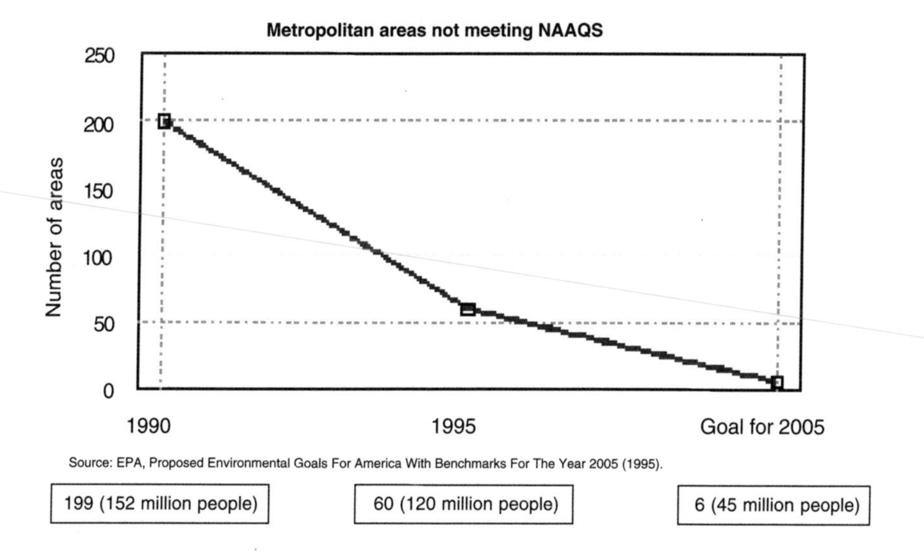

C. Population Living Where Air Quality Fails to Meet National Ambient Air Quality Standards (NAAQS). Graphic from the EPA, OPPE, ESID, Compendium of Selected National Environmental Statistics in the U.S. Government (Draft: 8/24/94 with updates by agency).

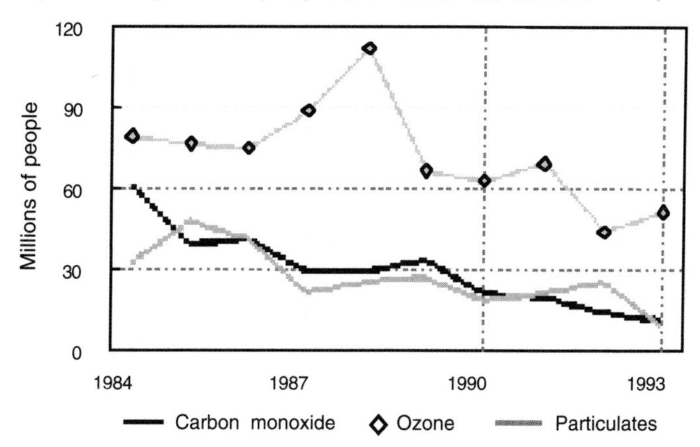

Source: EPA, OAQPS. National Air Quality and Emissions Reports.

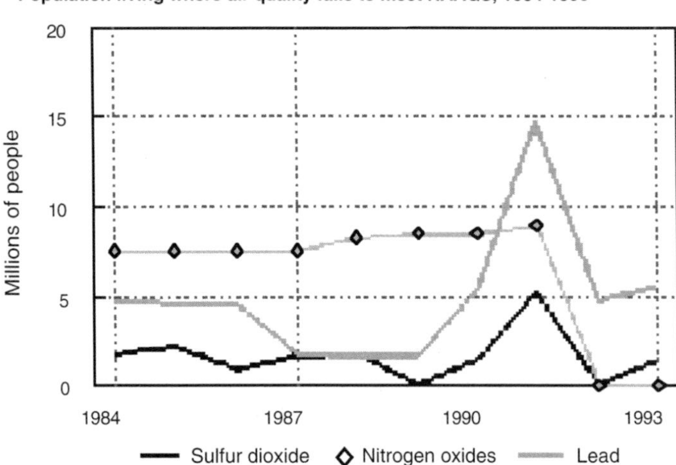

Source: EPA, OAQPS. National Air Quality and Emissions Reports.

II. Examples of Graphics Used as Indicators of Stratospheric Ozone Conditions

A. U.S. Production of Selected Ozone-Depleting Chemicals. Graphic from EPA, OPPE & OAR and World Resources Institute, Protection of the Ozone Layer (Draft Environmental Indicators Bulletin, May 1995).

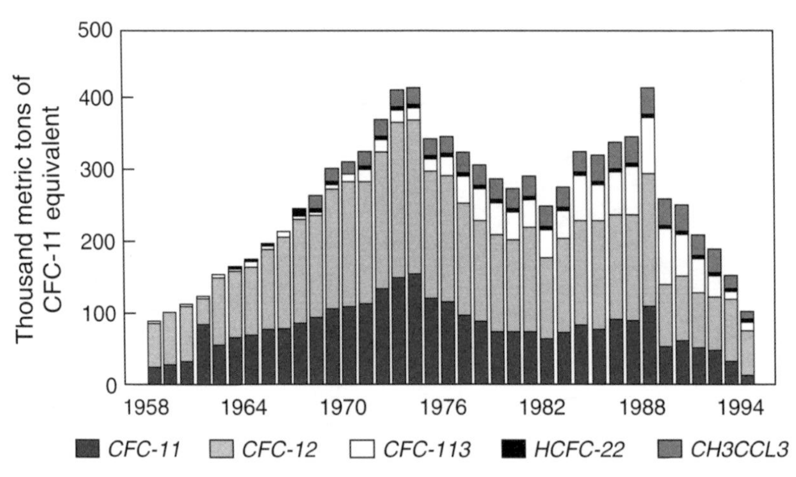

Source: U.S. International Trade Commission.

B. Cumulative CFC Production for the United States and the Rest of the World. Graphic from EPA, OPPE & OAR and World Resources Institute, Protection of the Ozone Layer (Draft Environmental Indicators Bulletin, May 1995).

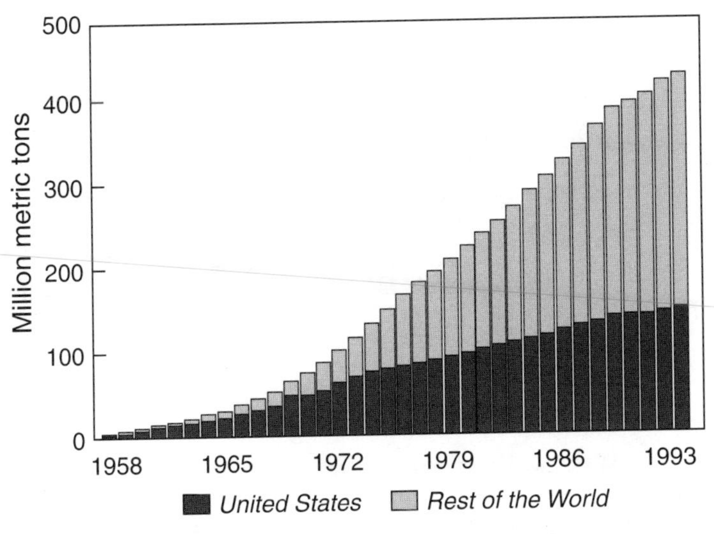

Source: U.S. International Trade Commission and AFEAS.

C. Stratospheric Ozone Decline and Recovery. Graphic from EPA, OPPE, Proposed Environmental Goals For America With Benchmarks For The Year 2005: Summary (Draft For Government Agencies' Review, February 1995).

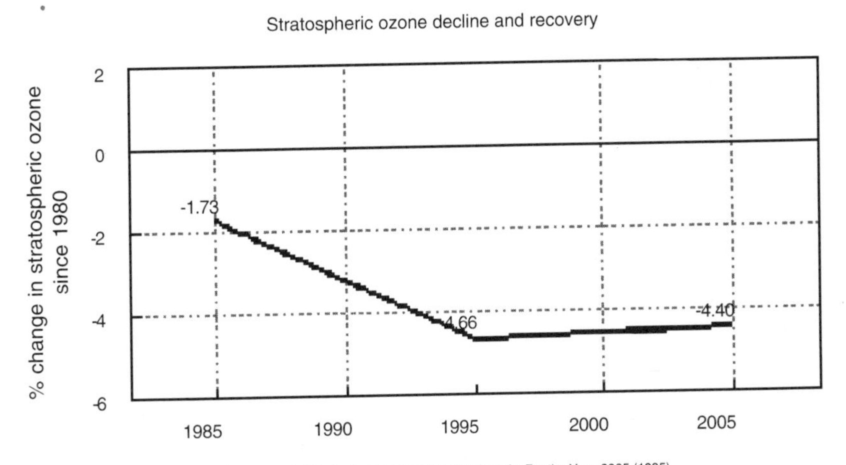

Source: EPA, Proposed Environmental Goals For America With Benchmarks For the Year 2005 (1995).

III. Examples of Graphics Used in Water Quality Status and Trends Reports

A. Stream Water Quality, by Pollutant Concentration Indicator. Graphic from the EPA, OPPE, ESID, Compendium of Selected National Environmental Statistics in the U.S. Government (Draft: 8/24/94 with updates by agency).

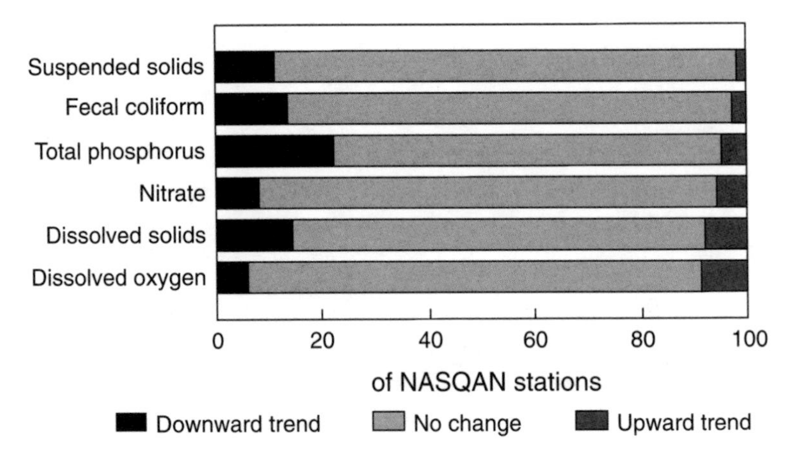

Source: DOI, USGS, Stream Water Quality in the Conterminous United States.

B. National Trends in River and Stream Water Quality. Graphics from Council on Environmental Quality, Environmental Quality 1993 (1995).

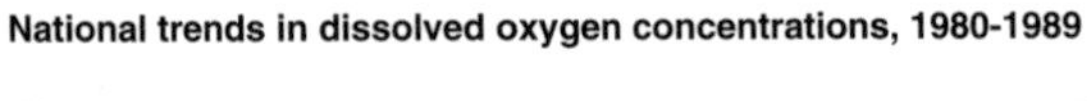

National trends in dissolved oxygen concentrations, 1980-1989

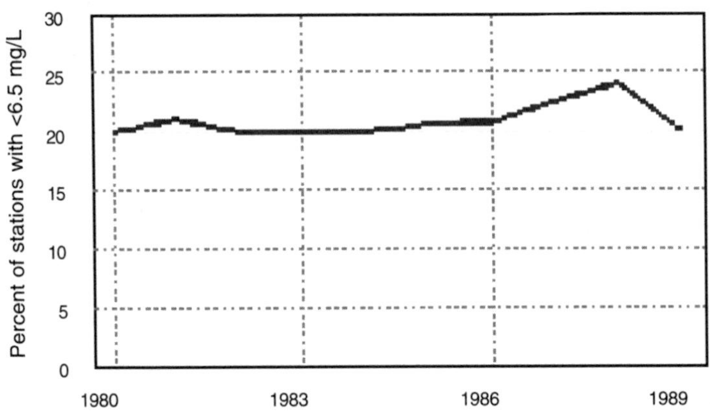

Source: USGS, National Water Summary (1993).

B. National Trends in River and Stream Water Quality—*continued*

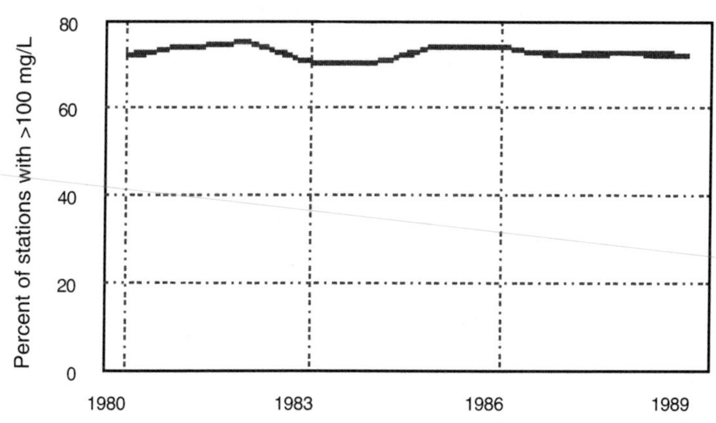

Source: USGS, National Water Summary (1993).

National trends in coliform bacteria concentrations, 1980-1989

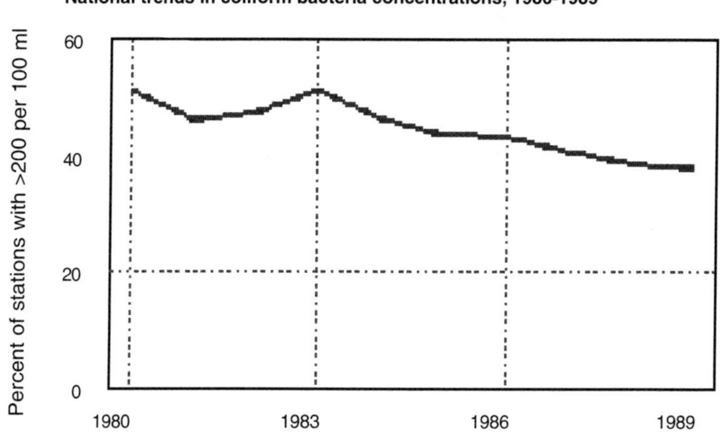

Source: USGS, National Water Summary (1993).

B. National Trends in River and Stream Water Quality—*continued*

National trends in total phosphorus concentrations, 1980-1989

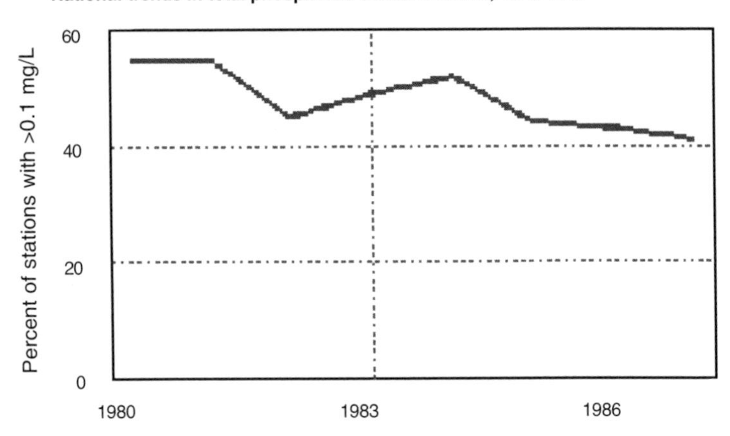

Source: USGS, National Water Summary (1993).

National trends in suspended sediment concentrations, 1980-1989

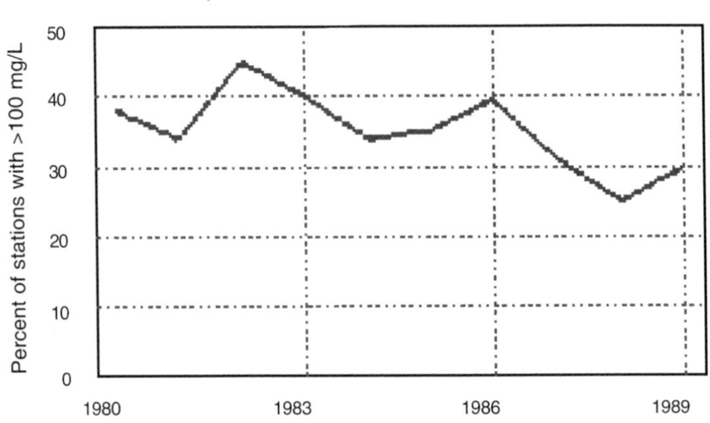

Source: USGS, National Water Summary (1993).

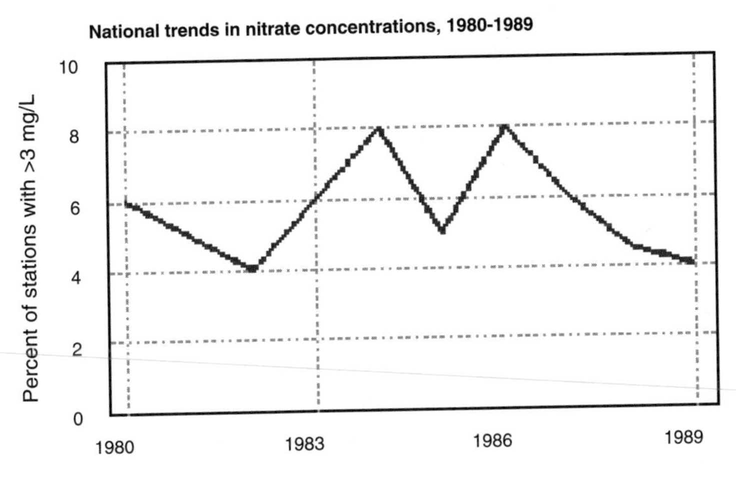

Source: USGS, National Water Summary (1993).

C. U.S. Waters Supporting Healthy and Diverse Aquatic Life. Graphic from EPA, OPPE, Proposed Environmental Goals For America With Benchmarks For The Year 2005: Summary (Draft For Government Agencies' Review, February 1995).

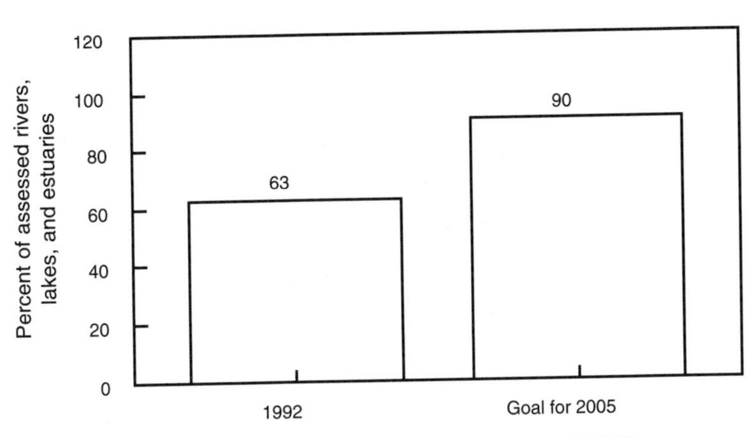

Source: EPA, Proposed Environmental Goals for America with Benchmarks for the Year 2005 (1995).

D. Contaminant Levels in Herring Gull Eggs From Great Lakes Colonies. Graphics from the EPA, OPPE, ESID, Compendium of Selected National Environmental Statistics in the U.S. Government (Draft: 8/24/94 with updates by agency).

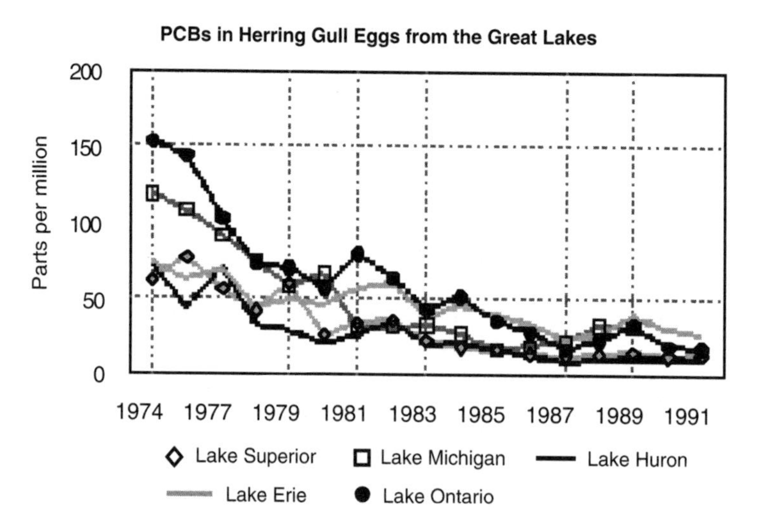

Environment Canada, Canadian Wildlife Service, Canada Center for Inland Waters.

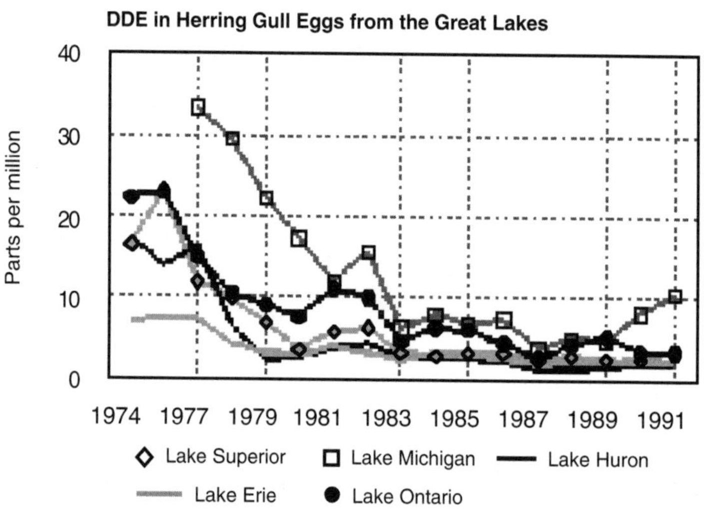

Environment Canada, Canadian Wildlife Service, Canada Center for Inland Waters.

E. Rate of Wetlands Loss. Graphic from EPA, OPPE, Proposed Environmental Goals For America With Benchmarks For The Year 2005: Summary (Draft For Government Agencies' Review, February 1995).

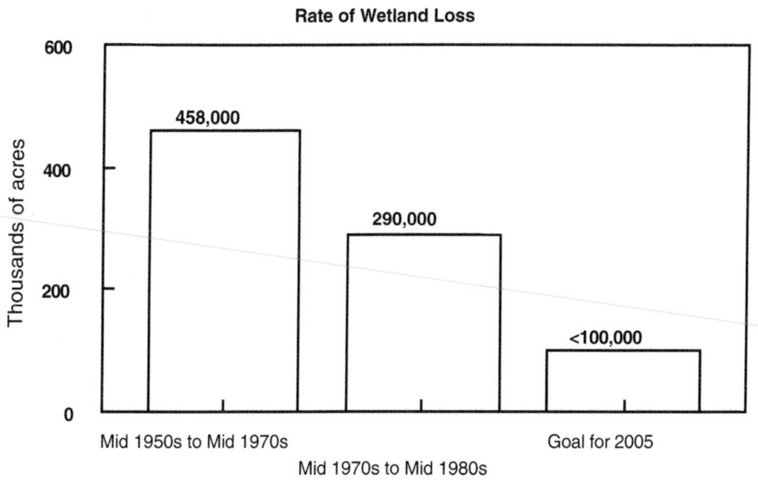

Source: EPA, Proposed Environmental Goals for America with Benchmarks for the Year 2005 (1995).

F. Changes in Wetlands, by Type. Graphic from the EPA, OPPE, ESID, Compendium of Selected National Environmental Statistics in the U.S. Government (Draft: 8/24/94 with updates by agency).

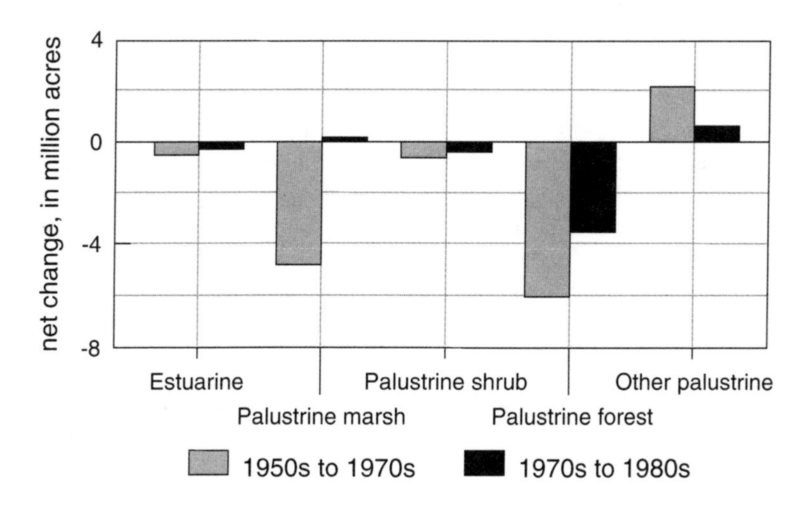

IV. Examples of Graphics from the Compendium of Environmental Statistics That Have Been Updated and Augmented with Metadata and Interpretive Text in a LOTUS NOTES DATABASE (Under Development). The LOTUS NOTES Entries Have Been Converted to Wordperfect Files for the Purpose of this Demonstration. Two Pressure Entries That Are Related to Air Quality Are Presented.

A. Energy Consumption by End-Use Sector

Environmental Statistics and Information Division
Compendium of Environmental Statistics

Data Entry Author: Carroll Curtis **Date Created:** 07/27/95 02:58:40 PM
Topic: Energy Consumption by End-Use Sector **Environment Media:** Energy

GRAPH

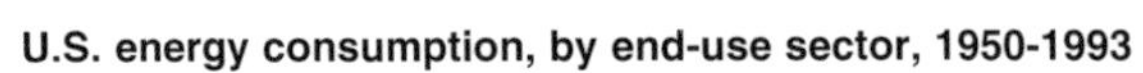

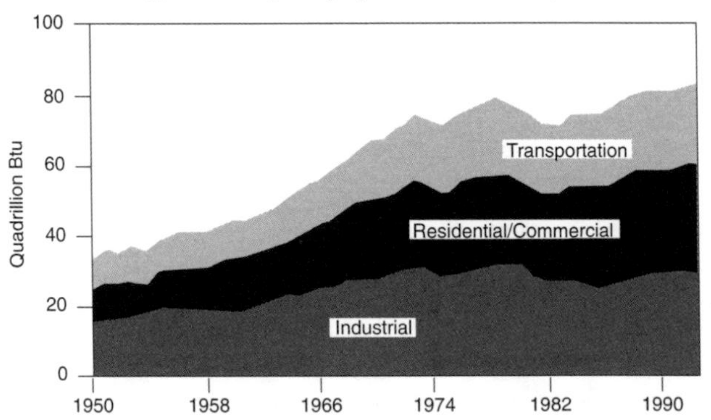

Source: DOE, EIA, Annual Energy Review 1993 (1994).

Technical Notes: Totals include fossil fuels consumed directly in the sector, electricity sales to the sector, and energy losses in the generation and transmission of electricity (allocated in proportion to electricity sales per sector). Due to a lack of consistent historical data, some consumption of renewable energy resources is not included. For example, in 1992, 3.0 quadrillion Btu of renewable energy consumed by U.S. electric utilities to generate electricity for distribution is included, but an estimated 3.0 quadrillion Btu of renewable energy used by other sectors in the United States is not included.

Publication Name(s): U.S. Department of Energy, Energy Information Administration. Annual Energy Review 1993 [DOE/EIA-0384(93)], Table 2.1, p. 39 (Washington, DC: DOE, EIA, July 1994).

Last Updated by ESID: 06/27/95 CNC

INTERPRETIVE TEXT

Energy consumption by the industrial sector increased throughout the 1960s and in 1973 reached 32 quadrillion Btu. Of the three end-use sectors, the industrial sector proved to be the most responsive to the turmoil in energy markets after the 1973-1974 Arab oil embargo. In 1979, industrial consumption of energy peaked at 33 quadrillion Btu. In the 1980s, a stagnant economy restrained industrial consumption, which declined to a 16-year low of 26 quadrillion Btu. In 1988 and 1989, economic growth spurred demand for energy in the industrial sector, and industrial energy consumption in 1989 rose to 29 quadrillion Btu. Despite slow economic growth in the 1990s, industrial energy consumption trended upward. In 1993, industrial consumption energy reached 31 quadrillion Btu, the highest level in 14 years.

Much of the growth in energy consumption during the 1950-through-1993 period occurred in the residential and commercial sector. Residential and commercial consumption leveled off in response to higher energy prices in the late 1970s and early 1980s, but lower prices in the 1986-through-1993 period played a role in boosting residential and commercial energy consumption to the record level of 30 quadrillion Btu in 1993.

Energy consumption by the transportation sector was primarily petroleum consumption. Over the 44-year period, the transportation sector's consumption of petroleum more than tripled, but growth was slower during the 1980s than in previous decades. In 1993, consumption of petroleum in the transportation sector totaled 23 quadrillion Btu, up 1.6 percent from the 1992 level.

Publication Name (if different from Table &/or Graph Publication Name):

Last Updated by ESID: 06/27/95 CNC

TABLE U.S. energy consumption, by end-use sector, 1950–1993 (quadrillion Btu)

Year	Industrial	Residential & commercial	Transportation	Total
1950	15.71	8.87	8.49	33.08
1951	17.13	9.30	9.04	35.47
1952	16.76	9.54	9.00	35.30
1953	17.65	9.50	9.12	36.27
1954	16.58	9.78	8.90	35.27
1955	18.86	10.41	9.55	38.82
1956	19.55	10.96	9.86	40.38
1957	19.60	10.98	9.90	40.48
1958	18.70	11.64	10.00	40.35
1959	19.64	12.15	10.35	42.14
1960	20.16	13.04	10.60	43.80
1961	20.25	13.44	10.77	44.46
1962	21.04	14.27	11.23	46.53
1963	21.95	14.71	11.66	48.32
1964	23.27	15.23	12.00	50.50
1965	24.22	16.03	12.43	52.68
1966	25.50	17.06	13.10	55.66
1967	25.72	18.10	13.75	57.57
1968	26.90	19.23	14.86	61.00
1969	28.10	20.59	15.50	64.19
1970	28.63	21.71	16.09	66.43
1971	28.57	22.59	16.72	67.89
1972	29.86	23.69	17.71	71.26
1973	31.53	24.14	18.60	74.28
1974	30.70	23.72	18.12	72.54
1975	28.40	23.90	18.25	70.55
1976	30.24	25.02	19.10	74.36
1977	31.08	25.39	19.82	76.29
1978	31.39	26.09	20.61	78.09
1979	32.61	25.81	20.47	78.90
1980	30.61	25.65	19.69	75.96
1981	29.24	25.24	19.51	73.99
1982	26.14	25.63	19.07	70.85
1983	25.75	25.63	19.13	70.52
1984	27.86	26.48	19.80	74.40
1985	27.22	26.70	20.07	73.98
1986	26.63	26.85	20.81	74.30
1987	27.83	27.62	21.45	76.89
1988	28.99	28.92	22.30	80.22
1989	29.35	29.40	22.58	81.33
1990	29.93	28.79	22.54	81.26
1991	29.57	29.42	22.12	81.12
1992	30.58	29.10	22.46	82.14
1993	30.77	30.34	22.83	83.96

Publication Name(s): U.S. Department of Energy, Energy Information Administration. Annual Energy Review 1993 [DOE/EIA-0384(93)], Table 2.1, p. 39 (Washington, DC: DOE, EIA, July 1994).

Technical Notes: See Technical Notes above. Also see Compendium Database entry for Renewable Energy Resources. Totals may not equal sum of components due to independent rounding. Current-year data are preliminary and may be revised in future publications.

Administrative Notes: The Energy Information Administration publishes two sets of statistics on end-use energy consumption. The first set, based on surveys directed to suppliers and marketers, provides continuous series for the years 1949 through 1993 and allocates U.S. total energy consumption into one of three end-use sectors: industrial, residential and commercial, and transportation. The statistics from these surveys are presented above. The second set, based on surveys directed to end-users of energy, provides detailed information on the type of energy consumed and the energy-related characteristics of manufacturing establishments, commercial buildings, households, and household motor vehicles. For information on the second set of statistics, contact EIA specialist for the Manufacturing Energy Consumption Survey [John L. Preston (202/586-1128)], Residential Energy Consumption Survey [Wendel L. Thompson (202/586-1119)], Residential Transportation Energy Consumption Survey [Ronald Lambrecht (202/586-4962)], or Commercial Buildings Energy Consumption Survey [Martha M. Johnson (202/586-1135)].

Review Contact for Statistics (if different from Contact below):

Last Updated by ESID: 06/27/95 CNC

CONTACT INFORMATION

Agency/Department: U.S. Department of Energy **Bureau:** Energy Information Administration

Division/Office: Office of Energy Markets and End Use **Branch/Address:** Forrestal Building, EI-231

City: Washington **State:** DC **Zip Code:** 20585

Contact's Last Name: Brown **First Name:** Samuel E.

Phone #: (202) 586-5108 **Fax:** **Internet Mail Address:**

B. Aging of U.S. Vehicle Fleet

Environmental Statistics and Information Division
Compendium of Environmental Statistics

Data Entry Author: Carroll Curtis **Date Created:** 07/27/95 02:58:40 PM
Topic: Aging of the U.S. Vehicle Fleet **Environment Media:** Transportation

GRAPH

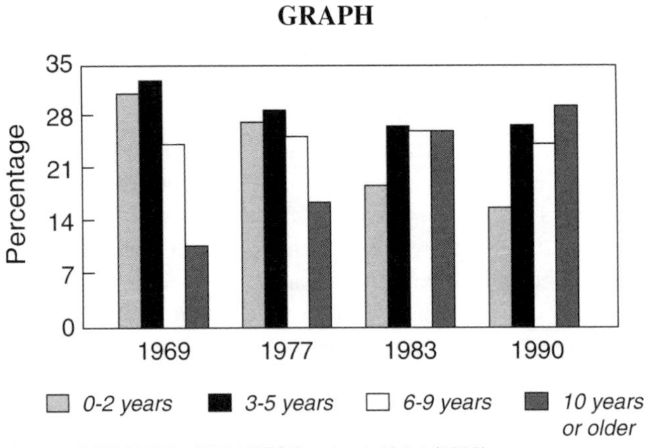

Source: DOT, FHWA, 1990 NPTS Databook, Vol. 1 (1993).

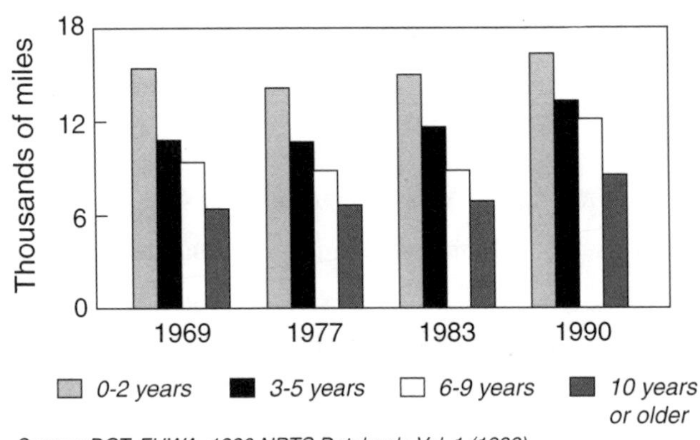

Source: DOT, FHWA, 1990 NPTS Databook, Vol. 1 (1993).

Technical Notes:

Publication Name(s): U.S. Department of Transportation, Federal Highway Administration. 1990 NPTS Databook: Nationwide Personal Transportation Study, Vol. I (Washington, DC: DOT, FHMA, November 1993).

Last Updated by ESID: 07/27/95 CNC

INTERPRETIVE TEXT

American households tend to keep their cars and trucks for a longer period of time. Over the period 1969-to-1990, the average age of household vehicles increased by 51 percent, from 5.1 years in age to 7.7 years. The percentage of household automobiles that are 10 or more years old increased from 10.8 in 1969 to 29.9 in 1990. While 31.4 percent of household automobiles in 1969 were less than two years old, this percentage decreased to 15.6 percent in 1990. Vehicles of all ages are driven more than in previous years, yet the rate of increase was more between 1969 and 1990 for older vehicles than younger ones. In 1990, between 20 percent and 26 percent of all vehicle trips were taken in vehicles 10 years and older, regardless of the number of vehicles available to the household.

The aging of the U.S. vehicle fleet has implications for energy consumption and air pollution issues, and the introduction of recent safety features into the household vehicle fleet. The fuel consumption characteristics of the older fleet clearly lag that of the newer fleet. From a level of 13.5 miles per gallon when the energy shock hit in the early 1970s, the fuel economy of the fleet has risen to a level approximating 21 miles per gallon in 1990. This suggests that for each mile of VMT occurring in older vehicles we pay a substantial energy penalty. The air pollution control consequences are probably even more pronounced. The year 1981 was a key turning point in the air quality control characteristics of the vehicle fleet. The differences in pollution per vehicle mile for vehicles pre- and post-1981 are extraordinary. In terms of safety there so many new safety features—anti-lock brakes, airbags, traction control, etc.—that will only slowly gain penetration into the fleet. On the other hand, one of the benefits of a fleet that lasts longer is for resource conservation and minimizing junk yards.

Publication Name (if different from Table &/or Graph Publication Name: U.S. Department of Transportation, Federal Highway Administration. 1990 NPTS Databook: Nationwide Personal Transportation Study, Vol. I (Washington, DC: DOT, FHMA, November 1993).

—, 1990 NPTS Report Series: Special Reports on Trip and Vehicle Attributes, Chapter 3. The Demography of the U.S. Vehicle Fleet, Table 1.2, p. 3-16 (Washington, DC: DOT, FHMA, February 1995).

Last Updated by ESID: 07/27/95 CNC

TABLE Aging of the U.S. fleet, 1969-1990

	Characteristic	1969	1977	1983	1990
Vehicles, by average age, all	age, in years	5.10	5.60	7.60	7.70
0-2 years	%	31.40	27.80	19.00	16.60
3-5 years	%	33.20	29.60	27.30	27.50
6-9 years	%	24.60	25.70	26.80	25.30
10 or more years	%	10.80	16.90	26.90	30.60
Annual VMT, by vehicle age, average	1,000 miles	11.60	10.68	10.32	12.46
0-2 years	1,000 miles	15.70	14.46	15.29	16.81
3-5 years	1,000 miles	11.20	11.07	11.90	13.71
6-9 years	1,000 miles	9.70	9.20	9.25	12.55
10 or more years	1,000 miles	6.50	6.76	7.02	9.18

Publication Name(s): U.S. Department of Transportation, Federal Highway Administration. 1990 NPTS Databook: Nationwide Personal Transportation Study, Vol. I, Table 3.24, p. 3-40 and Table 3.26, p. 3-43 (Washington, DC: DOT, FHMA, November 1993).

Technical Notes: The 1969 survey did not include pickups and other light trucks as household vehicles.

Administrative Notes:

Review Contact for Statistics (if different from Contact below):

Last Updated by ESID: 07/27/95 CNC

CONTACT INFORMATION

Agency/Department: U.S. Department of Transportation **Bureau:** Federal Highway Administration

Division/Office: Office of Highway Information Management **Branch/Address:** 400 Seventh Street, SW

City: Washington **State:** DC **Zip Code:** 20590

Contact's Last Name: **First Name:**

Phone #: (202) 366-0180 **Fax:** **Internet Mail Address:**

Attitudes Toward the Environment Twenty-Five Years After Earth Day[1]

KARLYN BOWMAN
American Enterprise Institute for Public Policy Research

CONTENTS

---

[1]This transcript of a presentation by Karlyn Bowman at the National Forum on Science and Technology Goals—No. 1: Environment, August 22, 1995 is under the copyright of the American Enterprise Institute for Public Policy Research. These remarks are a condensation of a monograph, *Attitudes Toward the Environment: Twenty-Five Years After Earth Day*, authored by Karlyn Bowman and Everett Carll Ladd and published by the American Enterprise Institute for Public Policy Research, The AEI Press, Washington, D.C., 1995.

I bring to the issues you are discussing only a layman's knowledge. My work and that of my colleague Everett Ladd is in public opinion. We conduct no surveys of our own; we rely on data in the public domain—the Gallup, Harris, network/print partnerships such as Gallup and CNN and *USA TODAY*—to write about public attitudes. We are by no means uncritical admirers of the survey device. It is a blunt instrument ill-suited to many tasks. Still, we believe polls are a useful device to understand a complex public.

We have recently completed a review of public opinion on the environment. Our study focused mostly on attitudes since Earth Day and that will be my focus today. We did, however, review briefly surveys conducted before that time. Our study is different from many we reviewed in preparation for the monograph we wrote because we put views on the environment into context by comparing them to views about other issues. It is useful to know, for example, that in the abstract, Americans say we are spending too little on the environment. But it is also important to know that they say we are spending too little halting the rising crime rate, improving the nation's education system, and improving and protecting the nation's health, and that these issues appear more urgent.

Our review revealed a central weakness of the huge collection of attitudinal data on the environment. Americans have been asked repeatedly in a wide variety of formulations to affirm a core value, in this case the importance of the environment. Each time, not surprisingly, they respond that a clean and healthful environment is important to them. These questions tell us little about what a society with many demands on it is willing to do to advance the value, what trade-offs the public is willing to make for it, or what happens when one important value clashes with another. The pollsters have missed an opportunity to advance our knowledge. Let me briefly outline attitudes today.

WHAT DO AMERICANS WANT?

In 1995 Americans are committed to a clean environment and to economic growth. What is more, we are optimistic about the country's ability to achieve both. We are confident about science and technology and about our ability to make environmental progress. We continue to believe that the federal government has an important role in meeting environmental objectives.

In 17 iterations of the question *Cambridge Reports* began to ask in 1976, pluralities or majorities have responded that we can combine material progress and a clean environment. But now, in the 1990s, that belief has substantially broader support than previously. In the latest *Cambridge Reports* survey, 67 percent agreed that we could have both.

The optimism this and other questions reflect is rooted in Americans' confidence in the nation's technical and scientific prowess. The number saying in 1994 that they have a great deal of confidence in those directing science is virtually identical to the number giving that answer twenty years ago, the first time the question was asked. Only those managing medical care received a higher vote of confidence in the surveys conducted by the National Opinion Research Center. Confidence in many other key players (including Congress and the executive branch) was much lower and has dropped sharply since the early 1970s.

Another reason for the belief that we can encourage economic development and protect the environment may stem from the belief that government is doing a good job in this area. In 1994, 46 percent told Roper Starch Worldwide interviewers that protecting the environment was a definite responsibility of the federal government, and 37 percent called it highly desirable. In the follow-up question, a solid majority felt that government was carrying out that responsibility fully or fairly well.

THE LABEL AND THE MOVEMENT

One way to assess support for the environment is to look at how strongly people identify with the movement. In 1987 and in 1994, in *Times Mirror* surveys, about a quarter of Americans placed themselves at points 9 and 10 on a scale to indicate that the term environmentalist was perfect or near perfect for them. Of the sixteen groups Times Mirror inquired about, only "a religious person" and "a supporter of the civil rights movement" got broader backing.

A *Times Mirror Magazine* group survey found that slightly over 20 percent called themselves "active environmentalists." A majority fell into the "sympathetic but not active" group. Almost no one said they were unsympathetic. Any way you slice it, "the environment" and hence "environmentalists" are popular.

HOW IMPORTANT IS THE ENVIRONMENT?

That large majorities feel positive about the environment as a social value is important. But politically the real question is how much Americans think should be done now and in the future to advance environmental goals and how much they are willing to do themselves. Answering these questions is not easy because most Americans do not get involved in specific policy issues, nor are they particularly attentive to the details of the debates surrounding them. They offer broad policy direction to legislators that reflects their belief about the importance of the environment.

Many of you are familiar with the work of Robert Cameron Mitchell. Mitchell, a thoughtful commentator on public attitudes toward the environment, has argued that in thinking about any issue, we need to distinguish between salience—"how much immediate personal interest people have" in it—and strength of opinion—"the degree to which people regard the issue as a matter of national concern and are committed to improving the situation or solving the problem." This formulation is useful, but it is not exhaustive.

Salience can be measured by looking at responses to "open-ended" questions, for example, What is the most important problem facing the country today? that people answer with any response they wish. By this salience measure, the environment ranks low. CBS News/*New York Times* asked its "most important problem" question ten times between January 1994 and January 1995, and in only one poll did the issue register at even 1 percent. In the early August 1995 question, the interviewers recorded mentions of 26 different issues as "the most important problem." The environment was not among them.

A poll done by Peter Hart Research for the National Wildlife Federation in December 1994 also shows that the environment has been eclipsed by other issues. Those who reported voting in 1994 were asked which two or three issues were important to them in making their choices. The environment was cited by just 6 percent—ranking behind crime (at 24 percent), health care (21 percent), the economy-recession (16 percent), high taxes (15 percent), unemployment-jobs (9 percent), education (8 percent), and efficient government (7 percent). Hart Research then asked those who had not cited the environment why they did not do so: 35 percent said other issues were more urgent; 23 percent, that the candidates did not discuss the environment; 10 percent, that the Clinton administration was doing a good job on the issue. Only 8 percent said that enough had been done for the environment already.

The property that Robert Cameron Mitchell contrasts to salience, the strength of public opinion, can be measured by poll questions asking how serious a problem is, how we view government efforts to regulate the activity, and how much we want to spend on it.

Roper Starch Worldwide asks Americans whether environmental pollution is a "very serious threat these days to citizens like yourself, a moderately serious threat, not much of a threat, or no threat at all." In May 1994, 47 percent called environmental pollution a very serious threat, 39 percent a moderately serious threat, 11 percent not much of a threat, and 2 percent none. A decade earlier, the percentages were roughly the same. This would seem to suggest enormous concern. In fact the poll results merely show that many people do not like pollution. To put the numbers in perspective, in 1994, 77 percent said crime was a very serious threat, 72 percent said illegal drugs were, and 60 percent said drunken drivers on the road were a very serious threat. Many polls like the ones above confirm the view that other problems are far more urgent than the environment for Americans today.

WHAT DO WE WANT GOVERNMENT TO DO?

Survey researchers naturally want to explore just how far the public is prepared to go in terms of taxing and spending and regulation to improve the environment, but it is extremely hard to get at this kind of information in the abstract. Hypothetical questions are almost always a problem in opinion research. Much of the public does not think hypothetically about policy choices. It posits broad values. Because the public does not put itself in this kind of a judgment situation, responses to questions vary. The wording of a particular question pulls the public in one direction, a change in a few words pulls it in another. Trade-off questions exacerbate the problem.

General questions about the proper level of government regulation show high commitment—once again—to the core value identified earlier, the importance of the environment. More specific questions show a concern about balancing costs and benefits. Since 1989, Roper Starch Worldwide has found that strong majorities or pluralities say that environmental laws and regulations have not gone far enough, with around 30 percent saying that we have struck the right balance. In the late 1970s and early 1980s, about two in ten said we had gone too far. That number has since dropped.

A *Cambridge Reports* question shows that since 1982, solid majorities or pluralities have said that in general there is too little government regulation and involvement in the area of environmental protection. The proportion saying that there is too much has moved up sharply from roughly 10 percent in the late 1980s to 31 percent in 1994.

Polls today show that Americans feel it is important to balance costs and benefits. A January 1995 question by Yankelovich Partners for *Time* and CNN illustrates the point. Two thirds suggested that "an environmental regulation which addresses a specific risk to people's health" should be "subject to an analysis to determine whether eliminating that risk justifies the cost." It is impossible to know how Americans would have answered that question ten or twenty years ago. These kinds of questions weren't asked then.

TRADE-OFFS

Because Americans offer broad values and general conclusions about directions to be pursued, asking them about specific policy choices when people do not really think much about them can produce misleading information. The problem is exacerbated when the nature of trade-offs to be considered is presented in an imprecise and misleading manner.

Consider a question asked by the National Opinion Research Center in 1994: And how willing would you be to accept cuts in your standard of living in order to protect the environment? What do the surveyors mean by "cuts"? How big a cut? In what area? What kind of environmental benefits would be achieved? Because these issues are left unresolved, it is impossible to know what to make of the answers given: 31 percent said they would be willing to accept cuts in their standard of living, while 45 percent said they would not be, with 23 percent on the fence. The numbers suggest that the strong backing for the environment is not unconditional. But we already knew that. This question does not tell us anything new. The actual proportions yielded by it are almost meaningless because of the imprecision and confusion about what is being asked.

Because the public affirms the general value of a clean environment and is not engaged in specific policy choices, the wording and timing of questions that ask people to engage in specific policy choices can be extraordinarily important. Four questions asked by CBS and the *New York Times* in 1992 and early 1993 illustrate the point. The public switched its view from May to September about whether protecting the environment was more important than stimulating the economy. Negative coverage of the economy in the fall of 1992 was making people more anxious about economic prospects, and that atmosphere probably explained the shift in responses.

The second pair of questions was asked in September 1992 and in March 1993. People were asked whether we must protect the environment even if it means jobs in your community are lost because of it. The public split evenly in September, 45 to 45 percent; in March 1993, 60 percent agreed with the statement. Had people changed their minds about anything related to the environment over the ten months those four questions were asked? Of course not. An election campaign took place, and a new president was inaugurated. The political context in which the questions were asked shifted from negative coverage of the economy to more positive coverage of a new president. Americans use polls to send messages to their governors—to indicate subtle changes in moods, as they did over these ten months.

SPENDING

When a value such as the importance of the environment occupies a substantial position in public thinking opinion researchers need to take care not to attribute specific conclusions to it mistakenly. Consider the familiar question about whether we are spending too little, too much or the right amount on the environment. The National Opinion Research Center has asked Americans that question nineteen times over the past twenty years. Strong majorities have consistently told the pollsters that we are spending too little. The proportion saying we are spending too much is always small. Policy makers should not read these responses literally. They are expressions of a broad general commitment to a clean environment, not an endorsement of higher spending.

Public thinking about what spending is actually needed today is more complex. Americans say we are spending too little on many problems, and it appears from some questions that they think we can move a little more slowly in the environmental area. In January 1995, 53 percent of Americans told Yankelovich Partners that "given our other problems right now," it would be better to go slow in spending money to clean up the environment. Still 40 percent said we should go full speed ahead.

It is striking how few questions over the past twenty years have asked Americans to think about their tax burden when thinking about federal spending on the environment. In surveys conducted by General Electric from 1966 to 1981, people were asked about their concern about air and water pollution. They were then asked how much they would be willing to pay in taxes, utility rates, or other prices to do something about the problems. In the last year of the survey, a near majority said they would pay $5.00 to do something about air pollution, 27 percent indicated that they would be willing to pay $50.00, and only 6 percent said they would be willing to pay $500.00. The results were similar for water pollution. Since the question did not specify a yearly appropriation and seems to indicate as well a one-time assessment, the public was clearly setting very tight limits on environmental spending.

Throughout the health care debate, pollsters probed the dollar amount Americans would be willing to pay for health care for the uninsured. The responses varied considerably for something Americans care deeply about, but the overwhelming impression conveyed by these questions was much like the impressions gleaned from the GE data. We are not willing to spend much at all.

Americans are willing to pay only small amounts for things they care deeply about not because they are not generous. Incomes are modest and many people feel strapped, and there are so many problems to solve. Unless survey questions of this sort are carefully designed, asked over a long period, and provide comparative perspectives we will never know how much Americans are willing to spend.

LITE GREENS?

We tell the pollsters we are sympathetic to the environmental movement and a plethora of questions show that we remain committed to the environment even if it is less urgent than in the past. We are thinking twice about additional spending. Other types of questions exploring actual behavior on environment-related matters generally show a public not inclined to do much to advance the cause. We have become Lite Greens.

Recycling is a relatively painless exercise, and the number of Americans who say they have sorted newspapers or bottles for recycling has risen considerably since 1980. We are also recycling at work. Far fewer say they have reduced the amount of their driving or boycotted a company's products because of its environmental record.

Roper has asked the public twice since 1989 whether the respondent or someone in the household makes a real effort to do a list of things about the environment on a regular basis, does these from time to time when it is convenient, or does not bother about it. Solid majorities said they did not really bother about doing volunteer work for local environmental groups, writing letters, not patronizing restaurants that put take-out food in Styrofoam containers, not cutting down on the use of their car by using public transportation, etc. Majorities did a few things on a regular basis from time to time such as returning beer or soda bottles, recycling newspapers, sorting trash, buying products in pumps, and using biodegradable soaps.

Beyond this, Roper Starch Worldwide has repeated an extensive battery of questions asking people whether they would be willing to see a list of things happen in their communities to protect the environment. The results have been fairly stable over the three years Roper has asked the question. A majority was willing to ban the use of CFCs to reduce harm to the ozone layer even though it may mean that the prices of refrigerators might rise, and a bare majority would ban use of these chemicals even though it might mean home and car air conditioning won't cool as well. But majorities were not willing or did not know whether they would be willing to put a new burden on dry cleaners that might force some of them to go out of business and mean higher costs for consumers, see an increase in utility bills, make us more dependent on other nations, and so forth. Even though the items in this battery are very specific, most respondents have never thought about many of them. How are we to evaluate the response to the item about requiring those who use charcoal grills to use electric lighters rather than lighter fluid. Is the 35 percent figure of those willing to see this happen in their community high or low? Is the question asking whether electric starters might be a good idea or whether a categoric ban on lighter fluid should actually be imposed? The item gives little practical guidance.

The organization asked an interesting battery of questions that get to the heart of the issue about what we are willing to do. The exercise had two parts. The first question asked Americans whether they would favor laws that would help clean up the environment but make products more expensive. In the abstract, solid majorities favored laws to reduce pollution. The follow-up question asked Americans whether they would be willing to pay a lot or a moderate amount more for these products. Public resistance to paying more was considerable.

Young people are thought to be more committed to environmental improvement than their elders are. Alexander Astin at the University of California at Los Angeles has been probing the attitudes of young people entering college for nearly three decades. Astin finds that these people have consistently said that the government is not doing enough to protect the environment. But the number who said that being involved in environmental cleanup was essential or very important to them dropped from 43 percent in 1971 to 23 percent in 1994.

CURRENT CONTROVERSIES

When the public is not engaged in a specific policy issue, question wording influences their responses. A number of national pollsters have been in the field recently who claim to know what Americans think about reauthorization of the Endangered Species Act and how private property owners should be compensated when regulations cause a loss of value in their property. In many of these areas, the responses are clearly contradictory.

Peter Hart Associates, in its December 1994 survey for the National Wildlife Federation, asked a national sample which of two statements came closer to their opinion. Fifty-seven percent chose, "Some people say Congress should maintain strong requirements in the Endangered Species Act because certain plants and animals could become extinct if they are not protected, and some of these plants and animals might be used for medical cures or to develop disease resistant crops." About a third, 32 percent, chose, "Other people say Congress should relax certain requirements of the Endangered Species Act, because these requirements can slow down or even stop business growth and development in order to protect all endangered plants and animals." These alternatives are leading and contrived. It is not surprising that different kinds of questions come up with different findings. A question posed by Roper Starch Worldwide for *Time Mirror Magazines* illustrates. The question is set up this way: In 1972, Congress passed a law called the Endangered Species Act. This law requires the federal government to take whatever steps necessary to prevent any type of plant, animal, or insect species from becoming extinct, even at a cost to landowners, businesses, or the local economies where the species live. When asked what government policy should be, 63 percent said that the endangered species policy should take account of costs, while only 29 percent said that all species should be saved.

In areas with little public knowledge, responses—particularly hypothetical ones—can be highly misleading. Groups representing various positions in conflicts, such as the one over reauthorization of the Endangered Species Act, like to claim they have public opinion on their side. We simply don't know if they do.

POLITICS

I'll say just a word about the politics of the environmental issue. When the issue emerged, neither political party had a particularly strong advantage on handling the issue. Over the years, however, the Democrats have developed and maintained a strong lead over the Republicans. Yet although Democrats generally and Democratic candidates are seen as better able to protect the environment, election results seem to suggest that the issue is not a significant one for most voters. In most of the exit polls where pollsters have included the category "environment" in a list of most important problems, only a small number have said the issue was one of the most important to them in casting their votes. Those voters have pulled the lever for Democrats. In 1994, however, the issue appeared to work for the GOP in the West. The national exit polling consortium of the four networks and AP, Voter News Service, did not include the "environment" in their list of most important problems facing the nation. But VNS did ask voters in the West whether Clinton environmental or land use policies had helped their states, hurt their states, or had no effect. In eight of nine western states, voters said the policies had hurt their states. These voters voted for GOP candidates. Only in Colorado, where voters said the policies had had no effect, did voters vote for the Democratic gubernatorial candidate. When a public agrees on the ends we as a society should pursue, something we did in the 1970s about the environment, we tend to disengage from the debate over the means. That has happened at the national level, though the issue is now very potent at the state and local level. If the public perceives that the national consensus is threatened, as some felt Jim Watt and Anne Gorsuch Burford did, the possibility then exists that the issue will engage the public significantly again at the national level.

HOW MUCH PROGRESS?

The 1994 Hart poll for the National Wildlife Federation found that 62 percent rated the overall quality of the air, water, land, and wildlife where the respondent lived as excellent or good, 28 percent as only fair, and 8 percent as poor. The numbers are similar to those obtained by Roper Starch Worldwide. Reinforcing the impression of general satisfaction are the results of a question Hart and Teeter Research Companies posed for *Newsweek* in 1991. Surveyors asked about a number of problems where the respondent lived. Four in ten said that pollution was really not a problem, while 23 percent described it as just somewhat of a problem. Thirty five percent said it was very or fairly serious. Far more Americans thought that drugs, economic stagnation, the cost of housing, crime, the financial condition of local government, the cost of living—the list goes on—were very serious problems in their community than felt that way about pollution.

Questions asked about the environment nationally produce more pessimistic responses—reflecting a familiar pattern in polls. Americans are far more likely to say crime is a problem in the nation than they are to believe it is a problem in their own communities. Similarly, we tell the nation that education is a big problem for the nation, but that our schools are performing well. We rely more on survey data that asks Americans about things they can expected to have opinions about. Negative national assessments are often calls for leaders to perform better. We continue to want government to be vigilant, and business to be attentive to our concerns, but we are generally satisfied with progress on the environment on the home front.

SUMMARY

To conclude, on occasion, a value that was not politically salient or central comes to be seen as essential. The environment made this transition over the 1960s and 1970s. Large majorities of citizens across class and other social group lines are deeply committed to a safe, healthful environment and are prepared to support a variety of actions that seem reasonable in promoting those ends. The challenge opinion research faces when a general value occupies this substantial standing in public thinking is to ensure that specific conclusions are not mistakenly attributed to it. Questions about spending, for example, should be taken as expressions of genuine commitment to a clean environment. Publics assert general values. They do not engage in specific policy choices. It is extremely difficult to get information on how far the public wants to go in terms of taxing, spending, or regulation by asking abstract questions. The public points to ends, and does not think much about the means.

Unfortunately, the existing polling literature has not answered satisfactorily the question of whether the balance has shifted, even if subtly, such that groups saying, "We must protect the environment" find a more skeptical audience now than they did a decade ago. The issue is not whether Americans have soured on the environment or esteem a clean environment less as a central value. Clearly, the vast majority of our citizens are environmentalists. But we are now more inclined to think that for most Americans, the urgency has been removed, and the battle to protect the environment is being waged satisfactorily. Despite the many ambiguities, impressive evidence of such a shift exists.

Environmental Goals and Science Policy: A Review of Selected Countries

KONRAD VON MOLTKE
Institute on International Environmental Governance, Dartmouth College, and World Wildlife Fund

CONTENTS

Science and technology play an unusual role in environmental policy. Science makes the environment speak. Many environmental threats would be unknown without scientific research, or known too late to permit appropriate policy action. In other words, environmental policy rests on a foundation of scientific research without which it would not even exist.

This creates a unique and uneasy relationship between scientists and policy-makers. The ethics and process of scientific research are not geared to the needs of policy-making. Science seeks to prove or disprove hypotheses as a strategy for reaching enduring answers. Policy is limited in time and location: decisions must be made at a given time for a specific jurisdiction, so policy-makers are seeking the best possible answers to issues they have chosen not because they may be susceptible to being answered but because there is a constituency that requires an answer. Thus for policy-makers any answer is better than no answer.

Because environmental policy needs science to identify its objects but science is not normally organized to provide information that can be used in policy-making, most countries have developed specialized procedures for science assessment. This generally involves a process rooted in science and the values of scientific investigation but is not itself a scientific undertaking. It involves a review of available evidence and provides an assessment of what is known about a specific issue at a given point in time. Scientific assessments are not readily transferable from one jurisdiction to another because they already incorporate certain aspects of the policy environment for which the assessment is being undertaken.

Even while science and technology are at the heart of environmental policy, they are also widely perceived as being at the origin of the environmental crisis. Without many scientific developments of the past century, and their adaptation by technology to practical uses, even 10 billion humans would be incapable of threatening the natural fabric of the planet. The extraordinary magnification of the human presence through technologies, ranging from fossil fuel combustion to organic chemistry, from medical and biological interventions to electronics, is a precondition of threats to the environment that are qualitatively different from the historical impact of humans on the planet. The population explosion itself is fundamentally a product of simple technologies relating to sanitation and nutrition, which have expanded average human life expectancy beyond anything dreamed of but a century ago.

Finally, no solutions short of human catastrophe are conceivable without further resort to science and technology. The modern human condition has created a dependence on technology which implies that only technology holds the prospect of saving us from technology, a paradox often barely perceived and never resolved.

Each of these reasons would link science and technology closely with environmental policy. All three taken together cause them to be in an almost "symbiotic" relationship. The stakes are extraordinary high, ranging from the universal to the prospects of individual material benefit:

- Science policy must ensure research is undertaken that is essential to identify environmental threats in a timely fashion so as to avoid the kind of crisis caused by "unknown" natural events.
- Science policy must ensure that science itself respects limits that are defined by the environmental impacts of unknown applications of scientific discoveries.
- Science policy must contribute to developing technologies that redirect human efforts from environmentally damaging to environmentally benign activities.
- Science policy can give direction to future social and economic development, providing extraordinary comparative advantages to the individuals, corporations, and societies that take the "right" decisions earlier than others and thereby define the parameters of future development and reap the social and economic benefits associated with this paradigm shift that may follow.

While all countries face the need to articulate science and technology goals for environmental research and policy, each will tend to go about this process in characteristic ways.

Environmental policy is confronted by essentially the same agenda in all countries. In the temperate zone, the environment will be more forgiving than in extreme climates, but everywhere the basic need is to protect air, water, soil, fauna, and flora from the impacts of human interventions. Everywhere the extraction of natural resources, their transport and transformation, their use, and the wastes attendant upon these processes are the stuff of environmental policy. Despite these basic similarities, due to the universality of nature, environmental policies differ widely from one country to the next because they reflect specific environmental conditions, because differing social and economic priorities exist, and because they can only be expressed through the existing political and administrative culture of each country.

COMPARING ENVIRONMENTAL POLICY[1]

The Framework for Environmental Policy

Environmental policy represents a relatively recent development. In most Western industrialized countries, systematic attention was first given environmental management in the late sixties and early seventies. The problems were everywhere the same: economic growth had reached a stage where the consequences of emissions could be felt over large areas, affecting significant segments of the population. Public pressure increased to limit the risks associated with the practice of using the ambient environment for waste disposal. The responses were also everywhere quite similar: the adoption of laws regulating emissions to air and water, the establishment of procedures for environmental management, and legislation concerning the control of hazardous wastes and toxic substances. Table 1 shows the early pattern of regulation in selected countries. It is most remarkable for the overall symmetry of responses.

TABLE 1 Major National Environmental Laws in Some OECD Countries

	Type of Law					
	General	Water	Wastes	Air	Impact Statement	Other
Canada		1970			1973	1975[a]
United States	1970	1972	1965	1963	1969	1976[b]
		1977	1970	1970		
			1976	1977		
			1984			
Japan	1967	1958	1970	1962		1973[c]
	1970	1970		1968		1973[c]
Australia	1974				1974	
New Zealand		1967		1972	1972	
		1974			1977	
Austria		1959				1973
Belgium		1971	1974	1964		
Denmark	1973	1978	1978			1978[d]
		1980				1979[b]
		1982				
Finland		1961	1978	1982		1923[d]
		1979				1965[e]
France	1976	1964	1975	1974	1976	1977[b]
Germany		1957	1972	1974	1975	1976[d]
		1976				
Greece	1976	1977		1983		1977[l]
	1980	1978				1985[g]
Iceland						
Ireland	1976	1977		1977	1976	
Italy		1976		1966		
Luxembourg	1982	1961	1980	1976		1976[j]
Netherlands	1952	1969	1976	1970		1963[h]
	1979	1975	1977			1979
						1983[i]
Norway	1981					1977[b]
Portugal	1976	1977		1980		1976[d]
						1983[b]
	1983					1983[k]
Spain			1975	1972		
Sweden	1969	1969	1975	1969	1969[l]	1964[d]
	1981	1981		1981	1981	1973[b]
		1983				
Switzerland	1983	1971				1966[d]
						1969[b]
						1979[k]
Turkey	1983	1960				1983[d]
		1971				
United Kingdom	1974	1961	1974	1956		1974[j]
		1974		1968		1975[c]
				1974		1981[d]

continued

TABLE 1 *Continued*

SOURCE: Organization for Economic Co-operation and Development (OECD), *The State of the Environment 1985*. Paris: OECD, 1985, p. 242.

[a]Some federal countries such as Australia and Austria have laws at state level.
[b]Law on the general control of chemicals.
[c]Law on compensation.
[d]Law on nature conservancy.
[e]Law on public health.
[f]Law on protection of forests and forest areas.
[g]Law on noise and air pollution from motor vehicles.
[h]Law on nuclear energy.
[i]Law on soil.
[j]Law on noise.
[k]Law on territory planning.
[l]Essentially a specific administrative procedure.

Given the similarity of the problems and the symmetry of responses, it might be expected that environmental management is essentially the same in industrialized countries. In fact, it is difficult to compare environmental policies between countries because newly emerging environmental policies did not develop in a void. A number of important factors have created a framework that contributes to the specificity of responses.

Existing Related Legislation

In most countries, legislation concerning water supply and quality dates to the 19th century.[2] Initiated at widely differing times, ranging from the earliest such legislation in the Netherlands where water management represents an existential need, to the United Kingdom in the 19th century, to the United States, which—blessed with abundant resources—felt compelled to address this issue at a relatively late date. Similarly the regulation of industrial nuisances, essentially the impact of industries on their neighborhoods, follows the pattern of industrialization itself. The United Kingdom, having been the first to industrialize, first confronted industrial pollution. Germany followed at some distance, and many countries did not address this issue until the early 20th century. Countries also began to address issues of workplace safety and health. In the United States these issues were not tackled until a relatively late date.

Despite these differences, early linkages already existed between countries regarding these newly emerging policy areas, particularly through trade. Indeed, the United States first moved on pesticide safety in response to a trade ban for reasons of environment and public health: a British embargo imposed in 1925 on apples from the United States. Forced to comply with British requirements concerning arsenical residues on apples or lose an important export market, the United

States adopted the British standard for exported apples. It could hardly provide its own citizens with less protection than it afforded British consumers and proceeded to impose the standards on its own production, despite vigorous opposition from apple growers.[3]

Land use planning represents an important forerunner of environmental policy. Most European countries have known land use controls for decades, growing out of a limited concept of property in societies subject to monarchical rule. In the United States, federal land use planning does not exist and state programs are restricted by traditional reluctance to limit individual property rights and the Constitutional doctrine of "taking," which requires compensation for the diminution of rights through certain public actions.

The great surge of environmental legislation in the United States in the late sixties and early seventies is often seen as the beginning of modern environmental policy and is usually interpreted as a signal act of US leadership. Seen in a comparative perspective, the major acts—the Clean Water Act, the National Environmental Policy Act, the Toxic Substances Control Act, the Resource Conservation and Recovery Act, and the Clean Air Act Amendments of 1977—contain important legislative innovations. They have been used as benchmarks worldwide. At the same time, these acts filled a legislative void which had been allowed to continue longer than in most other developed countries and represent belated recognition of necessities that had been addressed elsewhere over longer periods of time. The United States had a strong tradition of conservation, a relatively short tradition of industrial safety regulation, and almost no tradition of land use planning. In the western United States, extraordinarily large areas of federally owned lands permitted the federal authorities to act directly in a manner that was impossible elsewhere. The resultant environmental legislation reflected these pre-existing circumstances. In particular the environmental assessment requirement of NEPA is comprehensible only in the context of a country that had neglected land use planning and consequently did not dispose of the basic data and decision-making structures that had existed elsewhere for decades. In this respect, the United States of 1965 resembled the countries of the developing world more than those of Western Europe or Japan.

Political and Administrative Culture

The policy areas that were ultimately to form a core of environmental management—water supply, neighborhood protection, worker safety, and public health, as well as land use planning—developed independently of each other. Each responded to a particular need, and while linkages may have been recognized, it did not seem essential to the success of each endeavor to undertake them jointly. As a consequence, each of these policy areas was typically housed in separate administrative units, often with incommensurate hierarchical structures. For example, water quality aimed at integrating river basins, worker safety was linked to economic policy-making, and land use planning needed a strong local base.

In addition to their independent development, these institutions tended to reflect political and administrative traditions of the respective country. Bureaucracies in different countries, while exhibiting well-known structural similarities, also reflect characteristic differences determined by history, the constitutional framework, and the educational system. As a result, essentially similar administrative procedures as basic as the issuance of identity papers or the description of factual information are undertaken in a distinctive manner in different countries. Permits with equivalent effect will tend to be structured differently, rendering comparison difficult.[4]

One of the major innovations inherent in the concept of "environmental policy" is the recognition of linkages that exist between seemingly disparate policy areas and of the fact that their joint management is a condition of success in each of them. This requires close coordination between the existing areas and new issues such as air pollution, toxic substances control, waste management, or global phenomena such as climate change.

In most countries, environmental agencies were formed in several stages, and certain aspects of environmental policy are frequently still managed outside the environmental agency. In the United States, for example, marine pollution is in the Commerce Department, nature protection in the Department of the Interior, and there are no land use planning functions at the federal and few at state level; in Germany, marine pollution is in the Ministry of Transport and new chemicals must be notified to a unit attached to the Ministry of Labor, while land use planning is the responsibility of a third ministry; in the Netherlands, water quality is handled by the environmental authorities but all other aspects of water management by the Ministry of Transport. In Japan, the Ministry for Industry and Trade (MITI) plays a central role in most aspects of environmental policy that concern industrial production. There exists no universally recognized definition of the responsibilities that need to be assigned to a ministry to qualify it as "environmental." Frequently, the name preceded the reality of administrative authority as it is easier to identify the issues that need attention than to reorganize the structure of government.

Environmental Conditions

The natural environment varies from region to region. To the extent that environmental policies are designed to achieve certain environmental outcomes, they may be expected to be different from one region to the next. In economic terms, these differences appear as elements of comparative advantage. In other words, a company producing in Ireland with emissions primarily to the open ocean should face less stringent environmental controls than a company producing in the Ruhr region whose emissions affect densely populated regions, sensitive ecosystems, and rivers utilized by others. In practice it is not possible to fine-tune policies to reflect all variations in each environmental medium. Moreover, few emissions have no environmental impact whatsoever and many can have long-range or long-term impacts that are difficult to identify. Moreover, governments must maintain consistent policies for all affected persons to avoid arbitrary decisions and ensure predictability.

Different policies may sometimes even be needed to achieve identical environmental outcomes. An example is the preservation of lobster stocks in the United States and Canada through size limitations.[5] In this case, a smaller size limitation in Canada achieved better conservation than larger US limits because lobsters mature faster (i.e., at smaller size) in Canadian waters, which are warmer on account of the Gulf Stream. Other examples concern the control of photochemical smog in Southern California, which experiences long periods of sunshine and limited atmospheric exchange. Western Europe has less sunshine, lies at more northerly latitudes and has greater wind movement, and consequently experiences different smog events.

In some instances the impact of environmental conditions can be observed in the phasing of environmental measures or in differences in priority-setting. The United Kingdom (like Japan but unlike the rest of Europe) is an island with short, swift-flowing rivers. Modest levels of water treatment can achieve dramatic improvements in water quality, as demonstrated by the Thames. In continental Europe, even vigorous water treatment can still result in limited water quality, as demonstrated by the Rhine.[6] Even though the lack of water treatment in the United Kingdom is essentially transferring pollution to the oceans, it has taken twenty years to demonstrate the impact of such policies and to induce a shift in priorities to more closely resemble those of the countries most affected by deterioration of the North Sea.

Certain chemical substances react differently under different ambient conditions. For example, pesticides will typically volatilize more rapidly under hot climatic conditions, requiring larger applications to achieve comparable levels of receptor exposure.

These differences in environmental conditions are frequently overlaid by long distance effects of emissions, which are hard to detect and have tended to be identified only when they reach crisis proportions, as in the case of acid rain in Europe or the transport of toxic substances into the Great Lakes region of North America, or when their presence is very unusual, as with particulates from the United Kingdom found in remote Swedish lakes in the late 1960s or industrial chemicals in the tissue of Arctic and Antarctic animals.

Past Emissions

Environmental policy does not originate in a pristine environment. In many countries, accumulated environmental impacts, some of which are reversible only over very long periods, severely limit the options available to policy-makers. The countries of Central and Eastern Europe are a dramatic demonstration of the manner in which environmental liabilities can destroy the capital base of enterprises that appear to be going concerns.

It is often said that current generations must not appropriate the possibilities of future generations to utilize environmental resources. In practice, the current generation already faces a situation where past practices have eliminated numerous options that might otherwise have been available. This may require appropriate investments to remediate environmental problems and sometimes can produce rapid results. It may also foreclose any possibility of using certain substances. For example, it is well known that the characteristic London smog that persisted throughout the first half of this century was attributable in large measure to the practice of burning coal in open fireplaces to heat private residences. The smog events were successfully curtailed by banning this practice. Less well known is the fact that the burning of coal also involved emissions of trace amounts of lead. Since lead is not very mobile once emitted into the environment, these lead emissions have accumulated in the soil and are to be found in dust. As a consequence, the London environment is intolerant to additional emissions of lead—for example, from gasoline—and the United Kingdom presses for a more rigorous elimination of lead from other products than is advocated by most other EC countries. Even while the United Kingdom began to confront the health hazards posed by this lead reservoir, Germany assessed the health risks of lead in the environment and concluded that they were not sufficient to push for the elimination of lead additives in gasoline; however, the desire to limit sulfur and nitrogen oxide emissions from automobiles to reduce acidification (and to permit forests to recover from the accumulated acidifying deposits) caused the German government to advocate the introduction of unleaded gasoline almost simultaneously with the British, but for other reasons. The United States failed to adopt the White Lead Convention when it was agreed to by the International Labour Office in 1921. As a result, paints laced with lead continued to be used in US dwellings for fifty more years, creating a huge reservoir of lead, which will be released into the urban environment over centuries and create conditions similarly intolerant of additional lead burdens, albeit for different reasons again.

Economic Conditions and Social Preferences

There has been a vigorous debate in Western countries concerning the association of strong environmental policies and elevated levels of economic activity, particularly as measured by gross domestic product. Some observers view environmental quality as a luxury good for which demand rises as disposable income rises.[7] Many of these observations are based on empirical data derived from the past twenty years of environmental policy showing that levels of sulfur dioxide emissions started to fall as GDP grew beyond a "threshold" of approximately

$5,000 per capita. The assumption is that increasing economic activity will have the effect of increasing demand for environmental quality and consequently lead more or less directly to environmental improvement.

These arguments are flawed for two principal reasons. The empirical data are derived from a period when environmental policies are known to have been inadequate. Consequently their interpretation is liable to be misleading since they reflect a period during which significant environmental costs continued to be deferred. Moreover, they reflect a period when environmental management and economic policies were inadequately integrated, thus increasing the cost of environmental measures.

On the other hand there exists a level of economic development at which no resources beyond those required to meet basic human needs are available—essentially a subsistence economy. At this level, the trade-off between environment and economic activities is akin to the farmer faced with the need to consume the next year's seed stock to assure immediate survival.

At all levels of economic activity, however, social preferences for environmental quality may differ. Most developed societies have come to tolerate certain levels of environmental risk and actual pollution. Social choices concerning these risks are liable to differ depending on a wide range of factors. Countries (and even jurisdictions within countries) must remain free to determine these preferences through their own processes of social choice, insofar as these decisions do not have impacts on others.

Differing Pressures on the Environment

The intensity of human pressures on the environment varies widely. Some regions, centers of population and economic activity for the most part, are the focus of intense pressures. Those who live and work there benefit from the advantages of their location and suffer its disadvantages. Other regions, rural areas not used for agriculture for the most part, are characterized by an absence of human pressures on the environment, again resulting in specific advantages and disadvantages. It is difficult to determine how these differences are to be taken into account when comparing environmental management.

One approach is to focus on the areas of greatest intensity of use, which generally are also considered the motors of economic activity in the country or region, and to compare policies and practices for these areas. In most countries, special rules apply to areas like Southern California or the US East Coast, the Tokyo region, or the Ruhr area, reflecting the special circumstances of these regions. This generally requires analysis that goes well beyond the national level and takes into account the role and the discretionary flexibility of local authorities and regional government.

Another approach is to focus on national rules, on the assumption that they represent an average or at least a minimum standard that must be universally respected. The result is a clearer comparison but less specificity for practical applications.

Criteria for Comparing Environmental Management

Until recently, environmental management represented a marginal activity of public policy. In many countries, environmental agencies did not have the status of a ministry or remained subunits of some ministry which handled matters that were considered more important—typically health or housing. From an economic perspective, environmental affairs were viewed as a minor issue, of concern mainly as a possible impediment to the priorities of economic growth in terms of GDP. The steady increase in the importance of environmental management as a government priority tended to occur as a result of unremitting public pressure, a forceful expression of changing societal priorities.

Until the late eighties, differences in environmental management between countries were a curiosity rather than a serious problem. It was perhaps useful to understand how other countries conducted their environmental affairs, since these may certainly have some indirect impact on the conduct of policy elsewhere, but it was not essential. In recent years, environmental issues have multiplied and their interrelationships have become more evident. Our understanding of the fragility of many ecosystems has increased. Environmental protection measures have become more extensive both because human activities have continued to increase in volume and because their environmental impacts have been better understood. How other countries manage "their" environment has become a matter of growing concern. Failure to control chlorofluorocarbons in New Zealand will deplete the stratospheric ozone layer over Frankfurt. Continuing deforestation will contribute to global warming, threatening Dacca and Rotterdam alike with inundation. Extensive use of DDT in Mexico will contaminate fish in the Great Lakes (through air transport), and the ocean dumping of sewage sludge threatens a food chain that ultimately reaches Europe, Japan, and North America alike. Moreover, economic actors constantly fear that their competitors elsewhere are advantaged by less rigorous regulatory requirements, and superficial analyses lend themselves to false conclusions.

In addressing the emerging environmental challenge, most countries sought to supplement an existing policy structure for water management, public health, worker safety, and land use planning. Consequently, the first phase of environmental management can be described as a patchwork approach to a common problem, again complicating comparisons between countries.

Environmental policy in the nineties is different from policies pursued in the previous decades. These differences frequently reflect perceived inadequacies of earlier policies. The new approach has several important characteristics that distinguish it both from the early and disjointed efforts to contain industrial pollution and the first stage of more systematic environmental management. Environmental policy is integrated in the double sense that it is sensitive to the ecological linkages that exist in the natural environment and reflects the complex linkages that exist between environmental management and other areas of public policy.[8] It is procedurally complex, using environmental assessment, freedom of information, and public participation to establish priorities and to ensure implementation of measures that have been decided. It seeks an equitable distribution of costs associated with environmental management by ensuring as far as possible the internalization of related environmental costs in all economic activities. Indeed, this process of internalization has increasingly been recognized as creating desirable economic incentives to find the most cost effective means of environmental management; this in turn leads to the development of new tools of environmental policy such as pollution charges, resource taxes, or tradable permits.

There are now two principal reasons to investigate the hypothesis that the outcomes of environmental management are broadly comparable in developed countries, even though the processes by which these results are achieved may differ significantly:

- The increasingly significant international dimension of environmental management implies that countries will need to know what other countries are doing to protect the environment, beginning with their neighbors, if they are to achieve their own policy goals.
- As environmental management becomes more complex, more comprehensive and more effective, significant differences in levels of environmental control or degrees of internalization of environmental costs can cause noticeable economic distortions that impact the relative competitive position of the countries concerned both positively and negatively.

The literature comparing environmental management is sparse and largely limited to Western Europe and North America. It remains unclear just what must be compared to adequately assess environmental management in different countries. At least five dimensions need to be kept in mind.

Comparing Legislation

Comparative studies are a well-established field of legal scholarship. Consequently several authors initially thought that a comparison of legal requirements in developed countries would provide important insight into the state of environmental policy in those countries. The results have generally been unsatisfactory, mainly for two reasons. Environmental regulations are but part of an extended process, which begins well before the adoption of legislation and continues long beyond it. The legislative stage is certainly a key way station since it codifies agreements reached up to that point and defines the framework within which the process is to continue. Focusing on existing law often fails to capture this dynamic. While this is a problem with most comparative legal studies, it is particularly acute in an area that has been changing rapidly, like environmental regulation. Furthermore, environmental policy is an indirect activity: it seeks to influence human behavior with the ultimate goal of changing environmental conditions. Because the environment responds to laws of nature and not to laws made by people, it has proven difficult to achieve satisfactory results through legal analysis alone.

A two-tiered implementation gap exists: laws are not adequately enforced, and even adequately enforced laws do not change environmental conditions sufficiently. As a result, environmental policy has proceeded in cycles as it has become increasingly clear that certain measures or standards do not achieve the desired result. This has inevitably increased the complexity of the policy structure. It is now manifest that single measures of environmental performance do not exist: measurement is necessary along the entire pathway of pollutants to ensure that the ultimate goal—environmental quality—is actually achieved, and policy must follow actual environmental conditions. An OECD study of the early eighties put it succinctly: ". . . in practice, there is no single control procedure which can provide a safe barrier to the spread of pollutants, and thus safe environmental protection."[9]

Legislation is needed until satisfactory environmental conditions are attained, and that can frequently require many iterations of the legislative process. Exclusive focus on legislation does not capture the actual nature of environmental management. It can, however, provide useful insights concerning possible regulatory tools to employ.

Comparing "Standards"

In the search for more readily comparable aspects of environmental management, attention has turned to "standards." Generally expressed in technical terms, standards appear to offer a comparable basis for evaluating environmental policies in different countries. However, two difficulties exist in comparing standards: variations in the definition of standards and in their application in practice.

Several distinct types of standards exist.[10] It is common to distinguish among product standards, process standards, and environmental quality standards, but in practice further variations exist, including emission standards, exposure standards, and biological standards. Figure 1 provides a schematic overview of the possible points on the pollutant pathway at which standards or objectives may be set.

Attempts to "harmonize" standards internationally have proven difficult. The experience of the European Community (which has undertaken more extensive international harmonization of environmental standards than any other organization) is instructive. It indicates how apparently simple issues such as determining blood lead levels in humans, harmonizing water quality management or defining product standards for automobiles can lead to major complications.[11]

The EC has experienced what can only be described as competitive standard-

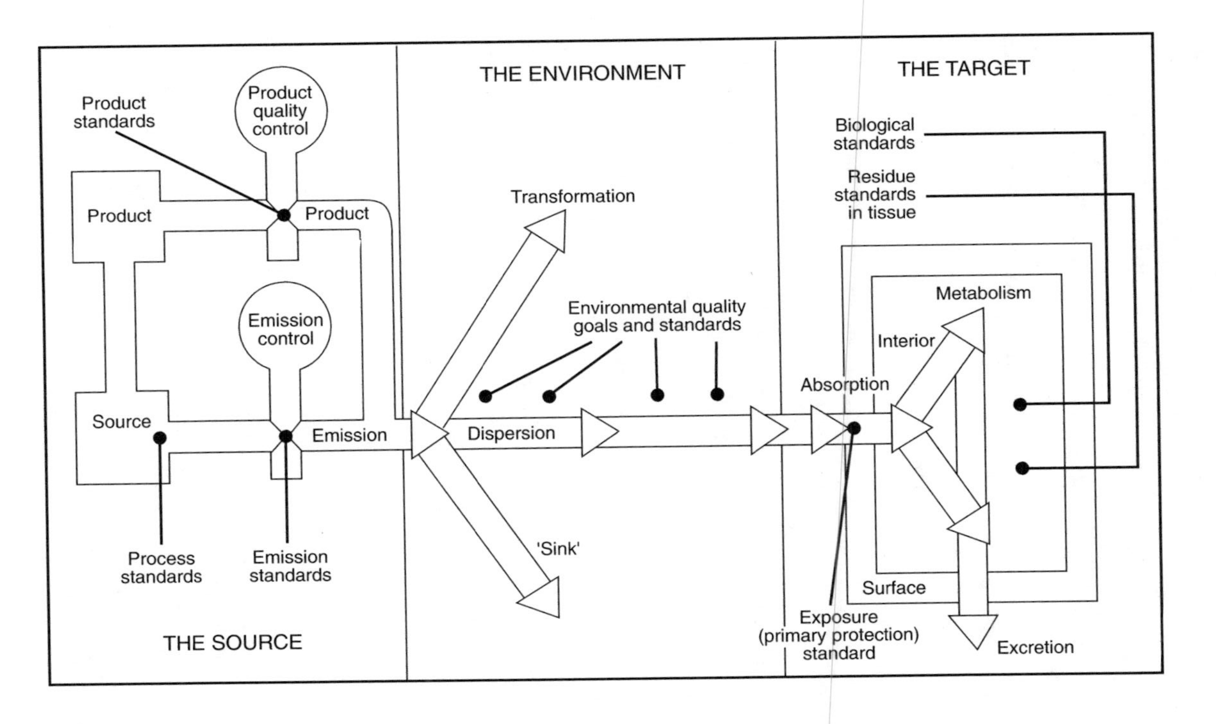

FIGURE 1 Possible points on the pollutant pathway at which standards may be set. SOURCE: Nigel Haigh, *Manual of Environmental Policy: The EC and Britain*. London: Longman (looseleaf), 3.2.

setting, as countries use the regulatory process to secure their position in the EC process and to protect the interests of domestic industries. For example, attempts to harmonize ambient air quality standards for sulfur dioxide were rendered difficult by German insistence on retaining its measuring system which was manufactured by a German company but was incompatible with the systems used by most other EC countries. The French government exercises close control over the regulatory process and has therefore been able to put in place domestic regulations rapidly when this appeared desirable to impel EC negotiations in particular direction. The pattern of French regulation with respect to environmental assessment, toxic substances control, waste oil collection and the management of packaging waste is suggestive in this regard.

The United States has also long promoted US environmental standards for adoption by other countries, with the purpose of improving environmental management and the incidental outcome of creating a market for US monitoring technologies. In practice, environmental standards can have a significant impact on the direction of scientific research and technological innovation, as public authorities first fund research to support the regulatory process and those affected by regulation subsequently move to implement requirements and to limit their economic impact through the introduction of new technologies.

What emerges from the EC experience is the difficulty of comparing, and consequently of harmonizing, environmental standards internationally. In some instances, the availability of a standard of comparison can be helpful. For example, the European debate on automobile emissions repeatedly used US emission standards as a benchmark for the evaluation of proposals. In other instances, US standards have been proposed as a guideline for international action without success. For example the ban on the use of chlorofluorocarbons in aerosols enacted in the United States in 1977 was pressed upon other countries as a reasonable first step.[12] Because such a ban would have had quite different impacts in different countries (some of which used fewer aerosols than the United States or had an industry structure based on smaller enterprises where changing technologies exacted higher penalties), it ultimately proved impossible to establish international action based on the US approach and negotiations only progressed when the United States embraced the EC approach of controlling production and use (which the EC had originally applied in a wholly unconvincing manner[13]). This illustrates the difference between national and international action. National standards can reflect the needs and criteria of a given country. International standards must be formulated in such a manner as to accommodate different control approaches while rendering their differences transparent and moving towards greater comparability in terms of outcomes. In the EC, this has made the directive the instrument of choice. Article 189 of the EC Treaties stipulates that "a directive shall be binding, as to the results to be achieved upon each Member State to which it is addressed, but shall leave to the national authority the choice of form and methods."[14]

EC Directives cover the entire range of environmental policy. They frequently provide a viable international standard of comparison since they are designed with this in mind. From this perspective it is hardly surprising that the EC approach underlies the Montreal Protocol, because it already reflects the needs of international action.

Comparing Procedures

In addition to standards, it is possible to compare procedures, that is, the rules and regulations governing the regulatory process. This assumes that the process has an impact on the outcomes that are achieved. It further assumes that the nature of the procedure to be followed influences the economic impact of the measures that are adopted as a result. It reflects the fact that environmental management has developed a substantial body of specialized procedural regulations.

The impacts of environmental degradation are difficult to predict. Traditional concepts to determine interested parties, based on property or membership in certain groups, do not apply without qualification. Geographic proximity can define affectedness, as can belonging to certain groups (for example, pregnant women or indigenous peoples whose diet includes a large proportion of fish from a single ecosystem). However, in other instances, users of certain products (automobiles) can be affected, or all those within a certain distance of major highways along which large numbers of automobiles may pass. Environmental phenomena such as stratospheric ozone depletion can affect every person capable of leaving the house. In many countries, environmental management has increasingly embraced the principle that everybody has an interest in environmental decisions, requiring procedures to allow anybody who feels affected, or potentially affected, to participate in decision-making as environmental regulations are elaborated. Some European countries and Japan continue to use traditional criteria (frequently relating to property ownership or neighborhood) to limit participation rights.

The need for public participation has engendered a number of aspects of environmental policy that are the proximate cause of its procedural complexity. Participation is not possible without an adequate information base that is publicly available. This informs the structure of many environmental assessment requirements and explains the central importance of freedom of information regulations to environmental management. Each of these can result in additional economic burdens on enterprises, which must comply with extensive procedural regulations prior to obtaining environmental permits and after these have been awarded.

Increasingly, the need for public participation across national boundaries is becoming apparent. Since environmental problems transcend national boundaries, there are no logical reasons why the participation rights of citizens must end at the border. However, this poses serious problems in relation to traditional legal systems. The European Community struggled with this problem in negotiating a directive on environmental assessment. After considering various forms of direct

TABLE 2 Comparison of Ambient Air Quality Standards

Country	SO_2 (ppm)	CO (ppm)	Suspended particulates (mg/m^3)	NO_2 (ppm)	Photochemical oxidants (ppm)
Japan	DMV 0.04 HV 0.1	DMV 10 8 HM 20	DMV 0.10 HV 0.20	DMV 0.04-0.06 or less	HV 0.06
USA	Primary EQS				
	AMV 0.03 DMV 0.14	AHV 9 1 HM 35	AMV 0.05 DMV 0.05	AMV 0.05	HV 0.12 (for Ozone)
	Secondary EQS				
	3 HM 0.5	the same as above	the same as above	the same as above	the same as above
Germany	AMV 0.05 30 MM 0.15	AMV 9 30 MM 6	AMV 0.15 30 MM 0.30	AMV 0.04 30 MM 0.11	
Canada	Max. Desirable Level				
	AMV 0.01 DMV 0.06 HV 0.17	8 HM 5 1 HM 13	AMV 0.06	AMV 0.03	DMV 0.02 HV 0.05
	Max. Acceptable Level				
	AMV 0.02 DMV 0.11 HV 0.34	8 HM 13 1 HM 31	AMV 0.07 DMV 0.12	AMV 0.05 DMV 0.11 HV 0.21	AMV 0.02 DMV 0.03 HV 0.08
	Max. Tolerable Level				
	DMV 0.31	8 HM 18	DMV 0.40	DMV 0.16 HV 0.53	HV 0.15

SOURCE: Japan Environment Agency (the JEA converted values where units of measurement differed from those used in Japan). Numbers are not strictly comparable because each country may use different instruments, analytical procedures, and requirements of the location of measurements.

Terms:

AMV: annual mean value; DMV: daily mean value; 8 HM: 8-hour mean; 3 HM: 3-hour mean; HV: hourly values; 30 MM: 30-minute mean.

Definitions used in each country:

Japan: the Environmental Quality Standards (EQS) are standards that should be maintained in order to protect human health.

USA: Primary EQS are standards necessary to protect human health. Secondary EQS are standards considered necessary to protect public welfare from known hazards.

Germany: The establishment of and operation of certain facilities must be subject to governmental approval. For approved facilities, neighboring residents cannot suspend operation of the facility for reasons of emission (hazards such as air pollution, noise, vibration, light, radiation, etc., affecting humans, plants, and animals), but they may claim for damages.

Canada: These are long-term targets and serve as a basis for long-term development of pollution prevention policies in non-polluted areas.

TABLE 3 Comparison of Water Quality Standards (for Health Purposes) [units: mg/l]

Country	Cadmium	Cyanide	Lead	Chromium (hexavalent)	Arsenic	Total Mercury
Japan	0.01	ND	0.1	0.05	0.05	0.0005
USA (in Ohio, for the Ohio R)	0.01	0.025	0.05 (dissolved Pb)	0.05	0.05	0.0002
Canada (Ontario)	0.0002	0.05	0.05-025	0.1 (total Cr)	0.10	0.0002
Holland	0.0025	NA	0.05	NA	0.05	0.005
China (draft)	0.01	0.05	0.1	0.05	0.04	0.001 (Inorg. Hg)

SOURCE: Japan Environment Agency. Numbers are not strictly comparable because each country may use different instruments, analytical procedures, and requirements of the location of measurements.

Terms:
ND: not detectable; NA: not available (either no standard, not published, or not convertible).

international public participation, the EC recognized that differences in procedural rights in bordering countries were sometimes irreconcilable. Consequently the directive contains only an obligation to provide the public authorities of a neighboring country with the information commonly provided the citizens of the country in which a project is located. Consultation of the affected public in the neighboring country is the responsibility of the authorities in that country, which are also charged with providing a response to the agency that is managing the assessment process. No general direct citizen participation proved possible. In several instances, projects close to the border have created tension between countries and caused demonstrations by citizens from one country who felt curtailed in their rights to be consulted.

Subsequent implementation of environmental regulations frequently involves broad participation as well, either because people find themselves affected by some practice with environmental impacts and make their concern known, or because potentially affected persons undertake informal monitoring activities that can help to set priorities for more formal enforcement efforts. Indeed, no country has implemented effective rules for environmental management without the direct participation of its citizens.

The bedrock of this procedural structure is a permitting system by which public authorities determine who may emit what amounts of which substances into which media at what times. The importance of such a system is identified by a number of European Community Directives which require member countries to bring these into existence and stipulate certain conditions for their operation (see Table 4). Given the number of possible variables, most permitting systems are highly complex. They have frequently grown over many years, with layer after layer of permits superimposed on each other.

Attempts are currently under way in several countries to develop more integrated systems of permitting that allow authorities to consider all environmental variables at the same time.[15]

Permitting systems reflect many variables, including the administrative culture of countries. Federal systems typically require more complex permitting structures, particularly when levels of government are not integrated (as in the United States), so that regulatory requirements and permits become primary means of communicating priorities between levels of government. Countries with a strong civil service tradition can rely on continuity in public administration to a greater extent than countries like the United States or the Netherlands where civil service is but a career station for many individuals. A strong civil service tradition generally results in more stable (some would say more rigid) administration. In countries with a strong civil service, comparable stability is frequently to be found in the staffing of private sector enterprises so that long-term relationships between administrators and enterprises can develop. Such relationships have both positive and negative results. All parties can afford to take a medium- to long-term perspective, relying on the continuity of the relationship to ensure that goals that have been postponed are also met. The need to deal again and again with the same people make commitments more readily enforceable since the alternative is to encounter difficulties at subsequent stages. On the other hand, lack of stability in relationships can lead to more aggressive pursuit of particular goals since delay is tantamount to failure. This fosters more confrontational relationships.

All of these factors in the permitting system will tend to impact the form and substance of environmental regulation. It is impossible in practice to determine the exact nature of these impacts since they can be both positive and negative.

Comparing Costs

The cost of environmental protection measures is an obvious measure of economic impact and appears to lend itself readily to comparison. Two problems arise, however. Comparing costs tends to obscure the related benefits, not only through environmental improvement but also through gains in efficiency. For example, the Canadian authorities moved in the mid eighties to significantly reduce acidifying emissions. This included major emission reductions at a nickel smelter in Sudbury, Ontario, the largest single source of sulfur dioxide emissions on the North American continent. The new requirements caused the operator of the smelter to rethink its smelting operations and undertake a far-reaching rede-

TABLE 4 European Community Directives with General Procedural Requirements

Directive	Title	Procedure
75/440/EEC	Drinking water	Plan of action
76/464/EEC	Dangerous substances in water	Authorization procedure
78/176/EEC	Titanium dioxide	Authorization procedure
78/659/EEC	Water for freshwater fish	Pollution reduction programmes
79/923/EEC	Shellfish waters	Pollution reduction programmes
91/676/EEC	Nitrates from agricultural sources	Action programme for vulnerable zones
77/795/EEC	Exchange of information—water	Monitoring; information exchange
75/442/EEC	Waste	Waste management plans; permits; record-keeping
78/319/EEC	Hazardous wastes	Permits; inspections and records; plans
84/631/EEC	Transfrontier shipment of toxic waste	Notification; record-keeping
76/403/EEC	Disposal of PCBs	Authorization
75/439/EEC	Waste oils	Permits; information campaigns; situation reports
85/339/EEC	Containers for liquids	Reduction programmes
86/278/EEC	Sewage sludge	Bans, authorizations
85/203/EEC	Air quality standards for NO_2	Consultation in border regions
84/360/EEC	Emissions from industrial plants	Authorization
88/609/EEC	Large combustion plants	Licensing; national programmes
594/91	Substances that deplete the ozone layer	Reports
82/459/EEC	Exchange of information—air	Monitoring; exchange of information
92/72/EEC	Air pollution by ozone	Monitoring
79/831/EEC	Sixth amendment	Testing; notification; inventory
82/501/EEC	Major accident hazards	Safety report; on-site emergency plan; off-site emergency plan; public information
2455/92	Export of chemicals	Notification (prior informed consent)
90/219/EEC	Genetically modified micro-organisms	Risk assessment
90/220/EEC	Genetically modified organisms—release	Notification
91/414/EEC	Authorization and marketing pesticides	Authorization
79/409/EEC	Birds and their habitats	Control of hunting; restriction on sale
3626/82	Trade in endangered species	Permitting; certification
3245/91	Fur from leghold traps	Certification
92/43/EEC	Habitats and species conservation	Establishment of a coherent European ecological network
85/337/EEC	Environmental impact assessment	Environmental assessment; public information
85/338/EEC	Information on the state of environment	Work programme; information gathering
90/313/EEC	Freedom of information	Access to information

sign, leading to both emission reductions and an increase in operational efficiency, which offset most of the costs. Since then, the design has been licensed to other companies, leading to environmental improvements elsewhere and increased profits for the originator of the design.

Environmental costs borne by others, for example as a result of emissions, are not paid by the polluter and consequently not properly accounted for in the price. Elimination of these costs consequently appears as an additional expense in relation to prior practice. In reality, environmental externalities are also indicators of internal inefficiencies since they reflect patterns of waste production. Consequently changes in internal efficiency can frequently contribute to the reduction of externalities without engendering additional costs. This basic fact of environmental management, which has not been analyzed systematically, explains why low- or no-cost solutions to environmental problems can frequently be found.

How are investments that improve internal efficiency while eliminating externalities to be accounted for? They were initiated by environmental considerations but their impact extends well beyond the domain of emission reduction. They may be economical by any standard, although alternate investments with a higher return may have been available. Depending on the interpretation of these questions, estimates of aggregate expenditures for environmental purposes can vary dramatically.

The Organization for Economic Cooperation and Development (OECD) has maintained a system of statistics comparing environmental protection expenditures in member countries (see Table 5). These figure are largely self-declared by the countries concerned. While there is no reason to assume that they are incorrect from each country's point of view, they will tend to reflect the variables outlined above, as well as significant differences in practices relating to the identification and recording of relevant data. Countries with more widely dispersed responsibility for environmental management and without a central statistical office that assesses these data independently of the self-reporting are liable to produce lower figures than countries with more concentrated authority and a continuing statistical exercise. OECD admonishes users of its data to "exercise great caution in interpreting" the table, which provides trends in pollution control expenditures by industry, and to "interpret carefully" the table, which identifies public research and development expenditures for environmental protection.[16] Despite these warnings, the data have been repeated frequently in the literature.

In a recent report on pollution control and abatement expenditures, the OECD explains: "It is difficult to assess the extent to which data are truly comparable over time or across countries. In some instances changes in definitions used, or the means of collecting data, lead to discontinuities in the time series for individual countries. These changes in the data collection procedures, or changes in the definitions used, pose more of a problem for some countries than others since, for practical purposes, the effect of such changes may, or may not, be substantial. Apart from typical 'end-of-pipe' installations, it is often very difficult to estimate

TABLE 5 Total (Private and Public) Pollution Control Expenditure as Percentage of GDP

Country	1972	1974	1976	1978	1980	1982	1984	1986
United States	1.22	1.44	1.57	1.57	1.62	1.47	1.44	1.47
Incl. households	1.50	1.76	1.92	1.92	1.99	1.85	1.85	1.88
Austria	...	...	1.09	1.10	...	...	...	...
Finland	...	...	...	...	1.31	1.24	1.10	1.16
France	...	...	...	...	...	0.86	0.84	0.89
Incl. households	...	...	...	...	...	...	...	1.15
Germany	...	...	1.26	1.33	1.45	1.45	1.37	...
Netherlands	...	...	...	...	1.11	1.18	...	...
United Kingdom[a]	...	...	1.66	...	1.57	...	...	...

SOURCE: OECD 1990, Table 2, p. 40.

[a]1977 and 1981 figures.

the part of expenditure that is really a result of environmental control. It is equally difficult, if not impossible, to measure increased expenditure caused by product control, as for example, when a pesticide or other chemical is banned and disappears from the market. . . . it should be noted that some data are survey-based, others are budgeted figures and others are estimates. Furthermore, some countries have provided data on a calendar year basis whilst others have provided information on a fiscal year basis. In a few cases data have been provided for combined calendar and fiscal years. Most of the data are provided on a current price basis, but a few figures are given on a constant price basis. Finally, the composition of the expenditure categories varies widely between countries, especially for the different environmental media."[17] This warning effectively implies that the OECD data should be used for broad comparisons only rather than as a precise indicator of costs.

Environmental Quality

The ultimate goal of environmental policy is environmental quality. It can be argued that policies should achieve comparable environmental quality, given comparable environmental conditions to begin with. In effect this approach compares ambient quality standards or other measures of environmental quality. It must face the fact that countries with lower standards, lesser procedures, and smaller costs may achieve better environmental quality than countries with high standards, rigorous procedures, and high costs. For example, the Rhine valley in Northern Germany holds an extraordinary concentration of population and activity. It is also an area through which two of the main transportation arteries of Europe pass. Consequently its environment is heavily burdened. At the same time, the Rhine Valley represents an area of high attraction for economic activity because it has relatively inexpensive energy resources (in particular lignite-based electricity), superb infrastructure, proximity to one of the most important markets in the world, and a profusion of vital services. It is not unreasonable to expect those who benefit from such locational advantages to accept higher burdens for environmental protection so as to maintain basic levels of environmental quality. Some states in North America show similar characteristics, providing one explanation why states considered to have high levels of environmental control have continued to attract significant investment. This is particularly true of an area such as Southern California where environmental characteristics (weather, proximity to the ocean, and accessibility of relatively untouched natural areas) represent critical elements of the quality of life and their preservation an important source of welfare. Regions reputed to be highly polluted do not normally attract the kind of economic activity that characterizes the most successful economies.[18] Thus comparison of outcomes represents a reasonable criterion, even if it is in practice difficult to apply.

In Europe, this approach has led to the development of the concept of "critical loads," defined by the 1988 Protocol to the 1979 Convention on Long Range Transboundary Air Pollution as "a quantitative estimate of the exposure to one or more pollutants below which significant harmful effects on specified sensitive elements of the environment do not occur according to present knowledge." Critical loads are currently being defined in several countries and at an international level for a range of common pollutants in a number of ecosystems. They allow a much more precise determination of environmental quality than the ambient standards previously applied since they explicitly take ecosystem stability as their point of reference and recognize that some ecosystems are significantly more fragile than others. The message from the study of critical loads has been unambiguous. Even countries with vigorous environmental policy and apparently quite unaffected ecosystems are experiencing cumulative ecosystem changes that will ultimately lead to broader environmental change. In other words, it remains necessary to reduce current environmental loadings by a sizable percentage, ranging from 60-90 percent for most common pollutants, if sensitive ecosystems are to be protected. The Netherlands has developed an explicit policy which recognizes that even the most vigorous environmental protection will make it impossible to protect certain ecosystems because these are affected by emissions originating in several countries and the Dutch government is unable to project the reduced levels of emissions for those countries that protection of the most sensitive ecosystems would require. Thus Dutch policy recognizes that these ecosystems will degrade progressively over the coming decades and that there is nothing the Dutch government can do to arrest this process.[19] This may be a reflection of the intensive research, assessment, and consultation that characterizes Dutch environmental policy.[20]

The Netherlands has taken the concept of critical loads one step further, developing a range of environmental policy performance indicators that apply to a number of themes such as climate change, acidification, or eutrophication, and to target groups such as agriculture, traffic and transport, refineries, or consumers.[21]

Implications for Science and Technology Policy

The above discussion suggests that international comparisons of environmental policy have become increasingly important just as they are becoming more difficult to undertake. Since environmental policy itself has become highly complex, comparisons of environmental policy will necessarily also need to be complex. These difficulties in comparing environmental policy have permitted public officials (and affected business interests) everywhere to claim that their policies are the most advanced, the most stringent, and the most effective. This litany is repeated by spokespersons from most OECD countries, and by picking the times and the areas where a country has been active it is even possible to provide proof for these mutually exclusive statements. For example, for decades, the United States lagged behind all other industrialized countries in its land use planning, worker protection and industrial safety measures. When it finally turned to remedying this situation in the late sixties, it adopted a series of remarkable environmental laws that appeared highly progressive, as long as the peculiar circumstances of prior US neglect were forgotten.

The role of science and technology is a constant among the diverse criteria for comparison of environmental policy. Significant environmental degradation is always linked to the introduction of "Western" technologies, which help to magnify the impact of humans on the environment. These technologies are everywhere the same, ranging from simple rules of conduct for sanitation and nutrition, which lead to a population explosion, to cutting edge electronic and biological technologies. Consequently appropriate control and the creation of necessary incentives to science and technology are a consistent need for the development of environmental policies. In general, only countries with a strong scientific community can participate actively in the definition, development, and implementation of policies concerned with the environment and sustainability because only these countries have the means to participate actively in the required processes.

It is generally assumed that countries undertaking particularly vigorous forms of environmental protection will be at a disadvantage relative to countries with lesser levels of protection. In practice the opposite may prove to be the case, but only if scientific research and technological innovation move in a direction that is consonant with environmental policy goals.

The assessment of environmental quality requires a long-term effort to establish appropriate monitoring facilities and to study and evaluate the results. This creates an important field of scientific research and technological development as the "interface" between society and the environment becomes better defined. It is unique in its consistent requirement for broad interdisciplinary cooperation, including not only various disciplines in a common area of activity—for example, the natural sciences or the humanities—but also interdisciplinary collaboration between the natural and the social sciences.

THE ROLE OF SCIENCE AND TECHNOLOGY POLICIES IN SELECTED COUNTRIES

Science is international in its self-image and generally international in practice. Scientific results should be capable of validation anywhere. This goal has proven illusive for the social sciences, which remain, for better or for worse, embedded in their own social and political environments. But even the natural sciences show surprising national variations, reflecting broader values that become part of the training of scientists and influence the kinds of questions they are liable to pose and the strategies they will pursue to find answers. It cannot be fortuitous that modern organic chemistry has its roots in Germany, that modern physics represents the outcome of an international dialogue within a small group of Europeans, that modern biology is rooted in the United States, or that France has played a special role in certain fields of human health research. Science policy is even more clearly a product of national circumstances, the result of subtle differences in the research community of each country and their cause. Countries have organized their research endeavors in strikingly different ways and have created different structures to determine the levels of funding available for science and technology and how it is to be spent. Comparing these structures is interesting. Why is it important? A number of reasons underpin the effort to compare science and technology goals for environmental policy and research.

- Environmental policy is international by its very nature. What one country does or does not do to protect its environment can affect the environment of other countries.
- Most environmental policy areas by now require some form of international coordination of measures. To develop equitable approaches it is important to be able to compare national measures.
- Countries must draw on all available scientific information when they undertake science assessments. Knowledge of research strategies and parallel assessments is important in avoiding mistakes and misunderstanding.
- Most countries do not have the resources to undertake research in all areas of environmental concern; some countries cannot undertake any such research. These countries must rely on information available elsewhere to base their policies.
- The existence of an active research community directly impacts the ability of a country to address an issue and in some cases limits the ability to even recognize its importance. Lack of independent domestic research on stratospheric ozone depletion was one of the reasons why European countries were particularly slow in responding to the emerging threats and resisting the self-serving information being circulated by affected industrial interests.

A comparison of science and technology goals for environmental research and policy is an uncertain venture. It requires an understanding of the strengths and weaknesses of environmental policy in a particular country, the structure of its research community, the manner in which science policy decisions are traditionally made, and the manner in which the specific linkage of science and policy has been resolved with respect to the environment. This paper will outline some of the processes and some of the implications of these interlocking issues for Canada, selected European countries, the European Community, and Japan.

Canada

Environmental Policy

Canadian environmental policy is characterized by the complex division of labor between federal government and provinces. The latter control practical environmental policy to a greater degree than the federal units of any other OECD country. This division of responsibilities has rendered the participation of Canada in international environmental agreements particularly difficult. On the one hand, Canadian federal authorities have shown great enthusiasm and support for international environmental activities—not least because their constitutional authority for foreign affairs is broader than in most other areas of policy. On the other hand, the provinces, which control most of the policies that need to be adopted to meet international obligations, have not generally been willing to have their hand forced by international agreements negotiated by the federal authorities.

Canada is a very large, extremely sparsely populated country. Consequently it has tended to confront the environmental impacts of human activity and modern technology at a relatively late date because visible degradation of the environment remains rare. For example, Canada adopted controls on acidifying emissions well after most other OECD countries because it argued that its own emissions were largely dissipated over very large areas and were not the direct cause of some of the phenomena that were being attributed to acidification. Only the need to match its pressure on the United States with comparable deeds of its own moved the Canadian provinces to adopt more stringent measures. Canada's size has also been a major factor in causing the country to have a greenhouse gas emissions profile that is highly disadvantageous—and relatively difficult to change. Long distances—and a cold climate—imply high transport and domestic energy requirements, which can be reduced only through major innovations and extensive investments in public and private infrastructure.

Canada is a country whose economy continues to depend in large measure on the production of commodities—primary economic goods extracted from the environment. Commodity markets provide less room for the recovery of environmental costs than most other markets and have been subject to long-term downward price pressures. At the same time, much commodity production is again relatively energy intensive, certainly when measured in terms of energy input per unit of GDP.

For all of the above reasons, Canada has engaged in a more active debate on issues pertaining to sustainability than many other OECD countries. In the area of general economic policy, the relationship between federal government and provinces is more balanced and an economy that depends on commodity production is more likely to benefit from more sustainable economic systems. The Canadian debate is characterized by heavy reliance on consultative procedures. The country pioneered the use of Roundtables to bring together government, industry, and environmental interests in the search for consensual approaches to the broader issues of environmental policy.

The Research Community

The Canadian research community is spread widely across the country, with extremely long distances creating major burdens for the exchange of information. Canada embraced the information highway with more enthusiasm than most other countries outside the United States. The distribution of research centers has implications beyond geographic spread as funding decisions at the national level are inevitably influenced by considerations relating to geographic distribution (particularly insofar as French speaking Quebec is concerned), which overlays more traditional scientific considerations.

Because of its dependence on commodity production, and forest products in particular, and because of its location as the largest Arctic country after Russia, Canada has developed particular strengths with regard to forestry research and research on extreme climates.

Science Policy

Canada has recently created a new Department of Industry, which encompasses several previously dispersed science and technology functions, in particular of the precursor Department of Industry, Science and Technology.

The Natural Sciences and Engineering Research Council, the Medical Research Council, and the Social Sciences and Humanities Research Council support university research and training. Their combined budgets totaled C$806 million ($520 million). Government R&D spending represents about 2.6 percent of the federal budget, the lowest figure for any of the G7 countries. However, provincial budgets include significant additional resources for science and technology. The Council of Science and Technology Ministers involves cabinet level representatives from the federal and provincial governments and is designed to provide a forum for addressing policy issues of common interest.

A significant proportion of the government's research funds is channeled to the Networks of Centres of Excellence programme, which was initiated in 1988 with a budget commitment of C$240 million ($156 million) for five years and recently renewed with additional funds. The NCE are required to include leading researchers from across Canada and to involve an active collaboration between researchers and the potential users of new technologies, industry for the most part or government agencies.

Environmental Considerations

Environmental research has not traditionally been the focus of a special institution. Consequently environmental activities have had to be funded through the traditional avenues of research support.

Environment Canada is a ministry with limited executive functions. Consequently funding of environmental research to support Environment Canada's mission represents one of its most important activities.

The Government's Green Plan, announced in 1990, is a national strategy to take a step towards sustainable development in Canada. Twenty-five of the Plan's initiatives have a significant science and technology content, amounting to C86.6 million ($56 million) in 1992-1993. Among these are Global Warming Science Program (C$4.8 million), Technology for Environmental Solutions (C$2 million), and Eco-Research (C$3.3 million). The latter program encourages cross-disciplinary research and training on environmental issues.

The Networks of Centres of Excellence included two environmental topics in its 1994 call for proposals but only one was funded, a project on sustainable forest management. No network was funded to address the linked issues of trade, competitiveness, and sustainability, which had also been identified as a priority in the call for proposals.

The peculiar distribution of authority between the federal government and the provinces makes the provision of funding by the federal government (essentially subsidies) an important instrument to leverage desired outcomes, either from provincial governments or from industry. These subsidies frequently support specific research and development efforts designed to permit the more rapid or more efficient adjustment of policies or enterprises to the demands of federal government environmental priorities. An example of this process was the promise of subsidies to a major mining and smelting operation in Sudbury, Ontario to facilitate reductions in what was at the time the largest single source of sulfur dioxide emissions in North America. The result of this effort was the development of new smelting technology, which not only reduced emissions dramatically but proved to be economically superior to other available technologies and can therefore successfully be commercialized and sold to other companies. In this case, the promise of subsidies alone proved an effective tool, since the profitability of the new technology obviated most elements of subsidy.

France

Environmental Policy

French environmental policy has been dominated by the national administration, which has more extensive authority and control than its counterpart in most developed countries. France has frequently been able to legislate new measures rapidly, well in advance of other countries, largely because the French administration is capable of ensuring passage of most laws it deems important. The counterpoint to this forward looking action is the almost total inaction when confronted with issues the administration is not convinced are of major importance. For example, France was one of the last countries with significant levels of production of ozone depleting substances to remove representatives of the affected industry from its delegations and to act decisively to control the production of the most important ozone depleters. Only massive citizen protests were able to slow down a process designed to regulate the Loire, the last major free-flowing river in France. Because higher echelons of the environmental administration are frequently recruited from the École des Ponts et Chaussées, essentially a national training institutions for civil engineers, there is a tendency to believe that engineering solutions are available for most environmental issues.

France has maintained an unwavering commitment to the development and extensive utilization of nuclear power generation. While this commitment is presumably linked to a similar commitment to the development of its own nuclear arsenal, the result has been an unparalleled investment in nuclear power, occasionally even over the protests of local populations. France is one of the few countries to operate a nuclear fuel reprocessing facility.

The Research Community

France has one of the highest proportions of researchers in the work force, with 5.2 per 1,000.

Like much else in France, the research community is centered on Paris. Despite continuing attempts to promote decentralization, most leading scientific institutions—and all important decision-making positions—are in Paris. This high degree of centralization has enabled the creation of a number of strong research institutions at the national level.

French public administration is characterized by the existence of a limited number of highly selective "grandes écoles," institutions to train an administrative elite. The grandes écoles are not, however, major centers of research, which is typically conducted in specialized institutions that are sometimes attached to universities. Graduates of the grandes écoles will tend to "colonize" certain government functions. For example, a preponderance of engineers from the School for Bridges and Roads is to be found in the Environment Ministry. In consequence the perspective of a given school can predominate in the government's approach to issues such as the environment.

A significant portion of French industry is under the influence of public authorities, and leadership of many of the major industrial enterprises is in the hands of persons from the grandes écoles who will frequently also have served in administrative positions at some time. The result is particularly close cooperation between government and industry.

Establishing closer linkages between research and universities is one of the priorities of current policy: 990 of the 1,370 research units receiving support from the National Research Council are formally associated with universities. However, research and teaching functions remain relatively separate, and the universities as such play a limited role in the determination of research priorities.

Science Policy

France devotes a larger proportion of its national budget than most countries to research (5.99 percent). However, this figure is somewhat distorted by the absence of any significant sources of public research funding at other levels of government. Over the past decade, research and development expenditures have risen steadily as a percentage of GDP, from 2.01 percent in 1981 to 2.36 percent in 1992. These increases stabilized in 1990.[22]

Following parliamentary elections in March 1993, responsibility for science and technology policy were assigned to a new Ministry for Higher Education and Research. Until that date, research and higher education functions had generally been separate. The National Research Council provides cross-cutting coordination of research policy and is the avenue for providing government funds to centers of research excellence.

In comparison to other countries, at 50 percent, government funds represent a particularly large proportion of total research and development funds. This implies that government resources also play an important role in funding pre-competitive industrial research.

Environmental Considerations

The French structure of policy-making and research is not particularly well suited to topics that are interdisciplinary and require the cooperation of large numbers of research institutions. In many instances, French researchers have used opportunities for international cooperation as a structure to complement the fairly distinctive French structure. The absence of independent policy research institutions—due to the absence of funding for this kind of work from sources outside government—leads to a comparative paucity of independent scientific research to prepare policy. Much research is either purely academic in orientation or directly related to technological applications. Again this represents a hurdle for addressing environmental issues.

The lower status of environmental research is reflected in the priorities articulated for the 1995 civil research and development budget, which lists three first tier priorities (medical and biological research, civil aviation, and scientific employment and training through research) and five second tier priorities, among them the environment.

The exceptionally powerful position occupied by the Presidency and the French Administration render parliament a relatively weak institution. While major policy decisions are submitted to parliament and are subject to debate in the Assembly and Senate, it is unusual for this process to result in significant changes.

Germany

Environmental Policy

German environmental policy can be divided into a pre-*Waldsterben* and a post-*Waldsterben* phase (*Waldsterben* being the German term for forest dieback, which was widely observed in the early eighties and attributed to acidification, even though the link has never been conclusively established). In the pre-*Waldsterben* phase, German policy was largely technocratic and skeptical of claims to extensive and pervasive environmental degradation. It placed significant faith in traditional approaches to industrial safety, land use planning, and the permitting process. Emblematic of this approach was the powerful rearguard action fought by German negotiators against adoption of the Convention on Long Range Transboundary Air Pollution and the undermining of work in the OECD on transboundary pollution. Water management was strictly oriented to state of technology considerations, and technology forcing policies were considered undesirable. With a powerful chemical industry, German authorities were unable to legislate domestic toxic substances control legislation without a strong mandate from the European Community, which largely overrode resistance from domestic interests. In procedural terms, German policy rejected legal standing for environmental groups and restricted the citizens' rights to participation to the property rights. Foreigners who were not German property owners had no rights to participate, even in administrative proceedings.

Since 1982, these policies have largely been modified with initial pressure coming from citizen activist groups and, surprisingly, from core conservative constituencies such as foresters concerned about *Waldsterben*. The experience of Chernobyl reinforced the transition to environmental policies based on the precautionary principle, a concept that lay dormant in German environmental law until the eighties when it was used to justify the dramatic shift in government policies without having to disavow previous measures. The German government has not shied away from technology forcing legislation, for example, in the areas of waste reduction and packaging.

German environmental policy is characterized by the unique distribution of functions between the federal government and the Länder. Federal environmental legislation is implemented by the Länder. There exists no federal environmental agency with implementing authority. Under these circumstances, federal legislation and regulation are typically highly detailed and the federal government must take recourse to indirect means of achieving its policy objectives, not least of which are research funding and the extensive use of subsidies to promote environmental management.

The Research Community

As with many other aspects of German life, the research community is characterized by the experience of political division and unification in the postwar era and the strong role of the Länder in funding higher education and research.

German universities have traditionally been institutions devoted to research, with teaching a dependent function of the research effort. In the nineteenth century, based on models developed in Göttingen and the Humboldt University in Berlin, Germany pioneered the development of research-based higher education, resulting in dramatic scientific discovery, including many of the fundamental developments that underpin the chemical industry. German universities are funded and managed by the Länder. A federal role in higher education evolved in the sixties as the expanding needs of higher education outstripped the availability of funds at the level of the Länder. The need to attract federal funds forced the Länder to cede some authority to the federal authorities, particularly in the area of research.

The chance occurrence that the Berlin Academy of Sciences, the leading institution of its kind until World War II, was located in the part of Germany that came under Soviet control, forced the development of new science policy institutions independent of the academies and their traditional functions in postwar Western Germany. These institutions have generally been transferred to the new Eastern Länder.

Similarly the headquarters of the former "Kaiser Wilhelm Institutes" lay in Eastern Germany, leading to the creation of a new network of research institutes named for Max Planck. In the German tradition, these institutes are built around individuals rather than around themes, at least in their initial definition of tasks. As a result, they have tended to be strong in mature fields of research and relatively slow to take up emerging topics such as the environment. In general it can be said that German universities have adapted with great difficulty to the demands of the type of interdisciplinary research required for environmental policy.

Science Policy

In 1992, Germany spent 2.53 percent of its GDP on science and technological development activities. Of this amount, 22.1 percent was provided by the federal government (representing 4.31 percent of federal budget outlays), 16.1 percent by the Länder, and 59.5 percent by industry. The latter figure decreased from 70.1 percent in 1987 and 65.6 percent in 1990 but is still high in comparison to other EC Member States.[23]

The distribution of responsibilities between the Länder, which are mainly responsible for R&D in the universities, and the federal government, which is mainly responsible for non-university R&D, is characteristic for Germany.

Current research priorities are to reconstitute and complete the research system in the new Länder (former German Democratic Republic); to assure a high level of basic research (20 percent of total R&D expenditure); to promote strategic technologies in the precompetitive field (in particular, information technologies, miniaturization of electronic and mechanical systems, biotechnology, research on advanced materials, research for transport, energy, and concentration on interdisciplinary research; to improve the innovation capabilities of small and medium enterprises; to continue preventive research (in particular, ecology, health and social problems, space, and polar research); to strengthen international cooperation in research and technology development; to continue public long-term programs (fusion and space research).

Environmental Considerations

The federal Ministry of the Environment is responsible for a Federal Environment Agency, which undertakes information and research management tasks for environmental policy development. All other environmental research is funded through the federal Ministry for the Future.

Germany has a highly developed structure of advice and consultation that permits the participation of a wide range of interest groups in the formulation of public policy. Two institutions are particularly relevant in the environmental field. The Council of Experts on the Environment is a group of academic experts who provide research-based advice to the federal government on topics of importance to environmental policy. Over the years, the Council has produced numerous reports that can be viewed as way stations in the process of structuring public debate on major environmental issues. In recent years, the Bundestag has established several Study Commissions to address environmental issues, utilizing a peculiar instrument of the German parliamentary system that has proven particularly suited to environmental ends. Study Commissions are composed of an equal number of members of parliament and outside experts (nominated by a political process) and have no decision-making authority. Over the past two parliamentary sessions, one Study Commission has addressed "Preventive Measures to Protect the Earth's Atmosphere."[24] Its recommendations played an important role in the development of German policy on stratospheric ozone depletion and climate change. During the late parliamentary session, a further Study Commission considered the issue of material flows without coming to specific recommendations. Both Commissions are being continued in the current session.

United Kingdom

Environmental Policy

British environmental policy has long been a matter of controversy in Europe, with one commentator going so far as to call Britain the "Dirty Man of Europe." This diatribe was recently returned by the British environment minister who accused Denmark of hypocrisy in the Brent Spar incident, claiming the Scandinavian country was in fact "the dirty man of Europe." Certainly Britain has taken an approach to environmental management more closely informed by (British) scientific research and less inclined to take a precautionary approach when confronted with inevitable scientific uncertainty.

The United Kingdom invented pollution along with the industrial revolution. It was confronted in the 19th century with the practical problems of managing untrammeled economic growth based on technological innovation. Its early measures to limit damage to neighbors of industrial facilities through HM Alkali Inspectorate and its work on water pollution, the provision of safe drinking water, and the development of wastewater treatment technologies were pioneer achievements of environmental management. At the same time, Britain's location on an island with short, swift-flowing rivers and few occurrences of poor air circulation created powerful incentives to distribute and dilute emissions with a view to avoiding harmful impacts on the immediate environment. Relatively modest efforts brought dramatic improvements in environmental quality, and for several decades, Britain was a staunch defender of environmental quality standards as the measure of success in environmental management.

It was not until the eighties that Britain came to accept its contribution to long-range pollution, including acidification in Scandinavia and pollution of the North Sea from land-based sources.

Britain has a strong tradition of conservation and humane concern for animal welfare, which has marked some of its policies in the area of agriculture and the environment.

Hardly any country has a stronger record of fulfilling international commitments once entered into. This has tended to make the British government particularly cautious when negotiating international commitments. Frequently it has taken direct participation of British researchers in international environmental research to convince the government of the need for action. In the early stages of the dispute about stratospheric ozone depletion, the lack of active participation of British atmospheric researchers contributed to the deeply skeptical appraisal of the available evidence by the British authorities. Even discovery of the Antarctic ozone "hole" by a British team did not fundamentally alter these perceptions.

Similarly it took a joint project of the British Royal Academy and the Norwegian and Swedish Royal Societies to create the scientific record needed to induce government action on acid rain. A comparable role was played by the Water Research Center (at the time still a government research establishment, since then privatized) with respect to North Sea pollution. British scientists have long played a prominent role in global change research.

Increasingly British environmental researchers are part of international cooperative research efforts, frequently funded by the European Community and involving partners from other European countries. These cooperative research programs tend to focus on large, complex interdisciplinary research tasks, typically like those required for environmental management.[25]

The Research Community

The number of scientists and engineers engaged in research and development in the United Kingdom is 4.5 per 1,000 of the labor force, high in comparison with most European countries.[26] A major portion of the research community is affiliated with higher education, which expanded substantially in the sixties and seventies. In addition, numerous government departments maintain their own research laboratories, in particular the Department of Trade and Industry; the Ministry of Defense; and the Ministry of Agriculture, Forestry and Fisheries. The Research Councils also directly support a number of laboratories. The Department of the Environment does not have major dedicated laboratories, although the recently privatized Water Research Centre was in large measure concerned only with environmental research even though it transcended the sphere of responsibility of the Department of the Environment.

The British government has engaged in systematic scrutiny of all research establishments that depend on government funding with a view to rationalizing the structure or privatization.

Science Policy

In 1992, Britain spent 2.12 percent of its GDP on science and technological development activities. Of this amount, 35.4 percent was provided by the government (representing 3.01 percent of budget outlays) and 49.7 percent by industry. In 1993, defense research accounted for 45 percent of government research and technology development expenditures. Government R&D has fallen progressively in real terms from 6.8 billion ECU in 1992-1993 to 6.02 billion ECU in 1995-1996 (at 1993 prices and exchange rates).

Higher education institute funding is channeled primarily through the six Research Councils. A new Research Council structure went into effect in 1994 with stronger links to the central government, with a Director General based in the Office of Science and Technology. Government Department laboratories, such as those of the transport, agriculture and forestry, and defense departments, are being reviewed with a view to possible privatization. Support to industry, particularly to small and medium sized industry, through the Department of Trade and Industry now concentrates on technology transfer, consultancies, standards, awareness, and best practice and has moved away from the generation of technology, an area that now falls largely within industry responsibility.

Following publication of a White Paper, a Technology Foresight Programme was launched to identify priority market/technology sectors of most relevance to industrial users and to assist the formulation of government science and technology policy. Fifteen broad areas have been identified for further analysis: agriculture, natural resources and environment, chemicals, communications, construction, defense and aerospace, energy, financial services, food and drink, health and life sciences, information technologies and electronics, leisure and education, materials, manufacturing production and business processes, retail and distribution, and transport.

Responsibility within the government for research funding has shifted twice over the past three years. In 1992, science policy and funding was moved from the Department of Education to the Chancellor of the Duchy of Lancaster (the Office of the Prime Minister), and an Office of Science and Technology (OST) was created. This was widely viewed as enhancing the role of science and improving the prospects for funding. In the most recent change within government, the OST was removed from the cabinet office and transferred to the Department of Trade and Industry, a large and powerful ministry with interests of its own in relation to science policy.[27]

Environmental Considerations

One of the six British research councils is devoted to Environmental Science. This Environmental Science Research Council (ESRC) funds primarily research in institutions of higher education and supports a number of research centers. In addition, the Department of Environment has a research budget at its disposal to support research that is relevant to policy. That this structure can produce undesirable results is illustrated by a decision by the relevant body to cut funding for the Antarctic group led by Farmer in the years immediately following its crucial discovery of the "ozone hole" above Antarctica. This funding was ultimately replaced by US government sources.

There is no established procedure to ensure public participation in the formulation of government science policy as it pertains to the environment. A standing Royal Commission on Environmental Pollution provides one avenue to develop priorities for future policy development in relation to research funding. This Commission has traditionally been chaired by a prominent scientist although its membership is more widely representative of groups with an interest in environmental policy.

Parliament has not traditionally played an active role in the formulation of general policy guidance and the implementation of legislation, its principal role being legislative in the strict sense of the term. In the early eighties, the introduction of the select committee system created closer links between these committees and the government agencies they were to monitor. Select committees are not, however, a strong vehicle for public involvement.

A peculiarity of the British system is the existence of the House of Lords, composed of hereditary peers who are members by birthright and life peers who are members upon elevation by the Monarch based on a list provided by the Prime Minister. Life peerages are frequently awarded to prominent scientists so that some committees of the House of Lords have significant levels of expertise among their members. This is particularly true of the committees that deal with science policy and environmental affairs.

In general, British environmental policy is particularly sensitive to advice from British scientists. In several instances, the government has maintained environmental policy positions in the face of international pressure—for example, on acidification, pollution of the North Sea, or ozone depletion. However, once scientific advice changed, the government has generally been quite willing to shift its position.

European Community

Environmental Policy

The year 1972 was one in which international organizations needed to make their initial determination concerning the significance of the environmental agenda and their need to respond to it. Through the Stockholm Conference, the United Nations system concluded that the environment was marginal to its major priorities and could be entrusted to a newly created United Nations Environment Programme, which was given vast responsibility, few resources, and no authority.[28] UNEP was not integrated into the UN development system, which was emerging simultaneously, centered on the United Nations Development Programme.[29] The GATT established a Working Group on the environment, which was not convened for the following twenty years. The European Community launched its environmental activities with a political mandate from the newly constituted meeting of heads of state and government (which was later formalized as the European Council) but with no particular legal authority in the Treaties.

The fate of each of these three initiatives reflects the different character of the institutions involved. UNEP developed far beyond reasonable expectations in response to a pressing agenda of international environmental issues but failed to have a significant impact on the UN system. GATT was not confronted with the full range of environmental issues until the early nineties when these suddenly threatened to upset the delicate balance of an institution long accustomed to effective action based on an uncertain institutional mandate.[30] The environmental activities of the EC expanded so strongly that they were given a clear legal mandate in the Single European Act, which has been further elaborated in the Maastricht Treaty.[31]

EC environmental policy is implemented by means of more than 300 legal instruments (primarily Directives) adopted over the past twenty years.[32] Its development is marked by a series of five consecutive multi-annual Action Programmes. Beginning with the first Action Programme, which sought to give more specific form to a legally questionable political mandate, these documents have provided direction to EC environmental policy. Each of the Programmes has set out an ambitious agenda, and while the implementation of the details has been quite poor, the general thrust of action has indeed followed the directions indicated. Thus the recently adopted Fifth Action Programme, entitled "Towards Sustainability," can be taken as a strong indication of the direction of EC environmental policy even though its details are likely to prove difficult to implement.[33]

By now, EC environmental policy covers virtually every aspect of environmental management, ranging from water (12 major directives) to impact assessment and information (8 major instruments) and from waste (8 major directives), air (15 major directives), and harmful substances (16 major instruments) to wildlife and countryside protection (8 instruments) and climate change (5 instruments). Environmental policy of individual member states can no longer be adequately understood without incorporating the EC dimension of these policies.

Originally driven by sometimes hesitant recognition that the process of economic integration could not proceed without an accompanying programme of environmental management, EC environmental policy has developed a dynamic of its own—abetted by the existence of unambiguous authority in the EC Treaties following the changes introduced in 1986 by the Single European Act, including a new Title on the environment (Art. 130r-130t) and some other changes concerning the environment, in particular concerning harmonization.

It is not easy to identify the motors of EC environmental policy. In an initial phase, they were primarily economic, reflecting the view that the elimination of economic barriers between the Member States (six until 1972, nine until 1973, ten until 1981, and twelve since 1986) required measures to harmonize environmental policy. This view was reinforced by the need to draw on a narrow legal base until 1986, primarily Art. 100 (concerning the "approximation of such provisions laid down by law, regulation or administrative action in Member States as directly affect the establishment or functioning of the common market") and Art. 235 (a vague mandate, which permitted the EC to take unspecified measures necessary to achieve the goals set out in Art. 2). For the purpose of Art. 235, the 1957 mandate to achieve "harmonious" development was interpreted to imply attention to environmental issues beyond the simple harmonization of standards.

Formulation of a Title in the Treaties does not ensure action by the Community. The EC mandate for energy policy has been unambiguous from 1957 on, with special treaties for coal and nuclear power. Nevertheless no effective EC energy policy has emerged. Indeed it has taken the pressure of an environmental issue—climate change—to move energy policy forward. Similarly transport policy, theoretically a matter of eminent concern for a Community in which barriers are falling, did not develop effectively until the completion of the single market, despite a corresponding title in the Treaties since 1957. Thus it has been more than the simple logic of linking economic integration and the environment, or the expressed desire of governments, but the internal dynamic of environmental management itself that has impelled the EC to develop strong and frequently effective environmental policies. The provisions of the Single European Act concerning the environment were effective because they simply legitimized what was occurring anyhow. The need to develop environmental policies at EC level has been driven by the joint concerns of economic and political integration and the equally powerful pressure to find environmental measures at all levels at which they were needed, ranging from the local to the transnational. In the latter category, the EC represents a forum of convenience, the only international organization capable of undertaking systematic policy development.

The environmental provisions of the Maastricht Treaty build on the Single European Act although they are not only a development of its approach. While they reflect serious consideration of the need to reflect environmental concerns and the need to achieve greater sustainability, they also reflect some haste in drafting and relatively limited public discussion prior to their formal adoption. For example, while there is explicit though tortuous reference to sustainability in the aims of the EC,[34] this issue is not picked up in the operative articles concerning the environment.[35] This reinforces the impression that the reference to "sustainable growth" in the aims was largely declaratory and not meant to entail specific actions to operationalize it.

The Research Community

It is not possible to identify a specifically "EC" research community, apart from the approximately 7,000 persons employed by the European scientific facilities and organizations.[36] The EC draws on the researchers of its member states. These have widely differing levels of research intensity, ranging from a low of 1.2 researchers per 1,000 persons in the labor force (Portugal) to a high of 6 per 1,000 (Germany). Belgium, Denmark, France, Luxembourg, Ireland, and the United Kingdom also have research intensity ratios above 4 per 1,000.[37] Greece and Spain are below 3 per 1,000 and the Netherlands is 3.8.

Science Policy

Articles 130i and 130h were introduced by the Single European Act into the Treaty establishing the European Community as part of new Title VI concerning

Research and Technological Development. They provide for two complementary instruments for research and technological development at the Community level: a Framework Programme setting out the Community's research and technological development activities and the coordination of national and European research and technological development policies.[38] While activities under the Framework Programme are well established, the coordination function has remained dead letter. In a recent communication, the Commission proposes to "achieve better coordination by intensifying cooperation at the various stages of drafting and implementing RTD policy."[39]

The European Community's resources represent approximately 4 percent of the research and technological development resources available in its member states. Consequently it can have an impact on science policy only if it focuses attention on priority issues. The area in which its Framework Programme has had the most important impact concerns large projects requiring international and interdisciplinary cooperation. In these areas, the most appropriate partners are not always in a single country and the Community can provide the resources to ensure that cooperation occurs. At the same time, it can provide access to already established research networks for researchers in the countries with low research intensity. Indeed, cooperative research and technological development endeavors have become the hallmark of the EC research program (see Table 6). In addition to the various European scientific facilities and organizations, this program is conducted primarily through EUREKA (a program of collaborative R&D involving firms, universities, and research institutes, which seeks to increase European productivity and competitiveness through closer cooperation between firms and research institutes in advanced technologies, developing products, processes, and services with a world market potential) and the European Cooperation in the Field of Scientific and Technical Research (COST), a program that is to provide a framework for R&D cooperation. COST actions consist of precompetitive or basic research or activities of public utility.

Environmental Considerations

Environmental policy requires large-scale projects involving international and interdisciplinary cooperation. In this respect, it is ideally suited to the EC approach to research policy. While environmental research represents a priority of most European research programs, it does not emerge as a major area of activity in terms of total budget allocation.

While environmental issues crop up in many of the EC research and technological development programs, the principal activities are subsumed within the Environment and Climate 1994-1998 Workprogramme[40] (see Table 6). The total budget allocation for this work program is 482 million ECU for four years. Additional areas of work are included in the Marine Sciences and Technologies (MAST) program and the Joint Research Centre's program.[41]

TABLE 6 Scientific Content and Means of Implementation, EC Environment and Climate Workprogramme, 1994–1998

Theme 1	*Research into the Natural Environment, Environmental Quality, and Global Change*
Area 1.1	Climate change and impact on natural resources
1.1.1	Basic processes in the climate system
1.1.2	The climate system in the past
1.1.3	Climate variability, simulations of climate, and predictions of climate change
1.1.4	Impact of climate changes and other environmental factors on natural resources
1.1.4.1	European water resources
1.1.4.2	Agriculture, forestry, and the natural environment
1.1.4.3	Land resources and the threat of desertification and soil erosion in Europe
Area 1.2	Atmospheric physics and chemistry, interactions with the biosphere, and mechanisms of environmental change impacts
1.2.1	Atmospheric physics and chemistry
1.2.1.1	Stratospheric chemistry and depletion of the ozone layer
1.2.1.2	Tropospheric physics and chemistry
1.2.2	Biospheric processes
1.2.2.1	The functioning of ecosystems
1.2.2.2	Alterations of processes as a result of UV-B radiation
1.2.2.3	Biodiversity and environmental change
Theme 2	*Environmental Technologies*
Area 2.1	Instruments, techniques, and methods for monitoring the environment
Area 2.2	Technologies for assessing risks to, and protecting and rehabilitating, the environment
2.2.1	Methods of estimating and managing risks to the environment and to humans
2.2.1.1	Risks to human health
2.2.1.2	Risks to the environment
2.2.1.3	Industrial safety
2.2.2	Analysis of the life cycle of industrial and synthetic products
2.2.3	Technologies to protect and rehabilitate the environment
2.2.4	Technologies to protect and rehabilitate European cultural heritage
Area 2.3	Technologies to forecast, prevent, and reduce natural risks
2.3.1	Hydrological and hydrogeological risks
2.3.2	Seismic risk
2.3.3	Volcanic risk
2.3.4	Forest fires
Theme 3	*Space Techniques Applied to Environmental Monitoring and Research*
Area 3.1	Methodological research and pilot projects
3.1.1	Methodological research
3.1.2	Pilot projects
Area 3.2	Research and development work for potential future operational activities
Area 3.3	Centre for Earth Observation
Theme 4	*Human Dimensions of Environmental Change*
Area 4.1	Socio-economic causes and effects of environmental change
Area 4.2	Economic and social responses to environmental problems—towards Sustainable Development
Area 4.3	Integration of scientific knowledge and of economic and societal considerations into the formulation of environmental policies
Area 4.4	Sustainable development and technological change

The EC legislative process is largely controlled by high-level bureaucrats from the Member States. Because many decisions are taken with minimal public accountability, the EC legislative process is particularly prone to capture by interest groups. The European Parliament has a mainly advisory role, which can involve co-decision under certain complex circumstances.[42] These general observations also extend to the determination of environmental policy priorities and their linkage to science and technology policy.

Japan

Environmental Policy

Environmental policy in Japan reflects a characteristic interaction between private interests and public authorities at various levels. Binding codification of environmental norms generally only occurs after a lengthy period during which less formal (but still constraining) negotiations between public authorities and affected enterprises continue. Consequently in Japan, environmental law and published standards are only an incomplete reflection of environmental policy at any given moment in time.[43]

Major local government agencies such as the prefectures and metropolitan authorities have considerable autonomy.[44] The prefectures (and under certain circumstances the metropolitan authorities) are responsible for compliance with national regulations. In practice, however, they have the authority to apply their own standards, and this leads to a "three tier environmental control strategy."[45] The first level is a "recommendation" (also frequently known as administrative guidance). The Tokyo Metropolitan Authority currently expects industrial boilers to achieve 50 percent of the NO_x emissions permitted by national law. These recommendations are adjusted to the state of technology on a continuous basis. After several years of experience with a particular recommendation, the local authority will transform the recommendation into a local ordinance that is legally binding and enforced against all facilities. The third tier is a contractual agreement between the local authority and companies, under which the latter agree to achieve certain levels of environmental performance that may deviate substantially from the national norm. As a result, Japanese environmental policy is subject to a ratcheting process, much of which is not visible in the public domain.

At the national level, environmental authority is divided between many agencies, with the Ministry of International Trade and Industry (MITI) and the Environment Agency sharing responsibilities but MITI wielding far more power on account of its close ties to industry and its organizations. A recent observer summed up the process as follows: "In summary, environmental rule-making at the national level displays an interlocking set of processes in operation: the Environment Agency's technical hearing system, which ensures that all relevant information is considered prior to legislation and then formally repeats the process to determine the best time scale for implementation; a similar but less formal dialogue between MITI (and sometimes others) and industry, which allows MITI to determine the optimal time scales for implementation, where it has that responsibility; finally, the local authorities, again maintaining a detailed, open dialogue with industry and providing a direct route for the public's concerns to be fed back to firms as an obligation to continuously reduce their impacts on the environment."[46] This policy-making structure permits extraordinary levels of consultation and the integration of research, policy, and industrial development.

The Basic Law for Environmental Pollution of 1967 was replaced in November 1993 by a revised version. This places the concept of sustainability at the heart of Japanese environmental policy and for the first time incorporates significant elements of international and global responsibility. It also codifies the previous procedure by introducing an obligation to undertake voluntary pollution reduction beyond the requirements of the law: "Corporations are responsible for making voluntary efforts to conserve the environment such as reduction of the environmental loads in the course of their business activities."[47] The Environment Agency is required to draw up a Basic Environmental Plan (of course in consultation with other agencies, industry, and local authorities) that sets out measures to achieve the goals of the Basic Law.

In recent years, Japan has devoted increasing attention to the international dimension of environmental policy. Nevertheless, the strong influence of Japanese industry on the formulation of environmental policy and the absence of any significant commodity production in Japan itself lead to a strong emphasis on manufacturing industry and management of the waste cycle (emissions and other waste disposal) and a comparative disregard for environmental management issues associated with commodity production, the extraction cycle in which natural resources are turned into economic goods.

The Research Community

In comparison with other countries, research in higher education in Japan is weak, while research in government and industry scientific institutions is strong.

Science Policy

In early 1992, the Council for Science and Technology, an advisory body to the Prime Minister and the principal national deliberative body for science and technology policy in Japan, recommended a basic national science and technology policy. On this basis, the Japanese government established the "Basic Policy for Science and Technology" in April 1992. The fundamental goals of this policy are coexistence of humans in harmony with the Earth, expansion of the stock of knowledge, and construction of a society where people can live with peace of mind.

The government science budget has expanded by 5.3 to 6.2 percent in each of the years 1990-1993.[48] Total expenditures in FY 1993 were ¥ 2,266 billion, or 2.7 percent of general budget expenditure. Proportionately, this places Japan at the lower end of the spectrum of major OECD countries although comparisons are complicated by the existence of countries (such as Germany, Italy, or the United States) with significant levels of expenditure at the level of federal sub-units. The largest proportions of these funds are disbursed by the Ministry of Education, Culture and Science (almost 50 percent), the Science and Technology Agency (more than 25 percent), and MITI (13 percent). Funds are disbursed to a number of semi-independent institutions with program responsibilities.

Environmental Considerations

In recent years, environmental considerations have loomed large on the Japanese research and technology agenda. While the research budget of the Environment Agency remains modest (about 6 percent of the total government research budget in 1993), its growth rates have been above average in most of the past years. Because "coexistence of humans in harmony with the Earth" is the first of three priorities of the new Basic Policy for Science and Technology, environmental research is by now spread throughout the government's research budget.

There are no formal avenues for public participation at the national level in the determination of science policy priorities. Local authorities participate in this process and are considered to be representative of the interests of the general public. The Diet has a marginal role in this process as well. The driving interests are representatives of the research community and of government agencies and private institutions with ties to industry. The increased emphasis on environmental science and technology in all Japanese government programs reflects an assessment by these groups, in particular by industry, that taking account of environmental considerations will be a major factor of future production technologies and even constitutes an area of important competitive advantage.[49]

A recent review of Japanese environmental policy did not discuss science and technology policy issues.[50] This suggests that the Environment Agency plays a relatively marginal role in the determination of science and technology priorities, even when these are directly relevant to environmental policy. This task is largely undertaken by groups with close links to industry.

CONCLUSIONS

In each of the countries considered, a balance exists between similarities that derive from commonalities in the issues confronted by policy-makers and differences based on specific characteristics of the respective environmental policy, research communities, and science policy. Nevertheless a number of issues emerge.

Interdisciplinarity

Environmental science is a quintessentially interdisciplinary field of endeavor. This is universally recognized. Indeed, in several countries the promotion of more interdisciplinary research is seen as a goal of science policy, apart from the special needs of environmental policy. Nevertheless most countries struggle to realize these goals.

The difficulties in supporting interdisciplinary research are by now well known. The range of topics covered is typically wide so that peer groups are correspondingly small, with all the problems that entails. Academic organization, and the related rewards structure, are still prominently disciplinary in orientation. Training is still typically based on disciplinary specialization. Young researchers devote some of their most productive years to meeting the requirements of advancement rather than pursuing interdisciplinary topics. More senior researchers are then established in their disciplines and many see little advantage in taking the risks entailed in launching interdisciplinary work.

These problems all exist in relation to interdisciplinary work within broad fields of research such as the natural sciences or the social sciences. They become even more pronounced when dealing with issues that may require interdisciplinary approaches across natural and social science boundaries, as policy-related environmental research typically does.

No country has found a clear answer to these problems. However, increasing emphasis on cooperative research patterns may represent the most promising approach for science policy.

Network Formation

Many countries, particularly smaller ones or those with widely dispersed research communities, have been seeking to foster the development of wider networks of cooperation. In many instances, these are internationally oriented in that their membership is recruited internationally and the standards of excellence are based on international criteria. These networks respond to the difficulty in addressing complex environmental issues, because of both the dimensions of the needed research and the interdisciplinary nature of the work.

Networks have the added advantage that they can draw on the particular strengths of several institutions. No research institution, no matter how large and accomplished, is capable of covering all aspects of environmental research. Indeed no single research institution will have the capability to undertake high quality research on all aspects of a single environmental issue. On the one hand the issues are typically complex and require a range of skills not generally to be found in a single institution. On the other hand, even when these skills are all represented in a research institution, the standing of individual researchers within their peer group can vary widely. While some may be internationally recognized, others may have a lesser reputation.

Government Laboratories

Most countries have laboratories that were created at a time when scientific research was expected to expand in a predictable fashion. Many of these research establishments were government funded and designed to support a putative transition to nuclear energy or were engaged in military research. The collapse of the nuclear energy industry, combined with the end of the Cold War, has created a crisis for these institutions, many of which are very large and some of which have scientists with permanent contracts.

The maintenance of these laboratories represents a challenge for many governments, and the emergence of environmental issues represents a unique opportunity to redirect at least some of the research effort of these establishments. Experience in several countries indicates, however, that it is not easy to redirect the work of these laboratories. Many researchers are disinclined to change the focus of their professional activities, and many who may be inclined to do so bring a perspective defined by their past work, which can be inappropriate to the new agenda they are being asked to address.

The presence of these large scale facilities puts significant pressure on the available research funds, pressure that is felt acutely in periods of stagnating or declining funds or in periods when new issues—in this instance, environmental needs—require attention.

Policy Support or Technology Development

The research requirements of policy support and technology development are notably different. Essentially the former focuses on issues identification and specification while the latter seeks to develop responses to known issues. In some instances, research on issue identification can lead to the development of applicable new technologies but this is likely to be the exception rather than the rule.

The audiences for policy oriented and technology developing research are quite distinct. Increasingly, countries appear to be moving away from publicly financed technology development and increasing support for policy oriented environmental research. This reflects an assessment that the needs for policy oriented research remain pressing while the use of scarce public resources for technology development is less efficient than the use of private resources.

Industrial Research and Development

For many years, public authorities have felt a need to promote the development of environmental technologies. Industry long questioned the existence of cost effective solutions to many publicly mandated programs, and research was needed to prove out basic technologies and to support their initial application in practice. Countries appear to be reducing their involvement in industrial research and development (apart from the continuing provision of tax breaks, which are non-selective in character). This presumably reflects an assumption not only that industry is a better location for this kind of research but also that industry is now willing to invest in the development of environmentally sound technologies, either because this is the most cost effective way to meet public mandates or because this is viewed as a promising market in its own rights.

SOME PERSONAL OBSERVATIONS

This paper began by pointing out that science is the bedrock of environmental policy. There is no greater threat to the environment than our continuing ignorance about the consequences of human interventions. We know next to nothing about the species with which we cohabit on the planet. We remain remarkably ignorant about the effects of new chemical substances that we introduce into the environment, including their long term effects on humans. While conclusive proof of climate change remains elusive, it is hard to avoid the overwhelming impression that climate change is under way and we know little about the likely effects of the experiment we are undertaking with the climate system.

The success stories in the short history of environmental management—the control of organic emissions to water, reduction in emissions of acidifying compounds to the atmosphere, protection of the stratospheric ozone layer—all are based on a mixture of systematic research and happenstance. In particular the story of the ozone layer contains several fortuitous elements without which we would be emitting ozone depleting substances without limit: that the initial ozone depletion hypothesis was formulated by a scientist who pursued it beyond the limits of scientific etiquette, and that the British Antarctic Expedition chose to undertake (theoretically meaningless) ozone measurements that happened to uncover the "hole"—and that they had the courage to publish their results. It is not unreasonable to wonder whether there are phenomena that are not being pursued with comparable determination.

No country has yet confronted the systemic challenge implicit in the environmental imperative. What is most disturbing is that everywhere the incentives for research are environmentally insensitive. Interdisciplinary research remains the exception rather than the rule. In industry the development of environmentally benign technologies is evaluated by economic criteria that force innovation in one direction and render environmental benefits an added value rather than the principal outcome.

It needs to be recognized that industrial societies "underproduce" environmental quality and "underproduce" environmental innovation. When faced with such problems, society has previously developed creative approaches to changing the incentive structure, for example, by introducing protection of intellectual property as an inducement to innovation in general. What is needed is a structure that gives an environmental direction to the processes of scientific investigation, technological development, and economic change, which are currently driven by rewards structures that are environmentally insensitive.

NOTES

1. The following is based in part on: Konrad von Moltke, *Comparison of Regulatory Trends in the West and Central and Eastern Europe.* Report for the European Bank for Reconstruction and Development, London, 1993.

2. J.C. Day et al., "River Basin Development," in: Robert Kates and Ian Burton, eds., *Geography, Resources, and Environment* (vol. II: Themes from the Work of Gilbert White). Chicago: The University of Chicago Press, 1986.

3. James Wharton, *Before Silent Spring. Pesticides and Public Health in Pre-DDT America.* Princeton, N.J.: Princeton University Press, 1974, pp. 133-137.

4. Konrad von Moltke et al., "Rechtsvergleich deutsch-niederländischer Emissionsnormen zur Vermeidung von Luftverunreinigungen," Bonn: Institute für Europäische Umweltpolitik, 1985. Konrad von Moltke, *Handbuch für den grenzüberschreitenden Umweltschutz in der Euregio Maas-Rhein (Schriftenreihe Landes- und Stadtentwicklungsforschung des Landes Nordrhein-Westfalen—Landesentwicklung Band 1.045).* Dortmund: Institute für Landes- und Stadtentwicklungsforschung des Landes Nordrhein-Westfalen, 1987b.

5. Daniel C. Esty, *Greening the GATT. Trade, Environment, and the Future.* Washington, D.C.: Institute for International Economics, 1994, p. 272f.

6. Graham Bennett, *Dilemmas: Coping with Environmental Problems.* London: Earthscan Publications, 1992.

7. G.H. Grossman and A.B. Krueger, "Environmental Impacts of a North American Free Trade Agreement," paper prepared for a conference on the U.S.-Mexico Free Trade Agreement, Princeton University, October 1991.

8. The Conservation Foundation, *State of the Environment: A View toward the Nineties.* Washington, D.C.: The Conservation Foundation, 1987.

9. Organization for Economic Cooperation and Development (OECD), *Control Policies for Specific Water Pollutants.* Paris: OECD, 1982.

10. Nigel Haigh, *Manual of Environmental Policy: The EC and Britain.* London: Longman (looseleaf), 3.1-3.10.

11. Konrad von Moltke, *Possibilities for the Development of a Community Strategy for the Control of Lead.* Bonn: Institute for European Environmental Policy, 1987.

12. Richard Benedick, *Ozone Diplomacy: New Directions in Safeguarding the Planet.* Cambridge, MA: Harvard University Press, 1991.

13. Haigh, loose-leaf (see fn. 10), 6-12.2.

14. Office for Official Publications of the European Community, *Treaties Establishing the European Communities* (Abridged Edition). Luxembourg: EC, 1987, p. 282.

15. Nigel Haigh and Frances Irwin, eds., *Integrated Pollution Control in Europe and North America.* Washington, D.C.: The Conservation Foundation, 1990.

16. Organization for Economic Cooperation and Development (OECD), *OECD Environmental Data Compendium.* Paris: OECD, 1985, p. 285.

17. Organization for Economic Cooperation and Development (OECD), *Pollution Control and Abatement Expenditure in OECD Countries, A Statistical Compendium.* (OECD Environment Monographs No. 38). Paris: OECD, 1986, pp. 11-12.

18. Christopher J. Duerkson, *Environmental Regulation of Industrial Siting: How to Make It Work Better.* Washington, D.C.: The Conservation Foundation, 1982.

19. Ministry of Housing, Physical Planning and Environment (VROM), et al., *Interim Evaluation of Acidification Policy in the Netherlands* (VROM 80148/4088). The Hague: VROM, 1988. National

Institute of Public Health and Environmental Protection (RIVM), *National Environmental Outlook 1990-2010*. Bilthoven: RIVM, 1992.

20. See Second Chamber of the States General, *To Choose or to Loose. National Environmental Policy Plan*. (session 1988-1989, 21 137 no. 2).

21. Albert Adriaanse, *Environmental Policy Performance Indicators. A Study on the Development of Indicators for Environmental Policy in the Netherlands*. The Hague: Sdu Uitgeverij, 1993.

22. Commission of the European Communities, *Research and Technological Development. Achieving Coordination Through Cooperation. Communication from the Commission*. COM(94) 438 final, 19. 10. 1994, p. 58.

23. Commission of the European Communities, *Research and Technological Development. Achieving Coordination through Cooperation. Communication from the Commission*. COM(94) 438 final, 19. 10. 1994, p.55.

24. German Bundestag, ed., *Protecting the Earth's Atmosphere. An International Challenge*. Bonn: Deutscher Bundestag, 1989. Konrad von Moltke, "Three Reports on German Environmental Policy," in: *Environment* vol. 33 no. 7 (September 1991), pp. 25-29.

25. Kristy Hughes and Ian Christie, *UK and European Science Policy. The Role of Cooperative Research Networks*. London: Policy Studies Research Institute, n.d. (1995), pp. 50-80.

26. Commission of the European Communities, *Research and Technological Development. Achieving Coordination Through Cooperation. Communication from the Commission*. COM(94) 438 final, 19. 10. 1994, p. 58.

27. "Merger of Ministries Dismays Scientists," *Financial Times*, July 8/9, 1995, p. 4.

28. See Konrad von Moltke, "Why UNEP Matters," Paper prepared for the Sustainable Resources Use Program of WWF International.

29. Konrad von Moltke and Ginny Eckert, "The United Nations Development System and Environmental Management," *World Development* Vol. 20 No. 4 (1992), pp. 616-626.

30. Konrad von Moltke, "The Last Round: The General Agreement on Tariffs and Trade in Light of the Earth Summit," *Environmental Law* Vol. 23 (1993), pp. 51-531; Konrad von Moltke, The Multilateral Trade Organization: Its Implications for Sustainable Development," Paper for the Workshop on Enforcement of International Environmental Agreements, University of California, San Diego, Institute on Global Conflict and Cooperation, September 30, October 1-2, 1993.

31. Konrad von Moltke, *The Winnipeg Principles on Trade and Sustainable Development and the Maastricht Treaty*. Winnipeg: International Institute for Sustainable Development, 1995.

32. See Nigel Haigh, *Manual of Environmental Policy: The EC and Britain*. Harlow: Longman (looseleaf), 12.1. Cameron Keyes, *The European Community and Environmental Policy. An Introduction for Americans*. Washington, D.C.: World Wildlife Fund, 1991.

33. Commission of the European Communities, *Towards Sustainability. A European Community Programme of Policy and Action in Relation to Environment and Sustainable Development*. Brussels: Commission, 1992.

34. See above.

35. Articles 130r-t.

36. European Organization for Nuclear Research (CERN); European Molecular Biology Laboratory (EMBL); European Space Agency (ESA); European Southern Observatory (ESO); European Synchrotron Radiation Facility (ESRF); Institut Max von Laue-Paul Langevin (ILL); European Science Foundation (ESF); Joint Research Centre (JRC) in Ispra.

37. Figures for Austria, Finland, and Sweden are not yet available but these presumably also rank high.

38. Commission of the European Communities, *Research and Technological Development. Achieving Coordination Through Cooperation. Communication from the Commission*. COM(94) 438 final, 19. 10. 1994.

39. Commission of the European Community, *Research and Technological Development*, p. 2.

40. European Commission, Environment and Climate 1994-1998 Workprogramme. Edition 1994.

41. Commission of the European Community, *Proposal for a Council Decision for the JRC Programme*, COM(94) 68.

42. See Konrad von Moltke, *The Maastricht Treaty and the Winnipeg Principles on Trade and Sustainable Development*. Winnipeg: International Institute for Sustainable Development, 1995.

43. Konrad von Moltke, *Environmental Product Standards in the United States and Japan*. Report for the European Bank for Reconstruction and Development, London, 1993.

44. The following is based on David Wallace, *Environmental Policy and Industrial Innovation. Strategies in Europe, the US and Japan*. London: The Royal Institute of International Affairs, 1995, pp. 95-110.

45. Wallace p. 96.

46. Wallace, p. 103.

47. Cited in Williams, p. 104.

48. Organization for Economic Cooperation and Development (OECD), *Science and Technology Policy. Review and Outlook 1994*. Paris: OECD, 1994, p. 73.

49. Curtis Moore and Alan Miller, *Green Gold. Japan, Germany, the United States, and the Race for Environmental Technology*. Boston, MA: Beacon Press, 1994.

50. Organization for Economic Cooperation and Development (OECD), *OECD Environmental Performance Review: Japan*. Paris: OECD, 1993.

Can States Make a Market for Environmental Goals?

RICHARD A. MINARD, JR.
Center for Competitive, Sustainable Economies, National Academy of Public Administration

CONTENTS

TWO TRACKS

State governments and communities are experimenting with a wide array of approaches that they hope will improve both their environment *and* their economies. The advocates for environmental improvement and economic development are rarely one and the same however, and their approaches differ in many important respects. While some of these differences enhance their mutual effectiveness, the overall result appears to leave significant problems unaddressed. It appears that a more explicit connection between society's goals for the environment and for environmental science and technology is in order.

This paper will describe some of the state and local government efforts to make environmental policy and technology more forward-looking, more technically sophisticated, and more in touch with societal goals and expectations. The paper examines the roles of experts, elected officials, and the general public in these efforts. The paper is in four parts: (1) this introduction; (2) a look at efforts to set measurable goals for environmental quality and to define useful benchmarks or indicators of success; (3) a look at state science and technology (S&T) programs as they relate to environmental problems; and (4) a discussion of the relationships between environmental goals and goals for environmental S&T. The appendix includes excerpts from state and local publications on these topics.

A note of caution: this paper is the product of interviews with numerous leaders in the field, but it is not a survey of the 50 states and the numerous institutions within each state that are involved in environmental planning and science and technology development. Thus the paper does not touch on many of the exciting and innovative programs under way around the country.

The Carnegie Commission's report *Enabling the Future*[1] describes one of the nation's social and environmental dilemmas: a kind of massive market failure that inhibits the country from securing the environmental quality Americans want for the future. Individual firms have to focus on the next quarter's bottom line; governors and legislators must respond to today's crisis or political fad and risk the voters' ire if they spend money on problems that do not yield results before the next election. State regulators are slightly insulated from this pressure, but receive their funds from legislatures and many of their marching orders from the U.S. Congress or the U.S. Environmental Protection Agency. Even academic scientists' work is often driven by the availability of research funds, which in turn may reflect today's crisis. Most of society's incentives reward short-term fixes and leave many difficult and obscure long-term problems unaddressed.

Enabling the Future lists three contexts in which explicit goal-setting activities may help the nation: to respond to a crisis, such as a disease, military threat, or failure to remain economically competitive with other nations; to provide a more coherent and efficient direction for particularly complex issues such as energy policy; and "perhaps the most difficult to respond to . . . situations in which important needs or problems are clearly seen by some (for example, some part of the S&T community or a public interest group) but are not universally recognized, and there is no consensus on the seriousness of the problem or on how to address it. The current question of how to respond to predictions of global climate change may be an example of this."

Enabling the Future stresses the importance of linking technology goals to societal goals that go deeper than the pervasive goal of creating new jobs. A companion report, *Science, Technology, and the States,*[2] notes that this is difficult: "Partnership between government, industry and academia requires consensus about broad issues. Few states have a formal process for developing such views." Nevertheless, states and communities are using a variety of planning and goal-setting approaches to move public investments and policies in more thoughtful directions. All of these efforts start with the premise that strategic decisions based on some modicum of data, analysis, and thought will yield better results than would the political system if it were left alone.

STATES' GOALS FOR THE ENVIRONMENT

After more than two decades of federal dominance of the environmental policy agenda, states are beginning to reassert their role in setting goals and priorities for environmental quality within their borders. State environmental agencies have assumed greater responsibilities as they have developed additional technical and legal capacity.[3] Three approaches to setting priorities and implementing a strategic agenda are capturing state attention: (a) comparing the risks posed by environmental problems and comparing the efficacy of alternative strategies for risk reduction; (b) tracking environmental "indicators" and publishing them in annual "state-of-the-environment" reports; and (c) setting measurable environmental goals and tracking progress toward them.

Together, these three approaches have the potential to help states and cities focus on serious problems, track the problems and the jurisdiction's effectiveness at dealing with them, and provide the impetus for corrective actions that will keep the state or city moving in the chosen direction. Most of the environmental agencies using these approaches have integrated them into a public education/public involvement strategy. The reports and indicators are designed specifically to give voters the technical information they need to make more informed choices about environmental priorities and policies and to bolster the connection between the public and their agencies. Some states have tried one or two of these approaches; several states have linked all three. The combination of the three approaches could approximate the kind of consensus-building forum envisioned by the Carnegie reports, and could strengthen the incentives for the S&T community to address serious longer-term problems.

Comparative Risk Projects

The U.S. EPA has financed more than 40 state, local, and tribal comparative risk projects. In them, participants collect the best data available about a wide range of environmental problems, then draw conclusions about the problems' relative seriousness. The task builds on EPA's risk assessment methods and technical data bases. People familiar with EPA's *Unfinished Business* and *Reducing Risk*[4] projects may remember that the act of ranking environmental problems is problematic: the data are typically poor and participants must make difficult value judgments when comparing the seriousness of dissimilar risks, such as the effects of exposure to lead paint, the effects of exposure to ground-level ozone, and the potential effects of global climate change. EPA's original comparative risk projects were conducted largely by technical staff for internal consumption. States and cities have transformed the process into externally focused partnerships engaging scientists and non-scientists alike.

A typical comparative risk project today includes one or more technical committees composed of state or city agency staff people, private-sector scientists, and academics. The technical teams typically do the homework for the projects: collecting data and analyzing the risks posed by specific problems. The technical teams may rank the problems or they may turn their findings over to an executive-level committee to rank. A public advisory committee or steering committee is usually composed of senior government managers and the leaders of essential stakeholder groups: representatives of business and industry, the Farm Bureau, environmental coalitions, other civic organizations, and elected officials. These multi-disciplinary committees are essentially a hard-working, well-read surrogate for the public at large: a diverse group willing to take the time to work through more technical material than public debates usually surface. The committees are designed to strengthen the technical quality of the product, the public legitimacy of the results, and the political impact of the change recommendations.

One product of the comparative risk projects has been a ranked list of environmental problems. In general, the states have tried to turn that list into the basis for strategic plans and budget choices. Some states have developed specific short-term and long-term strategies to address high-risk problems. Others have tried to use the risk information and estimates of the costs of various policy options to select the most cost-effective strategies for reducing risk. The Carnegie Commission, EPA's Science Advisory Board, the National Academy of Public Administration, and others have endorsed this process of priority-setting in a time of scarce resources.

The comparative risk projects have demonstrated that the nation has not yet found or implemented effective tools for addressing serious long-term problems such as climate change, habitat destruction, and indoor air pollution. Perhaps just as significantly, the projects have refocused policy-makers' attention on the external environment rather than on internal bureaucratic functions. When analysts and agency staff members had to decide which problems were most serious, they discovered how meager their data were, how little most people really knew about environmental quality, and how little the technical staff knew about how the public valued different aspects of the environment. The embarrassing extent of these "data gaps," as practitioners call them, has inspired more rigorous attempts to measure environmental quality and trends.

Environmental Indicators

As government agencies have tried in the last few years to focus on measurable results, the demand has grown for more useful environmental data, particularly data that could help analysts, policy-makers, and the general public assess the quality of their environment. The ideal would be a relatively small number of easy to measure conditions that would indicate the overall health of the environment. Although the ideal remains elusive, so many states and municipalities are compiling collections of information they find important that the approach is gaining sophistication and credibility—and possibly more credibility than it yet deserves. EPA has helped sponsor state efforts to establish environmental indicators; 25 states now have a formal environmental indicator project either in the planning stages or under way.[5]

Two states, Florida and Illinois, explicitly use their indicator data in policy and budget decisions. The Florida Department of Environmental Protection's "Strategic Assessment of Florida's Environment"[6] (SAFE) project is now more than four years old and was one of the first state indicator initiatives. Updated in 1994, the SAFE system now has 87 indicators grouped in categories related to the 13 environmental problem areas evaluated in the Florida comparative risk project. (Sample pages from the SAFE report are included in the Appendix.) Illinois's "Critical Trends Assessment Project" is a large ongoing effort to map and track changes in the state's environment. The trends data have been compiled in a seven-volume technical report and made more accessible in a handsome 90-page summary.[7]

Eleven states have published or are drafting "state-of-the-environment" reports, which present to the public collections of environmental data and analysis, usually focusing on the significance of trends revealed in various indicators. Notable state-of-the-environment reports have came out of Washington, Florida, Maine, Kentucky, and Vermont. These reports are designed to convey information that should help policy-makers and voters understand broad issues and begin to think critically about priorities.

The state-of-the-environment reports are generally products of state pollution-control agencies, though many also include natural resource and habitat information. The process of selecting the indicators to report is usually managed by agency staff and thus rests on the expertise and values of the scientists, engineers, lawyers, planners, public information specialists, and political appointees who comprise most agencies. The selection process involves many trade-offs and pitfalls because of the limits of technical understanding and data quality, and because of the value-laden aspects involved in deciding what is important enough to measure or report and at what level of aggregation.

Given the ambiguity in defining "environmental quality" and the limited understanding of the relationships among changes in environmental conditions, a vast amount of information is potentially relevant, or potentially misleading. Some state-of-the-environment reports are filled with data tables of significance primarily to experts. To meet an agency's goal of providing an informative document for the lay reader, however, many reports simplify the indicators, focusing either on only a few that are relatively easy to understand, or compressing numerous measures into a few aggregated indexes. The target audience for many of these reports is the state legislature, journalists, and the heads of the constituency groups who influence public opinion.

The state and federal employees working on environmental indicator projects are wrestling with the competing demands for technical integrity, objectivity, simplicity, and impact. Occasionally, a single indicator can meet all of these criteria, as does the famous "Bernie Fowler Sneaker Index," named for the Maryland senator who annually leads crowds of waders into Chesapeake Bay to measure the clarity of the water, an indicator of its nutrient loading. The illustration below (Figure 1) shows how EPA's Chesapeake Bay Program used that indicator as an educational tool to explain why nutrients are a problem. The illustration also shows how an indicator can be used as the basis for defining a measurable environmental goal: in this case, making it possible for Bernie to see his sneakers in chest-deep water.

Environmental Goals and Benchmarks: Minnesota's "Milestones"

States and EPA are embracing the idea of adding a target or goal to the trend lines featured in the state-of-the-environment reports. These measurable environmental goals are gaining popularity as tools to help guide state policy. The first table in the appendix titled "State Activities: Comparative Risk, Indicators, and Goals," based on a table compiled by the Florida Center for Public Management under cooperative agreement with EPA,[8] shows which states have started or completed these initiatives. The appendix also includes several pages from state reports showing indicators and goals at work.

Two state projects in particular have become models for numerous initiatives around the country: Oregon's "Benchmarks," and Minnesota's "Milestones." Minnesota Governor Arne H. Carlson initiated the Milestones project in 1991 with the assertion that "defining a shared vision, setting goals and measuring results will lead to a better future for Minnesota's people."[9] According to the project's 1992 report, hundreds of Minnesotans contributed to the project's vi-

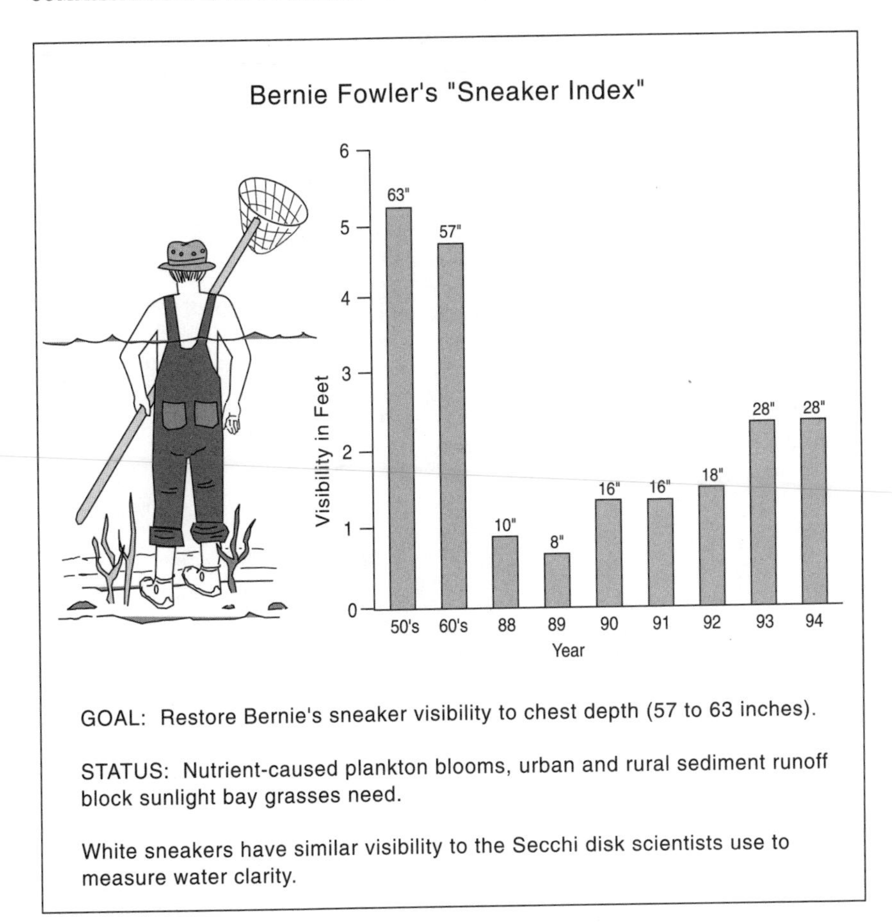

FIGURE 1 Bernie Fowler's Sneaker Index.[10]

sion statement and more than 10,000 participated in some aspect of the process. An advisory committee appointed by the governor played a significant role, but the process was essentially a populist one, designed to capture the public's values and ambitions and then provide a framework for motivating progress toward the goals.

The project adopted 20 broad goals on topics ranging from family stability to public participation in government. Each goal was defined in more specific terms by "milestones" or measures of progress toward the goal. There are 79 milestones in all. The goals and milestones that relate to the environment are listed in Figure 2. The text of the Minnesota report explains each of the milestones in a few informative paragraphs and presents specific numbers to show where the milestone had been in 1980 and 1990, and where the commission wants the milestone

FIGURE 2 Minnesota Milestones: Environmental Goals and Measures

Goal: Minnesota will have sustained, above-average, strong economic growth that is consistent with environmental protection.

Milestone 37: Minnesota per capita gross state product as a percentage of U.S. per capita gross national product.

Goal: Minnesotans will act to protect and enhance their environment.

Milestone 55: Average annual energy use per person
Milestone 56: Highway litter (bags collected per mile)
Milestone 57: Total water use (billions of gallons per day)
Milestone 58: Solid waste produced and recycled (in million tons)
Milestone 59: Percentage of students passing an environmental education test

Goal: We will improve the quality of the air, water and earth.

Milestone 60: Air pollutants emitted from stationary sources (thousands of tons)
Milestone 61: Number of days per year that air quality standards are not met
Milestone 62: Percentage of river miles and lake acres that meet fishable and swimmable standards
Milestone 63: Percentage of monitored wells showing ground-water contamination
Milestone 64: Soil erosion per acre of cropland (in tons)
Milestone 65: Toxic chemicals released or transferred (millions of pounds per year)
Milestone 66: Quantity of hazardous waste generated
Milestone 67: Number of Superfund sites identified and cleaned up

Goal: Minnesota's environment will support a rich diversity of plant and animal life.

Milestone 68: Diversity of songbirds
Milestone 69: Number of threatened, endangered or special-concern native wildlife and plant species
Milestone 70: Acres of natural and restored wetlands
Milestone 71: Acres of forest land
Milestone 72: Land area in parks and wildlife refuges

SOURCE: Minnesota Planning. *Minnesota Milestones: A Report Card for the Future*. St. Paul, MN. December 1992.

to be in 1995, 2000, 2010, and 2020. For example, the targets for Milestone 55, average annual energy use per person, decline from 300 BTUs in 1990 to 234 BTUs in 2020.

These milestones reveal a range of scientific or technical bases. Number 61, the number of days that air quality standards are not met (and counterparts in other states, often expressed as a percentage of the population living in EPA-designated non-attainment zones), is based on National Ambient Air Quality Standards set by the EPA and derived from the technical resources available to the federal government. The NAAQS are ostensibly based primarily on the health effects of criteria air pollutants on humans (with no overt consideration given to the cost of attaining the standards), so the milestone comes as close as any to being science based. And the goal that the milestone defines—reducing the number of days not meeting air quality standards to zero—could be called a risk-based goal derived from a social desire for healthy lives. The volume of litter on the highways, in contrast, defines a social goal that has no technical component, but expresses a desire for beautiful surroundings and shared respect for the environment.

Milestones that fall between these two extremes include energy use per person and diversity of songbirds. The significance of energy use, the *Milestones* report explains, is its contribution to acid rain, climate change, and nuclear waste. The statement of the indicator in personal terms (energy use per capita) is an implicit reminder that individuals have a role in meeting the targets. Diversity of songbirds, many of which are neotropical migrants, rests not only on habitat protection and chemical uses in the state but also throughout the hemisphere. The selection of songbirds as important indicators of change in biological diversity seems designed to touch a positive emotional chord in the public, and perhaps to illustrate the broader environmental connections between Minnesota and the rest of the continent. Thus each of these indicators attempts to connect the immediate concerns of Minnesota's citizens with scientists' understanding of local and global environmental problems.

The milestones define goals but they are silent on how the state or individuals should go about meeting them. For that, Minnesota has drafted other planning documents.

Soon after the publication of *Minnesota Milestones*, Governor Carlson and the state's Environmental Quality Board launched a new comprehensive planning process, the Minnesota Sustainable Development Initiative, which, in March 1994, produced *Redefining Progress: Working Toward a Sustainable Future.*[11] Although a product of state government, the project was essentially in the hands of teams of non-government stakeholders and the general public. The initiative drafted reports on each of seven sectors important to Minnesota's economy, including forestry, energy, settlement, and manufacturing. The introduction to the report notes that the vision embraced in the *Milestones* report "served as a beginning point for the teams," though there are few other references to the milestones in the text. The goals in the report do not relate directly to the milestones. The energy sector report, for example, sets out a variety of strategies for improving energy efficiency and increasing the use of renewable energy supplies, but the text does not mention the *Milestone's* target for BTUs per capita.

Redefining Progress did something the *Milestones* report did not, however: it spelled out strategies for achieving its goals. The text called for developing "sustainable manufacturing" in the state, using techniques similar to those in other state science and technology plans. The report called for strategic alliances among government, business, and consumers, and pointed to several research and development needs. For example, the report recommended that government and business should "improve the ability to measure and quantify the cost of environmental effects to appropriately cost products," and "redirect public dollars to conduct research on sustainable manufacturing processes and products and to identify other opportunities where needs can be met at lower social costs."

Staff members from several state agencies provided technical assistance to each of the sector teams, though neither the executive nor legislative branches was strongly represented on the teams. Some of the agency participants felt underused in the process.

Thus, it is perhaps no surprise that the Minnesota Pollution Control Agency has drafted its own strategic plan, *Strategies for Protecting Minnesota's Environment.*[12] The November 1994 document is an internally produced document that lists four goals for the agency: fishable and swimmable lakes and rivers; clean and clear air; uncontaminated groundwater and land; and sustainable ecosystems.

The plan focuses on eight "strategic directions," including two for immediate action: (1) "developing indicators and risk-based priorities"; and (2) developing a "geographically based approach to environmental management (with emphasis on nonpoint sources of pollution)." The document lays out specific operational plans for achieving these two directions. Although the Pollution Control Agency's plan does not mention the *Milestones* report or adopt anything like the specific milestones as targets, both of the agency's priorities are designed to strengthen the type of performance-based management approach that *Milestones* envisioned.

(The agency's strategic directions also include "environmentally sustainable economic development"—an unusual statement in a state regulatory agency publication—and "partnerships and intergovernmental coordination.")

Unlike the *Milestones* and *Redefining Progress* reports, the agency's strategic plan faced an immediate test in the legislature, which through the budget process did in fact support the agency's proposals for short-term work, including a proposal to conduct a somewhat abbreviated comparative risk project.

These snapshots of Minnesota's efforts to bring coherence, clarity, and continuity to its public policies would look familiar in many states. Even the multiplicity of reports coming out of Minnesota is familiar: three sets of actors produced three reports—within three years and under a single administration—with three formats, three sets of goals, and three sets of recommendations.

The *Minnesota Milestones* approach to goals and indicators is being adapted by many states; the attempt to bring stakeholders and experts together in the public eye to draft a consensus document about priorities and plans is now a standard approach employed by executive agencies, non-government organizations, and even some legislative branch institutions. The sponsors hope to overcome partisan conflicts, bring more technical knowledge into public deliberations than legislative hearings typically can, boost government's accountability to the public, motivate bureaucracies, and generate some genuine learning and agreement.

Public managers hope that by engaging the public in goal-setting exercises and then tracking progress toward the goals (through the indicators) the public will feel reconnected to the enterprises of environmental protection and be better able to send useful signals to legislators about spending and priorities. State agency officials also hope that the goals will encourage more consistency in policy and spending over the years, moderating the typical swings in policy that develop from daily crises and political transitions. Governors seem to like to initiate big goals projects because they appear to offer some hope of leaving a lasting imprint on the state.

EPA's Office of Policy, Planning and Evaluation is trying to do the same thing with the national environmental goals project and for roughly the same reasons. The project has many of the strengths and weaknesses of the state goals and indicator projects.

As Minnesota's experience showed, it is possible for a state to create a thoughtful, dynamic, and expansive forum for reaching consensus on environmental goals. Yet, even when thousands of people participate, many more thousands do not, adding to the difficulty of maintaining sufficient public attachment to the goals for them to influence decisions over time. That challenge is heightened when the forum attacks a broad range of issues and treats each of the resulting goals or milestones with equal importance.

Setting priorities among goals—and acknowledging that some may be mutually exclusive—is difficult, in part because goals are abstract until they are reduced to the specific investments, regulations, or activities that will actually lead to their attainment. In the goals and indicators projects described above, it is only at this last step that the legislatures have become fully involved. When real money is at stake and when decisions get close to home, the strength of the information-based, consensus-building process is put to the test.

Government gets closest to home at the municipal level, and several cities have developed some experience with environmental goal-setting.

Seattle: Environmental Goals at the Community Level

Seattle has had one of the nation's strongest environmental planning programs over the last decade. Its experiences illuminate some of the strengths and limits of the endeavor.

The City Planning Department conducted a ground-breaking comparative risk project in 1991. Technical advisory committees composed of city officials and others from business, non-governmental organizations, and other government agencies analyzed the city's environmental problems and ranked them in terms of the risk they posed and in terms of the priority the city government ought to place on addressing them. (See the last table in the appendix.) The mayor and city council used these rankings and the underlying data when crafting subsequent budgets. The city incorporated much of this priority-setting work in its current comprehensive plan.

J. Gary Lawrence, who for five years was head of the Seattle Office of Long-Range Planning and who now teaches at the University of Washington, said that the comparative risk process was extremely valuable because it brought together agencies, citizens, and organizations to focus on the city's problems and to agree on what matters, what goals the city wanted to achieve. The mayor and city council adopted some of the recommendations in subsequent city budgets. The technical committees agreed that the number-one risk and the number-one priority was the automobile's contribution to both air pollution and runoff to surface waters. The people of Seattle took that ranking seriously enough to invest in planning an $8 billion, three-county commuter rail system. After lengthy debates about the technical merits of the proposed system, voters recently rejected the proposal.

The story is a reminder that it is possible for a fairly representative group to agree on the nature of a problem, agree on a general goal, and disagree on specific strategies to achieve the goal, even when the strategies appear perfectly rational and well founded. Public opinion—and hence, public policy—will never be rational, although both can be better informed. The conflicts that arise when a public process frames problems carefully can foster useful public learning, though the line from goal-setting to strategy and implementation will not be straight or simple.

Out of the Seattle comparative risk project flowed an ambitious comprehensive planning process, and more recently, "Sustainable Seattle," an entirely volunteer effort to strengthen the community from the inside. Sustainable Seattle has adopted indicators of success similar to Minnesota's, though with an additional community-building aspect: one indicator of the city's sustainability is how many neighbors each individual knows by name. The participants in Sustainable Seattle include many of the people, companies, and organizations who were involved in the formal city government efforts. Here, as in Minnesota, the personal connections and learning that these processes establish endure beyond the process and probably contribute more to decision-making and the incorporation of technical expertise into policy than do the committees' formal plans and reports.

STATE INITIATIVES IN ENVIRONMENTAL SCIENCE AND TECHNOLOGY

"Green Business"

Across town, as it were, states have been increasingly active in setting goals and policies for the cooperative development of environmental technologies and focused environmental research. As noted in *Enabling the Future*, solving the nation's long-term environmental problems will require a long-term effort to understand the problems and to develop appropriate technological tools to mitigate, eliminate, or solve them. Several states have responded to that challenge as a market opportunity: a way to create jobs while contributing to the general well-being of the nation and the world. Their efforts have been more opportunistic than goal driven, however. States appear to be setting up processes to identify and promote technological "winners," rather than focusing on the problems that most need to be solved.

This environmental focus is a relatively new and small twist to a well-established set of public-private institutions that have attempted to stimulate economic development—more jobs and more income—through partnerships involving universities and research institutions, financial institutions, public agencies, public funding sources, and existing companies. Although relatively few states have specific *environmental* science and technology programs, every state has some kind of technology program. These are catalogued in some depth in a 641-page volume published in 1995 by Battelle, *Partnerships: A Compendium of State and Federal Cooperative Technology Programs*. In 1994, the 50 states "reported 390 discrete technology-based development initiatives, with annual expenditures of $385 million."[13]

The *Compendium's* criteria for including a program among its pages are significant: "To be included, programs needed to demonstrate that there was an ongoing connection with state government . . . , that they involved either a government-industry or government-industry-university partnership, and that the programs had as a primary goal the use of technology to enhance economic growth." The last criterion is a reminder that the environmental S&T work has been pursued, in most states, by economic development offices or technology offices, often with little connection to the environmental regulatory agency leaders.

The dual promise of economic prosperity and environmental gains has inspired states to add environmental S&T to their technology programs. The National Governors' Association published a useful report on the topic, *Cultivating Green Businesses,*[14] in 1994, which prominently repeated estimates that by the year 2000 the global market for environmental technology would total between $300 billion and $400 billion. States, particularly those with both a relatively green tradition and a high-technology infrastructure, had already begun to develop strategies for securing a piece of that market.

States have typically used the approaches recommended in the NGA report, and they bear striking differences from the approaches recommended in *Enabling the Future*. NGA recommends that states start by inventorying their existing environmental technology companies, determining what barriers inhibit their expansion, and identifying barriers to innovation. States then should respond accordingly with programs to help these firms grow. As the cases below illustrate, states typically work with their major traditional industries to promote innovation that may have secondary environmental benefits. In contrast, *Enabling the Future* recommends starting with a definition of an environmental problem that needs to be solved and then refining and implementing a long-term S&T goal to address it.

Massachusetts, for example, focused its attention on the potential for environmental S&T development in April 1993 when the Executive Office of Environmental Affairs published *Green Businesses: A Profile of the Massachusetts*

Environmental Industry. The report's recommendations did not suggest that the state effort should focus on any particular environmental problems. Rather, the report recommended steps the state could take to facilitate the expansion of the whole industry and accelerate the transfer of ideas from the state's extraordinary research institutions to the marketplace. These steps were similar to those the state would use in any other aspect of technology promotion: improve access to financing and markets; expedite and simplify the permit process to reduce delays and uncertainty; train the work force and bolster public education more generally; and foster the growth of trade associations to promote common interests. To follow through on the original report, state agencies and the University of Massachusetts established the Strategic Environmental Partnership in the fall of 1993, and Governor William Weld established the Massachusetts Envirotechnology Commission.[15]

Meanwhile, the state's Forum for Innovative and Alternative Technologies recommended the adoption of a strategic plan that would, among other things, lead to the establishment of "envirotechnology centers" at universities. Such centers have been major components of state S&T programs for many years. In 1984, New Jersey created the Hazardous Substance Management Research Center, which brings together several universities and some 34 industrial sponsors to research projects in six areas: incineration; biological and chemical treatment; physical treatment; site assessment and remedial action; health-effects assessment; and public policy and education.[16]

Barriers and Opportunities

The New York State Science and Technology Foundation (established in 1963) has helped to finance numerous university-based Centers for Advanced Technology, but none in fields directly related to environmental protection. The *Compendium* lists 13 foundation-sponsored centers, including centers for ceramic technology, materials processing, biotechnology, and automation and robots. As befits a rapidly changing economy, the list changes over time. Funding for each center periodically sunsets and must be renewed. In the process, the foundation has canceled those that fail to perform. The foundation invites universities to compete for the funds to establish these centers.

The foundation's former executive director, Graham Jones, said in an interview that he had tried to encourage applications for a center to focus on pollution prevention and environmental mitigation through innovative industrial chemical processes. He found no takers, despite the apparent social need and commercial value of advances in the area. He speculated that a fear of becoming encumbered with government requirements may be particularly strong among the private-sector experts in industrial chemistry. The decades-old divide between the regulated community and its regulators also discourages state agencies from taking mutually beneficial initiatives with firms. Graham said that a New York-based firm had created an etching process that eliminated the use of CFCs in some manufacturing work. The state's regulatory agency declined to help the foundation promote the firm's accomplishment for environmental technology because of the regulators' reluctance to show favoritism to particular firms.

A current spokesman for the New York S&T Foundation said he wasn't sure why the foundation had made no special focus on environmental S&T, since it had taken on other social problems including technology for the physically disabled. He theorized that the foundation was content to let the state regulatory agency drive environmental policy and technological investments. Graham also noted the foundation's belief that the regulatory agency had more money and the potential to have more of an impact on environmental technology than the foundation could.

The relationship between the foundation and the executive branch became particularly clear following the change of administrations in New York in early 1995. Despite the foundation's relative autonomy from state government, the governor's office remains an important source of goals and direction. As of July, the new administration had not yet set its course for science and technology—except to announce the likelihood of downsizing existing programs. An official in the New York Department of Environmental Conservation said that the new administration had scrapped an effort by the previous administration to have the department work cooperatively with the state's commerce agency on environmental goal-setting. The administration is now trying to establish its own process.

The nation's smaller states are rushing to establish technology programs and several have included environmental technology in their initial plans. Vermont, which only last year formed a formal Technology Council, now has a Science and Technology plan, which calls for the creation of four Centers of Excellence, including an environmental science and analysis center. The plan includes a mission statement for the proposed center, calling on it to focus on issues that might include pollution prevention systems and monitoring technologies; sustainable agriculture; recycling and recycled products; energy conservation; environmental services, including monitoring and analysis; and sustainable forest ecosystem management and forest product manufacturing.[17] Other small states, including Maine and Montana have developed similar approaches, often building on the personal and institutional connections established over their years during the states' participation in the National Science Foundation's Experimental Program to Stimulate Competitive Research (EPSCoR).[18]

California maintains an astonishing array of science and technology programs, partnerships, and research centers, many of which relate to environmental protection. Of them, the most relevant to this discussion may be the California Environmental Technology Partnership (CETP), which Governor Pete Wilson created with an executive order in 1992. The partnership is jointly managed by the California Environmental Protection Agency (Cal/EPA), the California Trade and Commerce Agency, and the Environmental Technology Council, an advisory committee made up of more than 70 representatives of industry, government laboratories, and government agencies, and academics. The advisory committee prepared a 10-year strategic plan in 1994 and an update in 1995,[19] which include the following goals:

- Provide for consistent testing and demonstration processes for new environmental technologies, and increase the certainty of acceptance in broader markets.
- Enhance the credibility, status and access to the marketplace of environmental technology companies through establishing strategic partnerships.
- Increase California's share of the national and international markets for environmental technologies, products, and services by augmenting domestic marketing efforts and expanding export assistance services.
- Develop and implement a comprehensive, integrated communication strategy for regularly providing California environmental companies with information that is advantageous to their business success.
- Establish California as the best place in the world to test and demonstrate environmental technologies.
- Encourage and support research and development of new California environmental technologies.

Again, these goals relate to the process of technology development and diffusion, not to specific societal goals for the environment or specific high-risk environmental problems. Cal/EPA completed the nation's most technically sophisticated statewide comparative risk project in 1994, but the project's findings do not appear to be reflected explicitly in CETP activities. The types of problems addressed in the comparative risk project, however, do feature prominently in the objectives and specific strategies listed in the strategic plan. For example, the plan calls for the completion of studies of new pesticide spray technologies that may reduce health risks to applicators. The plan also recommends that the state commit itself to purchase ultra-low emissions vehicles to help create a strong market for clean cars.

The plan identifies one issue as Cal/EPA's highest priority: implementing a first-in-the-nation technology certification program "to verify and certify performance of specific environmental technologies and to facilitate permitting." The certification program was authorized by legislation passed in 1993.

Cal/EPA's 1995 update of the CETP strategic plan was published in tandem with the *1994 Year-End Report* which tracked progress against the objectives published in the 1994 plan.[20]

CETP is just one of many California initiatives likely to have some impact on the environment. Indeed, CETP is not even mentioned in Battelle's *Compendium* section on California. The section does, however, describe the California Council on Science and Technology's "Project California," an attempt to "develop globally competitive, high-technology clusters of transportation businesses in California. Project California's primary mission is to create long-term, high-value-added jobs for Californians while enhancing the state's strong commitment to reducing environmental pollution and urban congestion." The quest for the ultra-clean car is also being motivated by regulatory agencies in the state, most notably the Southern California Air Quality Management District.

Meanwhile, an official in the California Governor's Office is focusing on a very different set of goals for environmental science and technology: natural resource issues and critical habitats. As part of the state's efforts to help regions and communities prevent conflicts between humans and endangered species, the state is developing CERES, the California Environmental Resources Evaluation System, which will put large amounts of place-specific environmental data on the Internet so that anyone can learn about local, regional, and statewide conditions. The information is intended to do what other state indicator programs hope to accomplish: make users more aware of trends and the choices individuals or communities can make to influence the trends.

The state's Natural Communities Conservation Planning effort has also established goals in an entirely different format from those described above. After conducting a biological survey of nearly 6,000 square miles of southern California, scientists have published maps showing critical habitat areas that ought to be managed to maintain the diversity of species. The maps translate the somewhat abstract environmental goal of maintaining biological diversity into a set of maps that can guide land acquisition programs and community growth management plans. Another of the goals of the project is to avoid the kind of rigid restrictions on land uses that may result from triggering the Endangered Species Act's protection.

When Goals Need Updating

The emphasis in virtually all of the technology plans described above is on building processes and networks that will be capable of launching an innovative idea into the marketplace as soon as the marketplace demands it. This capacity for opportunism requires systems that are dynamic and capable of change.

Pennsylvania oversees an environmental technology program in need of change because its original mission has lost its relevance in the market. In 1988, the Northeast was becoming increasingly concerned about a solid waste crisis: landfills were filling up and incinerators were being shut down. The Pennsylvania Legislature responded with a statute that required municipalities to recycle glass, paper, and yard waste. The statute also dedicated part of a solid waste tipping fee to an "Environmental Technology Research and Development Fund," which would provide grants to stimulate the recycling industry through the development of new recycling processes and markets. As part of the Ben Franklin Partnership Program, the grants were administered by the state's Office of Technology Development.

Bill Cook, chief of the research grants division, which manages the fund, said that the effort has produced some successes, but the number of good proposals for grants (up to $100,000 apiece for each of three years) has dropped in recent years because the fund can only be used to finance projects dealing with materials collected at curbside. The Department of Commerce wants to expand the list of materials to encourage more innovative proposals with potentially greater environmental and commercial benefits.

ISSUES AND THEMES

State Politics and Policies

States are often called the laboratories of democracy because state governments are generally more responsive to voters and hence more innovative than the federal government. The relative speed with which states have developed indicator programs, benchmarking systems, and environmental technology programs attests to the states' ability to build new systems quickly. Of course, the flip side of responsiveness is volatility: in innovation is instability.

With most changes of a governorship or the parties' control of a legislature comes pressure to throw out or reconstruct the predecessors' initiatives. Before leaving office, Washington Governor Booth Gardner described his administration's "Washington Environment 2010" risk-based priority-setting process as one of his proudest legacies. But his successor dismantled that effort and built something new to achieve similar objectives.[21] The new approach appears to be as strong and effective as any of its counterparts, though that is no assurance that it will—or should—last beyond the governor's term.

Because of the vicissitude of public opinion, government programs and priorities are often whipsawed by crises and fads. Depending on the type of program, managers may use different administrative structures and approaches to achieve some longer-term stability.

Environmental regulatory agencies face the challenge of keeping their resources focused on particular problems, which may go in and out of fashion with the legislators who control their annual appropriations. To foster stability in state environmental protection goals, many administrators have turned to collaborative processes like Minnesota's milestones and sustainable development efforts. Managers may hope to accomplish at least three things by engaging state employees, academics, business representatives, and the general public in a goal-setting process:

- to build a public constituency for the agency's goals;
- to effect a lasting change in the public's understanding of the environment; and
- to effect a lasting change in the agency's understanding of its challenge and of the public's values and goals.

The indicators and benchmarking programs are specifically intended to maintain links among the agency, the public, and the environment itself and thus foster continuity and continuous improvement.

Cooperative technology programs are generally more insulated from direct legislative control. Legislatures must approve investments in technology funds or in the research capacity of state universities, but these funds are then typically managed by committees or trustees with more policy discretion than is available to typical state bureaucrats. This insulation from detailed budget scrutiny ought to make it easier for technology programs to maintain a longer time horizon than a regulatory agency. Graham Jones of the New York Science and Technology Foundation noted the limits of that freedom. The foundation staff tried to do careful planning and to explain its goals to its board and relevant policy-makers, yet the foundation was nonetheless very much constrained by those programs passed and funded by the legislature.

Bill Cook of the Pennsylvania Office of Technology Development observed that other states' technology programs seem to be constantly starting over. It takes at least five years for a research program to show results and few are able to withstand the political pressure to produce quickly, he said. Cook's own dilemma with the recycling program shows another approach to achieving stability: financing a technology program through a special fund (a tipping fee), which can keep on generating cash without much annual review and take the pressure off the program to justify itself each year. (Similarly, the federal Superfund program has acquired much of its stability from its financial insulation from the rest of the Environmental Protection Agency's budget.)

In contrast to Cook's sense that most state S&T programs are politically tenuous, the authors of the *Compendium* reached a somewhat different conclusion:

> Despite the fact that there is no traditional, organized constituency for science and technology at the state level, state programs have nonetheless survived transitions in party control of statehouses and legislatures, state fiscal crises, external performance reviews, and changes in program management. Even in states that have made dramatic cuts in funding, the commitment to cooperative technology has proven tenacious. For example, in the early 1990s Illinois cut its $20-million-per-year program by 95 percent; today, at the insistence of the business community, it is weighing a plan to restore many of those services. Increasingly, states see their economic future in growth generated internally through higher education and technology, not just in recruitment of industry from outside.[22]

If legislatures, governors, and the public view the primary function of technology programs as producing jobs, then success depends not only on developing a great idea but also on finding willing buyers in the market. These buyers may not be there. Almost by definition, the underaddressed long-term problems identified in *Enabling the Future* are those for which markets for solutions may not exist today. The kind of market failures that create environmental problems inhibit their solution, absent some form of government regulation, intervention in the market, or public commitment. Unless an "environmental goal" or "environmental technology" goal is accompanied by regulatory action, the goal may remain elusive. This is why the apparent division between environmental regulatory agencies and environmental technology programs may be a problem.

Linking Goals, Regulation, and Action

Regulations can create markets. If the cost of disposing of wastes in accord with RCRA rules is high, firms will search for less expensive ways to treat, reduce, or prevent their waste. Cooperative technology programs favor emerging technologies that will fit into niches created by regulatory standards because the markets are well established. State technology programs have capitalized on these regulatory-driven markets even when they did not have technology programs specifically dedicated to environmental problems.

Different regulatory approaches will have different impacts on technological innovation. EPA's traditional regulations have often had the result of establishing a particular technology as the norm for many years while inhibiting further innovation. As many economists have maintained, market incentives of many types could reduce the aggregate costs of environmental controls while encouraging technological innovation. This message has not been lost on the states. Illinois has recently created a market for tradeable air pollution permits and expects to save industries millions of dollars while reducing pollution levels. Nevertheless, relatively few states have aggressively pursued market approaches to environmental risk reduction, and the topic rarely surfaces in the numerous technology plans consulted for this paper.

Many policy-makers[23] are calling on the federal and state governments to rely more on performance standards than technology design standards to control pollution. In a performance-based system, firms, cities, or even whole states would be told how much they may pollute and then given broader latitude than today in deciding how to achieve the standards.

EPA and state environmental agencies are using the same idea to redefine their oversight relationships. Under the "Performance Partnership" program adopted in the spring of 1995 by the states and EPA, EPA will back away from some of its more prescriptive process-oriented requirements and instead focus on the overall performance of the state in protecting the environment. The approach will require states and EPA to agree on performance measures—including environmental indicators as described above. Each state and the federal government will be negotiating on short-term environmental goals as defined by the indicators and performance measures they select. In these negotiations, the experts within the agencies will clearly dominate the selection process, though the growing reliance on indicators should gradually open the regulatory arrangements to public view.

In addition to these relatively obscure changes in environmental regulation, much of the nation at this writing is expecting the new Congress to pass regulatory reform legislation and to make significant changes in the nation's major environmental statutes. Some see this shift as one reason for significant weakness in the environmental technology industry. One trade paper concluded a story called "Envirotech Firms Can't Sit on Green Laurels" with the assertion:

> Another nail in the coffin of environmental technology firms has been the specter of changing environmental regulations, which fuels insecurity in the industry. Federal air pollution laws had called for certain cities with high pollution levels to start enhanced testing of cars' tailpipe emissions. But when local legislators decided they didn't want citizens who elected them to be forced to spend money testing their cars, the Environmental Protection Agency eased the rules and said localities could reduce pollution another way.[24]

This single paragraph captures the interactions of public opinion at the local and national level, regulation, political action, technological change, and the market. The example illustrates the tenuous nature of environmental goal-setting in America, which sometimes proceeded without sufficient public understanding or commitment to be sustained. In the case of automobile Inspection and Maintenance Programs, EPA has been unable to convince many angry car owners that mandatory tailpipe inspections are an economical way to reduce air pollution in their cities. Without the social commitment to submit to inspections, the technology of tailpipe monitoring was useless.

Experiences of states and communities suggest that setting goals for the environment is useful if it helps people focus on problems and discover a shared commitment to solving them. The very conditions that make explicit goals important—society's tendency to ignore problems and to delay action—make the goals difficult to attain. When the immediate steps required to attain a goal are unpleasant, expensive, or uncertain, individuals, firms, and governments can always find a good reason to avoid taking action. Almost every government effort to set goals will also suffer from the related dilemma of involvement: too few people will set the goals and too many will have to pay in some way to achieve them. The imbalance in that equation is one reason goals are difficult to attain.

The challenge of showing people what they are buying for their investment—what type of progress they are making toward their environmental goals—may be easier at the community and state level than at the federal level because the changes are closer to home. The people who live in the Chesapeake Bay watershed will feel a stronger sense of commitment to the Bay than would people who live elsewhere. Environmental goals and indicators set by collaborative processes at the state and local level thus have a chance of motivating action on the local and state levels. To address the bigger problems, however, the ones that cross boundaries and generations, higher levels of government will have to act.

In virtually all cases, the value of the goal rests not in its minimal potential to bind unwilling parties to honor a long-term commitment. Few goals have this kind of binding power. Rather, the value of goals appears to arise from the process that creates them. That process requires groups to learn about problems and agree on the desirability of change. As states and communities continue to experiment with benchmarking efforts and environmental performance measures, that learning should expand and deepen, influencing, in turn, the nature of goals for environmental science and technology.

NOTES

1. Carnegie Commission on Science, Technology, and Government. *Enabling The Future: Linking Science and Technology to Societal Goals.* New York. September 1992.

2. Carnegie Commission on Science, Technology, and Government. *Science, Technology, and the States in America's Third Century.* New York. September 1992.

3. National Academy of Public Administration. *Setting Priorities, Getting Results: A New Direction for EPA.* Washington, DC. April 1995.

4 U.S. EPA, Office of Policy Analysis, and Office of Policy, Planning and Evaluation. *Unfinished Business: A Comparative Assessment of Environmental Problems.* February 1987. U.S. EPA Science Advisory Board (SAB-EC-90-021). *Reducing Risk: Setting Priorities and Strategies for Environmental Protection.* Washington DC. 1990.

5. Florida Center for Public Management. "Information about Using Environmental Indicators in U.S. EPA/State Performance Partnerships," a report prepared by the State Environmental Goals and Indicators Project, a cooperative agreement between The Florida Center for Public Management at Florida State University and the US EPA's Regional and State Planning Division. Tallahassee, Fla. May 1995.

6. Florida Department of Environmental Protection. *Strategic Assessment of Florida's Environment (SAFE).* Tallahassee, Fla. November 1994. See also Florida Department of Environmental Regulation. *From Regulation to Protection: The Environmental Vision Plan (draft).* Tallahassee, Fla. July 1992. And The Florida Commission on Government Accountability to the People, Executive Office of the Governor. *The Florida Benchmarks Report (draft).* Tallahassee, Fla. March 1995.

7. Illinois Department of Energy and Natural Resources and The Nature of Illinois Foundation. *The Changing Illinois Environment: Critical Trends (Summary Report of the Critical Trends Assessment Project).* Springfield, IL. 1994.

8. Florida Center for Public Management. Op. cit.

9. Minnesota Planning. *Minnesota Milestones: A Report Card for the Future.* St. Paul, MN. December 1992.

10. U.S. EPA Chesapeake Bay Program. *Environmental Indicators: Measuring Our Progress,* Philadelphia. April 1995.

11. Minnesota Environmental Quality Board and Minnesota Planning. *Redefining Progress: Working Toward a Sustainable Future.* St. Paul, MN. February 1994.

12. Minnesota Pollution Control Agency. *Strategies for Protecting Minnesota's Environment.* St. Paul, MN. November 1994.

13. Coburn, Christopher, editor, and Dan Berglund. *Partnerships: A Compendium of State and Federal Cooperative Technology Programs.* Battelle Press. Columbus, Ohio. 1995.

14. Finegold, Allyn F. and Barbara B. Wells. *Cultivating Green Businesses.* National Governors' Association. Washington, DC. 1994.

15. Ibid. pp. 3-6.

16. Coburn. Op. cit. Page 308.

17. Vermont Technology Council. *Vermont Science and Technology Plan* (prepared for Governor Howard B. Dean, M.D., under Executive Order 14-93). South Burlington, Vt. December 1994.

18. See, for example:

The Maine Science and Technology Commission. *Maine's Science and Technology Plan: A First Step Toward a Productive Future.* Augusta, Maine. October 1992.

The Maine Economic Growth Council. *Goals for Growth: Progress 95.* Augusta, Maine. April 1995.

Montana Science and Technology Advisory Council. *The Montana Science and Technology Agenda.* Helena, Montana. October 1992.

Arkansas Science and Technology Authority. *FY 1993 Annual Report: The First Ten Years.* Little Rock, Ark. November 1993.

19. California Environmental Technology Partnership. *1995 Strategic Plan for Promoting California's Environmental Technology Industry.* Sacramento. April 1995.

20. California Environmental Technology Partnership. *1994 Year-End Report.* Sacramento. April 1995.

21. Minard, Richard A., Jr. "Comparative Risk Analysis and the States: History, Politics, and Results," in *Comparing Environmental Risks: Tools for Setting Government Priorities*, edited by J. Clarence Davies. Resources for the Future. Washington, DC. Forthcoming (late 1995).

22. Coburn. Op cit. Page 30.

23. National Academy of Public Administration. Op. cit.

24. Liz Skinner. "Envirotech Firms Can't Sit on Green Laurels" in *Washington Technology*, Vol 10, No. 6. Washington, DC. June 22, 1995.

APPENDIX

EXAMPLES FROM STATE AND LOCAL PROJECTS AND PUBLICATIONS

- States engaged in comparative risk projects, environmental indicators projects, and goal-setting or benchmarking projects, from "Information About Using Environmental Indicators in U.S. EPA/State Performance Partnerships," report prepared by the Florida Center for Public Management, Tallahassee, Fla., 1995.
- "Miles of Waterbodies Meeting or Exceeding Designated Uses," from *Strategic Assessment of Florida's Environment (SAFE)*, published by the Florida Department of Environmental Protection, Tallahassee, Florida, November 1994.
- "Vermont Indicators: Waste," from *Environment 1995: An Assessment of the Quality of Vermont's Environment*, published by the Vermont Agency of Natural Resources, Waterbury, Vermont, 1995.
- "Chapter 6: Prairies," from *The Changing Illinois Environment: Critical Trends, Summary Report of the Critical Trends Assessment Project*, published by the Illinois Department of Energy and Natural Resources and The Nature of Illinois Foundation, Springfield, Illinois, 1994.
- "Rankings of Air Issues," from *Environmental Risks in Seattle, A Comparative Assessment*, published by the City of Seattle, Office for Long-range Planning, Seattle, Washington, October 1991. (The report also included rankings for water issues, land issues, and cross-media issues.)

STATE AND LOCAL GOALS: APPENDIX

State Activities: Comparative Risk, Indicators, and Goals

State	Comparative Risk in planning (P) under way (U) or completed (C)	Indicator Project in planning (P) under way (U) or in action (A)	State of Environ. under way (U) or completed (C)	Goals/Benchmarks under way (U) or completed (C)
Alabama	U			
Alaska	P	P		
Arizona	U	P		
Arkansas	P			
California	C	U	C (1995)	
Colorado	C	U		
Connecticut		U		U
Delaware				
Florida	U	A		U
Georgia				
Hawaii	U	P		U
Idaho		U		
Illinois		A	C (1994)	
Indiana				
Iowa	P			
Kansas			U	
Kentucky	U	P	C (1992, 1994)	
Louisiana	C	U		
Maine	U	P	C (1994)	U
Maryland	U	U		
Massachusetts		P	U	
Michigan	C			
Minnesota	P	U		C
Mississippi	U	P		
Missouri	P	P		
Montana				
Nebraska				
Nevada				
New Hampshire	P	P		
New Jersey	P	P		
New Mexico				
New York	P			
North Carolina		P	U	
North Dakota	U			
Ohio	U			
Oklahoma				
Oregon				C
Pennsylvania				
Rhode Island				
South Carolina		P		
South Dakota				
Tennessee	U	P	C (1994)	
Texas	U			
Utah	C	P		
Vermont	C	U	C (1994, 1995)	
Virginia				
Washington	C	U	C (1995)	U
West Virgina				
Wisconsin	P	U	C (1995)	
Wyoming				

SOURCE: "The Comparative Risk Bulletin," Vol. 5, No. 5-6, May/June 1995. Northeast Center for Comparative Risk. South Royalton, Vermont; and "Information about Using Environmental Indicators in U.S. EPA/State Performance Partnerships." The State Environmental Goals and Indicator Project. The Florida Center for Public Management, Florida State University. Tallahassee. May 1995. Reprinted with permission.

MILES OF WATERBODIES MEETING OR EXCEEDING DESIGNATED USE*

Explanation of Indicator

Surface waterbodies are classified for their present and future most beneficial use by the Florida Environmental Regulation Commission under Chapter 62-302, Florida Administrative Code. Surface waters are classified as:

Class I — Potable Water Supplies;
Class II — Shellfish Propagation and Harvesting;
Class III — Recreation, Propagation and Maintenance of a Healthy, Well Balanced Population of Fish and Wildlife;
Class IV — Agricultural Water Supplies; and
Class V — Navigation, Utility and Industrial Use.

Each designated use has minimum criteria (both numeric and narrative) for various water quality parameters. Florida waters are sampled and various parameters tested to determine whether or not hey continue to maintain their designated use.

Source

The identification of waterbodies that meet, partially meet, and do not meet designated uses is in the *Florida Water Quality Assessment*, prepared to meet the requirements of Section 305(b) of the U.S. Clean Water Act. The data or the source of the data used to prepare the 305(b) report may be obtained by contacting Joe Hand, Florida Department of Environmental Protection, Bureau of Surface Water Management, at 2600 Blair Stone Road, Tallahassee, Florida 32399-2400, or at (904) 487-0505.

Data Characteristics

Data collection is ongoing and takes place at different frequencies, depending on the needs of the sampling program and on the responsible agency. The Florida Water Quality Assessment (CWA 305(b) Report) is produced every two years and provides a summary of waterbodies and maintenance of designate use. In 1994, 11,880 miles of streams were assessed (60 percent of these miles were assessed with recent, quantitative data and 40 percent were assessed with older data or qualitative information). The 5,648 square miles of lakes and estuaries were assessed and 70 percent of the miles were assessed with recent data with the

---

*Pages 268–270 reprinted with permission from the Florida Department of Environmental Protection.

balance assessed with qualitative data. Water quality data is sorted according to reach areas (USGS hydrologic units) for locations that are sampled and which have water quality analysis data entered into STORET. Generally, data is available for areas near the state's more urban regions. Data becomes less available in more rural areas. This data is available on a computer disk or hard copy format and maps are available in the 305(b) report. There are no additional costs of acquiring these data which are obtained as a result of conducting ambient water quality monitoring activities. However, there may be copy costs associated with data retrieval.

Overall Assessment

The information used to evaluate whether a water body meets its designated use is the most complete and voluminous set of surface water quality data available. It is the same information used to develop trophic state index, trends, and water quality index. With more agencies continuing to participate in entry of data in STORET, the utility of the data base and the quality of future assessments will continue to improve.

This indicator is useful in assessing the general condition of state waters. Although the majority of the information is developed for the state's urban areas, it represents most of the state affected by pollution.

Data may not be collected with sufficient frequency to provide a valid information base for all areas. Much of the information being used is actually an average of reach values gathered at irregular intervals rather than individual sites sampled at comparable time intervals for specific parameters. When insufficient information is available to make a determination, or where recorded values conflict with local knowledge, professional judgment is exercised to adjust the assessment. Therefore, the use of this indicator is limited to STORET and 305(b) data limitations.

Analysis of Indicator

The number of river miles and square miles of lakes and estuaries that meet, are threatened, partially meet and do not meet designated uses is shown below for four two-year intervals corresponding to the periods covered by the Florida Water Quality Assessment. The 1986 data is not totally comparable with the 1988, 1990, and 1992 data. Rivers and estuaries do not show significant changes in meeting their designated uses during the period of record available. Lakes have shown a significant decline during this period. This may partially be attributed to the reclassification of a part of Lake Okeechobee which, when reclassified went from meeting designated use to partially meeting designated use. Threatened is a designated reclassification between "partial" and "yes" and was not used in 1986 or 1988.

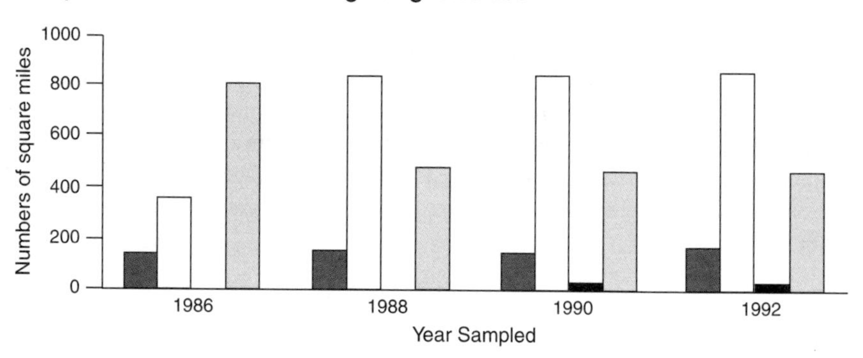
Square miles of lakes meeting designated use
Numbers of square miles
1000
800
600
400
200
0
1986
1988
1990
1992
Year Sampled

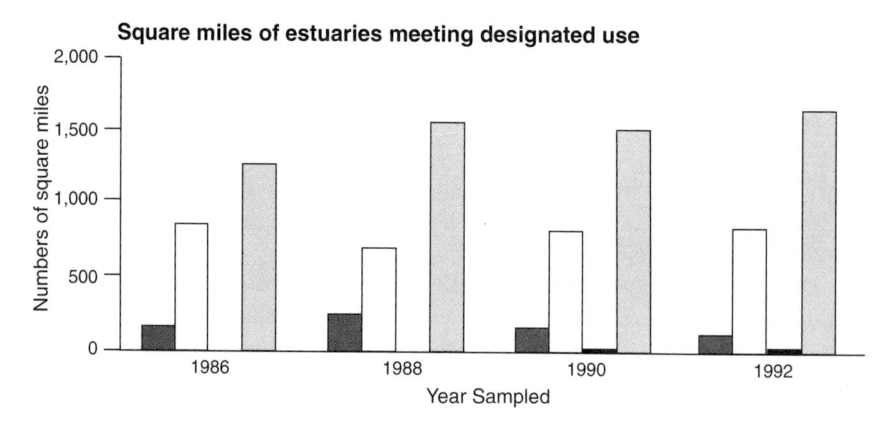
Square miles of estuaries meeting designated use
Numbers of square miles
2,000
1,500
1,000
500
0
1986
1988
1990
1992
Year Sampled

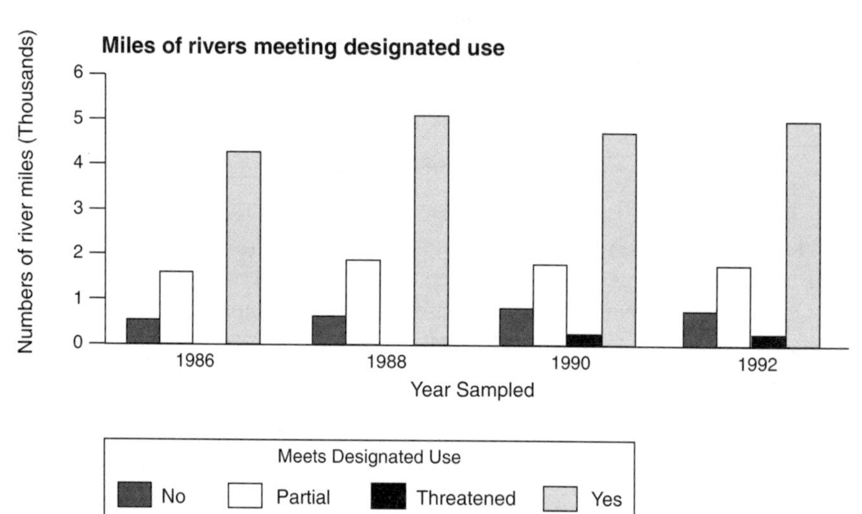
Miles of rivers meeting designated use
Numbers of river miles (Thousands)
6
5
4
3
2
1
0
1986
1988
1990
1992
Year Sampled
Meets Designated Use
No
Partial
Threatened
Yes

VERMONT INDICATORS*

WASTE

Waste is an inherent part of human activity. How much we produce and what we do with our waste, however, is very much under our control.

For most of this century, Vermont towns have operated dumps or landfills as centers for disposal of industrial as well as household waste. Over time, growing incidence of contaminated surface water, groundwater supply wells, and soils linked to these and other land disposal sites led to increased concern about waste management practices. In response, there are now stricter controls on the disposal of industrial and household hazardous wastes, and improved landfill design, construction, and operation standards.

While these changes have led to significant improvements in public health and environmental protection for Vermonters, solid and hazardous waste continues to be a serious problem facing the state. Recent increases in recycling and pollution prevention efforts by Vermont industries, governments, and citizens are promising responses to the state's waste problem.

Solid Waste

Over the last three decades, there have been significant changes in the way Vermont handles the solid waste it produces. Before 1968, open dumps and open air burning were common waste disposal practices. Unlined and uncovered, they posed significant public health risks and aesthetic degradation. The move to sanitary landfills in the 1970s reduced the most drastic health and aesthetic effects associated with open dumps, but still allow contamination of surrounding surface and groundwater.

Beginning in 1988, sanitary landfills were replaced with more protective lined landfills. Since 1991, 47 unlined landfills within Vermont have been closed. These closures, coupled with increased regional competition to provide solid waste disposal services for Vermont municipalities, have contributed to an increase in the amount of solid waste going to landfills or incinerators in neighboring states. The net effect has been a reduction in the amount of waste being disposed of in the older, less protective unlined landfills. In 1994, of the approximately 285,000 tons of waste disposed of in state landfills, only 20% (58,000 tons) were disposed of in unlined facilities.

As new landfills are constructed and operated by Solid Waste Management Districts in Vermont, the export trend is likely to reverse while continuing to offer significant improvements in public health and environmental protection to Vermonters.

---

*Pages 271–275 reprinted with permission from the Vermont Agency of Natural Resources.

Recycling continues to play a key role in reducing the amount of solid waste Vermonters send to disposal. Although there has not been a significant increase in the number of recycling facilities statewide, the number of towns adopting local recycling ordinances continues to rise. As a result, there has been a dramatic increase in the number of households and businesses participating in local recycling programs (Figure 1).

Hazardous Waste

In the 1950s, the town of Springfield opened its new landfill. Citizens were proud they had found a safe way to dispose of their industrial wastes. Today, that landfill is one of eight Superfund sites in Vermont on the National Priority List. Cleanup is currently under way with an estimated cost of $20 million.

The discovery of hazardous waste sites in Vermont has occurred steadily over the past 25 years. Many of these sites, such as the Springfield landfill, are the result of past disposal practices which, at the time, were considered safe and acceptable.

It was not until the early 1980s that Vermont become seriously concerned about the effects of hazardous waste releases. Since then, both the discovery and

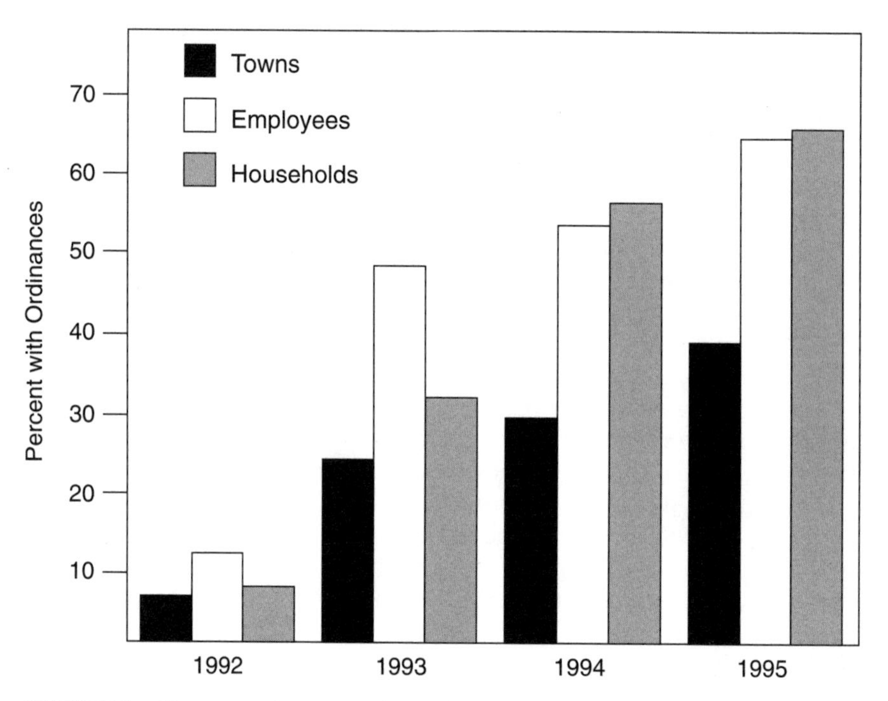

FIGURE 1 More towns have recycling ordinances.

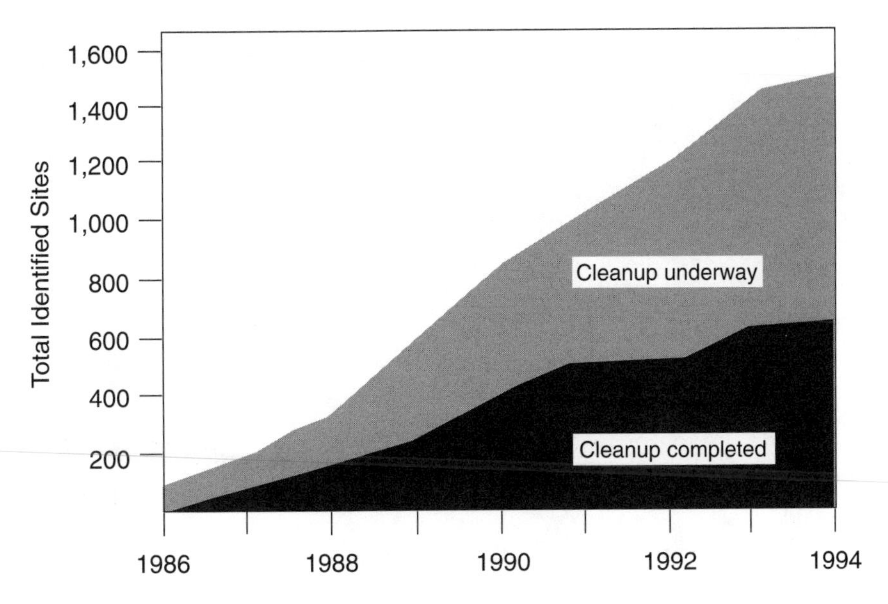

FIGURE 2 Waste sites in Vermont.

cleanup of hazardous waste sites in Vermont has steadily increased (Figure 2). By 1994, 1,499 sites had been identified in the state. Of these sites, 643 had completed cleanups with no further action necessary. An additional 856 sites are in various stages of investigation and cleanup.

Included in this total number of hazardous waste sites are eight sites (including six closed community landfills) that are on the National Priorities List as Superfund sites. Two of these sites have cleanups under way, three have proposed cleanup plans under sate and local review, and two have been found to not require cleanup under Superfund criteria.

A vast majority (74%) of hazardous waste sites in Vermont are from leaking underground storage tanks. The risk of an underground storage tank (UST) leaking is partly a function of its design characteristics. Older, single-walled USTs are more likely to leak and contaminate the surrounding soil and groundwater. The number of single-walled USTs has been steadily declining over the past eight years, dropping from 7,110 in 1986 to 3,367 in 1994-a reduction of over 50% (Figure 3). the number of safer double-walled USTs has increased over this same period from 37 to 1,245 as they replace the older, single-walled tanks. Overall, more than 2,500 USTs have been removed since 1986

Although progress continues to be made, a number of Vermonters are still affected by contaminated drinking water caused by these hazardous waste sites. Since 1987, contaminated drinking water has been found in 25 public wells and 228 private wells.

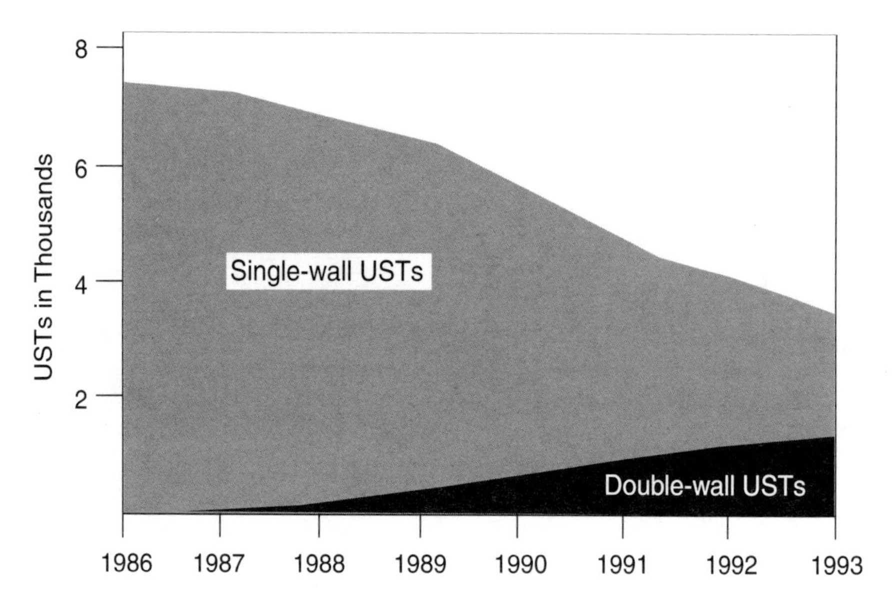

FIGURE 3 Fewer underground tanks.

Pollution Prevention Progress

Pollution prevention continues to be the preferred strategy to reduce the generation of hazardous wastes in Vermont. In Vermont, nearly 250 large and small quantity generators of hazardous waste are required to implement hazardous waste reduction strategies. In addition, beginning in 1995 Vermont companies using 1,000 pounds or more of a toxic substance per year must prepare toxic use reduction plans.

Data collected by the U.S. Environmental Protection Agency shows a significant downward trend in the release of toxic substances by Vermont facilities. The 1992 Toxics Release Inventory (TRI) indicates that since 1988, the 50 Vermont TRI reporters nearly halved the total amount of industrial toxic chemicals they released into the environment (Figure 4). In 1992, reporting facilities in Vermont reduced chemical releases by 13% from 1991 levels (compared with a national average decline of 6.6%). The 1992 TRI total ranks Vermont 52nd in the nation for toxic releases.

While the TRI data indicates that Vermont facilities are decreasing their release to the environment, it remains difficult to develop meaningful estimates of reductions in hazardous waste generation from toxics used and hazardous waste reduction plans submitted to date. The ANR is continuing its effort to encourage pollution prevention and to more accurately measure and monitor resulting toxics use and hazardous waste reduction.

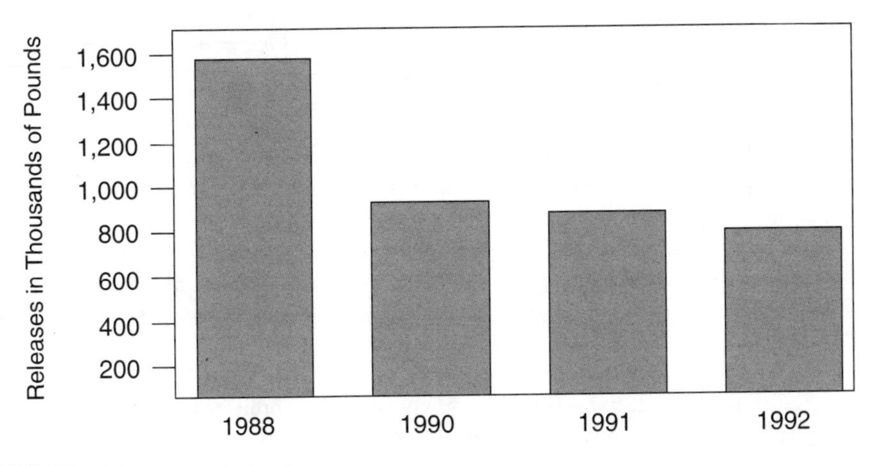

FIGURE 4 Toxic releases are down.

Challenges

Waste generation can often be influenced by the decisions of individual consumers. To help reduce the volume and negative consequences of both solid and hazardous waste generation, Vermonters may consider the following:

Practice Pollution Prevention. Buy durable goods that last longer, buy products with minimal or reusable packaging, give preference to products that are energy efficient and that are made from materials readily recycled in your community, and consider making deliberate choices to do without.

Make informed buying decision to reduce toxics. Read product labels to avoid hazardous household products and use less toxic alternatives to commonly available paints, pesticides, automotive products, chemical cleaners, and polishes.

CHAPTER 6*

PRAIRIES

What wasn't forest or open water in presettlement Illinois was prairie. The exact extent of these grasslands is disputed, but its safe to say that in 1820 at least 60% of Illinois' land area was grasslands of one type or another.

Modern scientists recognize six main subclasses of prairie in Illinois. These are distinguished mainly by differences in soils and topography; further subdivisions based on soil moisture produce a total of 23 distinct prairie types in the Prairie State.

Flat terrain and deep loess soils made most Illinois prairies ideal for agriculture. The breaking of the Illinois prairies began in earnest with the invention in 1837 of a self-scouring steel plow both strong enough to slice through the dense mat of prairie plant roots and slick enough to slip through sticky loam soils. Vast stretches of prairie were destroyed between about 1840 and 1900. According to one account, the 60-square-mile Fox Prairie in Richland County was reduced from more than 38,000 acres to 160 acres of prairie between 1871 and 1883. McLean County once had 669,800 acres of prairie; today it have five of high quality. Champaign County once had 592,300 acres of prairie; today it has one of high quality. (Figure 1) Prairie remnants in such counties probably have escaped natural areas surveyors (especially along railroad rights-of-ways) but including them would still leave Illinois with only a very small number of acres of surviving prairie.

The Illinois Natural Areas Inventory completed in 1978 found that only 1/100 of 1% (2,352 acres) of high-quality original prairie survives. Most sites of relict prairie occur on hilly land along the northern and western edges of the state and other places where plows and bulldozers can't reach, such as wetlands, cemeteries, and railroad rights-of-way. Four out of five of the state's prairie remnants are smaller than ten acres and one in three is smaller than one acre. Of the 253 prairie sites identified by the inventory, four out of five are not protected as dedicated nature preserves.

Prairie in the Prairie State Is Disappearing

- In 1820 at least 60% of Illinois' land area was grasslands of one type or another.
- The Illinois Natural Areas Inventory found that only 1/100 of 1% (2,352 acres) of high quality original prairie survives.
- Of the 253 prairie sites identified by the Illinois Natural Areas Inventory, four out of five are not protected as dedicated nature preserves.

*Pages 276–279 reprinted with permission from the Illinois Department of Energy and Natural Resources.

Illinois' Remaining Prairies Are Being Fragmented

- Four out of five of the state's 253 prairie remnants are smaller than ten acres and one in three is smaller than one acre—too small to function as self-sustaining ecosystems.

Illinois prairie remnants are often less than one acre in size, and the entire local population of some plant and animal species may by only a few individuals. The smaller such local populations are, the more vulnerable they are. Extremely isolated populations of plants and animals can develop so-called inbreeding depression, or an inability to produce, especially if they are not wind-pollinated species whose widely despersed seed gives them ample opportunities for cross-breeding with distant populations.

To date Illinois has few examples of inbreeding depression in its prairie preserves, although that may also reflect the lack of appropriate studies looking for it. Many prairie plants are long-lived, producing only a few generations per century, and thus are unlikely to quickly show the effects of inbreeding.

Because they tend to be inaccessible to the plow, hill prairies are among the last "living windows" into the presettlement Illinois ecology. Illinois hill prairies hold only half the acreage they did 50 years ago. Without periodic fires to check their growth, woody species invade hill prairies from adjacent lands. Comparing aerial photos from 1940 to the present shows that Revis Hill Prairie has decreased in size from 39.2 acres in 1939 to 17.4 acres in 1988.

The tallgrass prairie (where it survives) is a nitrogen-limited system, meaning that the exuberant growth of grasses and other plants consume most of the nitrogenous nutrients in the soil. This chronic nitrogen shortage helps prevent plant species not adapted to it from invading the grasslands. However, supplemental nutrients can enter prairie ecosystems in various ways and in various forms and are likely to alter their species composition.

Nitrate and ammonium compounds are delivered from the air by both wet and dry deposition. Total nitrogen deposition on Illinois soils through most of the 1980s ranged from 17 kilograms per hectare per year to less then ten in the Chicago area; nitrogen pollution from runoff and groundflow is locally even more concentrated than that from the air. Prairie plants have been shown to vary in their ability to capitalize on atmospheric carbon dioxide, another nutrient, which some experts expect to double during the next century.

Prairie Ecosystems Face Extirpation in Illinois, although Few Prairie Plants Do

- While 117 of the 497 plant species considered endangered or threatened in Illinois as of 1993 occur in prairies, only one occurs solely in prairies.

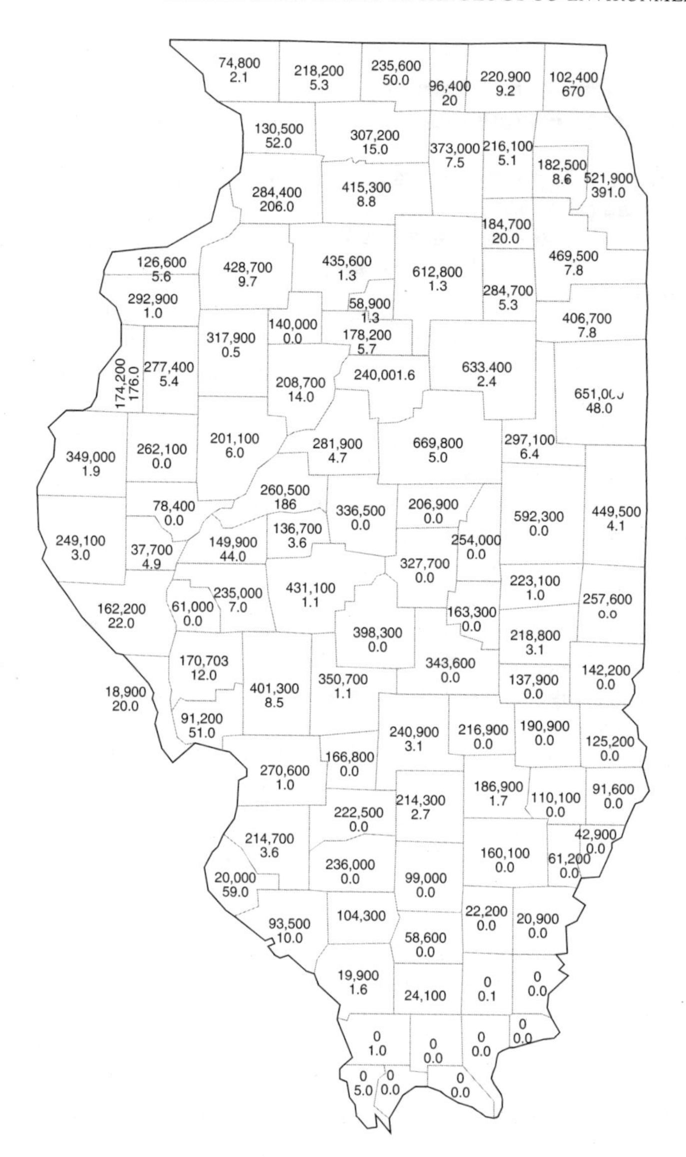

FIGURE 1 Changes in Illinois prairie acreage by county. The top figure is the number of acres of prairie in 1820; the bottom figure is the number of acres of high-quality prairie remaining in 1976. Source: *Ecological Resources,* Illinois Natural History Survey, 1994.

Plants

The number of "prairie species"—those plants capable of living at least part of their life cycle in that habitat—is quite large. The Illinois Plant Information Network counts 851 species of plant native to Illinois prairies. However, by no means do all of these occur in any one site. More than 100 species are seldom found in any one prairie, although (as is true of forest land) the larger patches are host not only to more plants, but to more kinds of plants.

Of the 497 plant and animal species considered endangered or threatened in Illinois as of 1993, 117 occur in prairies. Because nearly all species found in prairie occur in other states, or in habitats other than prairies, there are few species endemic to the Illinois prairie ecosystem.

Because of the peculiar conditions under which many Illinois prairie remnants survive, they are vulnerable to peculiar threats. Plants native to blacksoil prairies of the sort that flourish atop undisturbed country graveyards are being overtaken by non-native species planted by mourners as landscape ornamentals. Many exotic plants are little more than nuisances when they root in prairie patches, but species such as white sweet clover and giant teasel are very aggressive. Some can be eliminated by such approved management techniques as periodic burning.

Wildlife

The conversion of prairie to "secondary grasslands" in the form of hay fields and pastures actually enhanced Illinois' habitat for certain birds such as the dickcissel and the prairie chicken. But more recent changes in agricultural practice led to the decline of even these surrogate prairies. Three species of birds once common on the prairies have been extirpated in Illinois (the sandhill crane, once thought extirpated, recently reappeared in Illinois) and another thirteen species were endangered or threatened in the state as of 1993.

Insects have proven more adaptable, although some species may be struggling in Illinois. Typical is the Karner blue butterfly. A native of the Great Lakes and Northeast, the caterpillar of the Karner blue is uniquely adapted to feeding on the leaves of the wild lupine once common in northern Illinois savannas. Nationwide, populations of the Karner blue have declined by 99%, and in 1992 the U.S. Fish and Wildlife Service added it to the list of endangered species. The insect had not been seen for a century in Illinois until the summer of 1992, when five (perhaps a windblown "tourist" and its progeny) were spotted in Lake County.

The distribution of 255 of 640 prairie insect species surveyed since 1982 is restricted to prairie/savanna remnants. Perhaps one-fifth of these are found in only a very few, usually small, sites and must be considered imperiled. Among these is the loosestrife rootborer, which today is found at fewer than six sites; a handful of other species, such as the Dakota skipper, have not relocated to nonprairie habitat and are assumed to be extirpated in Illinois. Separate surveys of the protected prairie at Illinois Beach State Park have found that two butterfly, one moth, and a dozen leafhopper species that once inhabited it have not been seen there for many years.

RANKING OF AIR ISSUES[a]

ISSUE	RELATIVE RISK[b]	NEED FOR FURTHER ACTION[c]	CITY'S ABILITY TO FURTHER INFLUENCE[d]	OVERALL PRIORITY[e]
Transportation Sources	High	High	High	1
Wood Burning	High	High	High	2
Environmental Tobacco Smoke (ETS)	High	High	High	3
Other Indoor Air Pollution	Medium-High	High	Medium	4
Noise Pollution	Low	High	High	4
Fugitive Dust	Medium	Medium	High	4
Gas Stations	Medium	Low	High	4
Industrial Point Sources	Medium-High	Medium	Medium-Low	8
Centralia Power Plant	Medium	Medium	Low	8
Yard Burning	Low	Low	High	10
Other Non-point Sources	Low	Medium-Low	Medium-Low	10
Nonionizing Electromagnetic Radiation	Not Ranked	Not Ranked	Not Ranked	Not Ranked

SOURCE: "Rankings of Air Issues," from *Environmental Risks in Seattle, A Comparative Assessment*, published by the City of Seattle, Office for Long-range Planning, Seattle, Washington, October 1991. Reprinted with permission.

[a]Note that the Air Team chose not to rank nonionizing electromagnetic radiation, due to the high degree of scientific uncertainty surrounding the issue.

[b]Relative risks to the issue poses to human health, the environment, and quality of life.

[c]Need for action above and beyond existing efforts to reduce risks.

[d]City's ability to further influence the problem, given jurisdictional constraints and overall practicality.

[e]Overall priority for action, combining relative risk, need for further action, and City's ability to influence.

Setting Environmental Goals: The View from Industry

A Review of Practices from the 1960s to the Present

JOHN R. EHRENFELD* AND JENNIFER HOWARD†

Director, MIT Program on Technology, Business, and Environment, and Doctoral Candidate,† MIT Technology, Management, and Policy Program, Massachusetts Institute of Technology*

CONTENTS

In this paper, a model of a four-stage evolution of corporate environmental management is used to structure a discussion of changes in the environmental goals of industry. As companies move from treating environmental management as a case-by-case problem-solving issue, to a regulatory compliance issue, to an issue for proactive management, and finally, to a central element of business strategy, the environmental goals they set necessarily change. What trends are discernible in corporate goal-setting over the last 30 years? And, since most U.S. firms are currently at the compliance or proactive phases of environmental management, what goals should be set to help them move towards the final stage?

Prior to the 1980s, few firms actively set their own environmental goals. Typically, government regulations served as an embodiment of society's environmental expectations, and companies responded by complying with (or at least claiming to comply with) regulation. In the mid-1980s a number of high-profile incidents—the Bhopal toxic chemical release, the *Valdez* oil spill, and the assertion that ozone thinning was linked to CFC use—inextricably linked environmental degradation to the actions of industry. In response, companies took steps to articulate their own goals for environmental performance, which, in some cases, went beyond actions required by regulation. The internalization of responsibility for environmental protection demanded a significant shift in corporate attitudes and organization for some firms.

Early corporate goals that went beyond regulatory compliance primarily targeted waste and emissions reduction. Over several years, goals set by individual companies began to include process-oriented goals in addition to targets for desired "end-state" conditions. Goals began to include commitments to reduce waste at its source, rather than at the end of the pipe. Several leading companies publicly committed to targets (e.g., zero emissions) that were clearly beyond their current technical and organizational capabilities in order to push the internal development of environmentally sound alternatives.

Coordinated efforts on the part of industry groups and industry/government coalitions to define voluntary environmental codes of practice for their members emerged in the late 1980s. The goals set by such groups were different from individual company goals in that they aimed to set out new norms and values that would guide the actions of all firms in an industry. Rather than specifying particular performance targets or demanding strict compliance to specific standards, such codes allow companies to develop their own practices consistent with the guiding norms. While there is plenty of evidence that voluntary codes may reflect primarily a desire to legitimize the actions of an industry, or to preempt regulatory action by demonstrating the industry's leadership and commitment in environmental matters, the type of normative principles they espouse may lay the groundwork for a new approach to managing for the environment.

In order to reach the final stage of environmental management, a firm must begin to "manage for the environment" in the sense of bringing environmental concerns into every aspect of business strategy, product and process design, and supply chain management. Furthermore, goals at this stage must acknowledge broader issues of sustainability and orient the firm towards energy and resource conservation, dematerialization, and elimination of substances harmful to the global environment and human health. The challenge for companies, industry groups, and governments at this stage is to set goals that can be put into practice at the company level but aggregate in a consistent fashion to the multi-dimensional and complex goals of environmental sustainability. The Dutch Covenants are an example of such a coordinated attempt, while industrial ecology and design for environment are systematic approaches that can be taken by individual firms to reduce their environmental impact in a manner consistent with norms of sustainability.

Environmental goal-setting by industry, while it has changed drastically over the past 30 years, remains problematic. We may question the extent to which environmental concerns really can become a central strategic concern for business in the face of competitive and financial pressure. Furthermore, environmental goals are inherently difficult to set. There is no single "customer" who can specify demands (as in, say, the case of quality targets), there is little scientific agreement on the ecological consequences of many economic activities, and the prospect for a rational, technical solution to many complex environmental problems is dim. A broad, normative set of principles that embodies process considerations (i.e., emphasizing loop-closing or dematerialization) may be one way to guide change. Industry cannot generate such norms in a vacuum, however. It is the role of government, industry, and the public to jointly and continuously articulate norms for environmental practice.

INTRODUCTION

Historically, industry has played a minor role in setting broad environmental goals in the United States. Environmental policy in this country is based on a notion of the environment as a public good whose protection and development lies beyond the individual concerns of private business. U.S. industries have traditionally held a short-term and myopic view towards environmental goal-setting. In a sense, the laissez faire economic paradigm is antithetical to broad, long-term goal-setting at all. The last three decades, however, have witnessed a shift in corporate environmental goal-setting practices. Paralleling social sentiment, industry goals for environmental performance have become more aggressive, more explicit and more far-reaching, as the rise in environmental consciousness in the 1960s spawned current concerns over global sustainability. This paper traces changes in corporate environmental goal-setting in the context of the overall evolution of corporate environmental management. It examines characteristic patterns and examples in each stage and presents an assessment of the potential for industry to set broad environmental goals.

Stages of Evolution of Corporate Environmental Management

Distinctive, evolutionary patterns of corporate environmental behavior have been observed by many authors. Hunt and Auster presented an artifactual description of five stages of corporate environmental development starting with "beginner," and progressing through "fire fighter," "concerned citizen," "pragmatist," and "proactivist."[1] Shrivastava referred to this process of environmentally directed self-renewal as "greenewal." That process, he states, is initiated by a strategic threat from regulations, public pressures, public safety concerns, or social expectations. The embattled firm then forms ad hoc strategic programs, testing their competitive benefits and, if necessary, expanding the organizational systems through institutionalization and cultural changes.[2]

The five stages shown in Figure 1 have been used by researchers at MIT to classify observed and anticipated changes as corporate environmental management evolves. A brief description of the attitudes and practices is given for each stage. For consistency with the latter part of this paper, we combine the third and fourth stage into a single one, "Proactive Environmental Management," in the discussion below.[3]

Environmental Management as Problem-Solving

At this, the most basic level, the underlying assumption of the firm is that environmental protection is of little or no concern to corporate decision-making.

Environmental management as problem-solving

Environmental management as compliance

Environmental management as emissions reduction beyond compliance

Environmental management as source reduction

Managing for the environment: A transformation to the Green Company

Time

FIGURE 1 Historical evolution of environmental management.

Instead, it is viewed as an ancillary aspect of conducting business. Regulations are perceived as an unfamiliar nuisance. As a result, the organization is characterized by the lack of a permanent staff or budget for dealing with environmental issues. Very often, the plant engineering staff is called upon to handle environmental issues on an ad hoc basis. For example, when faced with the flood of environmental laws in the 1970s, Allied-Signal managers admitted to viewing the resulting expenditures as merely "a cost of the way we did business."[4] Few firms presently exist in this stage. Given the all-pervasive nature of today's environmental regulations, every company must have an understanding of environmental law, either in-house or through outside consulting services, in order to survive. Environmental goal-setting is virtually absent in this phase. Some smaller companies and larger laggards may still exhibit this type of behavior and management structure.

Environmental Management as Compliance

Advancing to this stage, the firm perceives environmental regulation as important enough to merit full-time attention. However, the firm views regulation, and not concern for the environment per se, as the motivator of new practices. Certain parts of the organization are altered, but the basic structure remains untouched. Dedicated compliance staffs labeled "government affairs" or "regulatory compliance" behave as buffers, limiting the collection of information and the impact that environmental regulation will have on the inner workings of the firm. These departments can exist on many levels, such as the operating level in the form of environmental engineering, the corporate level in the form of environmental counsel, or the political level through lobbyists who fight environmental statutes and regulations.

Most publicly held U.S. corporations can count themselves in this stage while also moving toward the next. For example, an Office of Technology Assessment study determined that the standard industrial thinking was to treat process wastes and emissions as separate and distinct from the process itself.[5] A few years ago, a Conference Board survey found that 65% of U.S. firms put resources into lobbying to change environmental regulations.[6] The survey also found that while some U.S. firms located their environmental affairs function in either manufacturing or engineering, the legal department was listed slightly more often. When rating the factors that reflected in environmental policy decisions, 69% of the companies were motivated in response to legal or regulatory requirements, 21% were motivated in response to liability pressures, and 32% were motivated by social responsibility. This emerging social responsibility may be what is driving companies on to the next stage of the greening process.

Proactive Environmental Management

At this stage, the firm believes that environmental protection has certain strategic advantages, as well as significant cost reduction opportunities. Requirements for public disclosure of emission levels forced companies to examine their environmental practices from the point of view of the public and other stakeholders. By reducing their emissions below permissible levels, companies could perhaps influence the attitudes held by the public and government regulators towards their products and operations. At this stage, the goals of the firm transcend mere compliance with government standards and encompass the voluntary establishment of stricter standards. Often such actions result in profit increases through lowered operating expenses. Organizationally, this shift is often accompanied by the establishment of an executive level individual who is responsible for matters concerning the environment, and the beginnings of the diffusion of responsibility for environmental protection throughout the company. Environmental management staffs move away from their alignment with regulatory compliance and begin to associate with quality control and corporate strategy. Generally, such moves are accompanied by public statements of broad corporate commitments to protecting the environment.

This is the transition through which market-based incentive regulations are attempting to push industry. By creating an economic value for pollution reduction, government, theoretically, will entice companies to reduce pollution below the stated requirements in an attempt to increase profits. As a result, many companies are beginning to enter this stage. According to a survey by Arthur D. Little of the top 100 companies in the Fortune 500 index, 49 had environmental Vice Presidents in 1991.[7] Several widely publicized stories, some of which are detailed below, tell of companies reducing pollution beyond compliance levels while at the same time increasing profits.[8]

Later in this stage, companies may expand the scope of their proactive environmental management beyond mere pollution prevention at the end of the pipe. They may consider further reductions through materials substitution, process improvements, product reformulations, and waste recycling. At this point, the objectives of pollution minimization become diffused throughout the organization, becoming a responsibility of each operating department, not just of a select environmental or corporate management staff. In essence, environmental management becomes integrated into corporate quality management plans.

Relatively few companies are at the stage of deeply examining their products and processes and making environmental considerations a priority at the design stage. It is difficult to determine whether a significant shift in corporate attitudes has occurred in those companies who do initiate such practices. Although corporate managers proclaim that "quality and pollution prevention are in the same boat," the corresponding underlying assumption appears to be lacking as evidenced by the reluctance of the operating staff to adopt this process change.[9] This example exemplifies the need to understand the dynamics of cultural change fully. Some argue that the transition to this stage will be a difficult hurdle, as evidenced by virtually all of the billions spent on environmental equipment and on services to industry and government going for pollution control rather than pollution prevention.[10]

Managing for the Environment, A Transformation to the Green Company

Since very few companies have made it to this point of managing for the environment, it is difficult to identify the corporate practices and beliefs of this stage through corporate example. This stage is reached when environmental concerns become a core strategic factor in corporate decision-making, infiltrating quality functions, financial measures, and performance criteria in a manner that leads rather than follows public policy. Companies will move beyond concern for the protection of the present environment, to considering the environmental impacts of their activities on future generations. Concerns about sustainable development are fundamentally different from the essentially compliance-related actions that characterize all of the earliest stages of corporate environmental management. Wise resource use and product life-cycle analysis become the instruments for firms to augment short-term goals with efforts at long-term sustainability through the design of new environmentally sound process and product technologies.

There is much uncertainty as to how far or fast companies will go in this final transition step. In many ways the transition is deeper and more wrenching than the shift to total quality management has been. In the TQM case, the field is still littered with firms that are struggling to move TQM into their core competence and culture. For most companies, the management of the greening process becomes a carefully controlled process, and requires continuous monitoring of changes in the overall industry and guiding their own actions accordingly. It is for these companies that trade associations play such an important role. By outlining the industry-wide definitions of what the environmentally responsible company ought to be, they are providing assurances that the industry as a whole is moving together. Each participating firm will expend comparable effort, face similar liabilities, and reach the new paradigm at roughly the same time and rate. Therefore, it is from these trade associations that we can see the beginnings of a forecast of the definition of a green company.

The Nature of Goals

Prior to discussing specific industry environmental goals, it is necessary to clarify what we mean by goals. Goals for industry may take the form of specific targets (to increase market share to X% by year Y) or general statements of intent (Hewlett-Packard's corporate objective on citizenship is a commitment to "honor our obligations to society by being an economic, intellectual, and social asset to each nation and each community in which we operate"[11]). In many cases the distinction between these more general "visions" of a desirable future and specific targets is dim or becomes ambiguous. The forthcoming ISO Environmental Management Standard (ISO 14001) makes a distinction between the two types that is worth mentioning here[12]

> environmental objective—overall environmental goal, arising from the environmental policy that an organization sets itself to achieve, and which is quantified whenever practicable.
>
> environmental target—detailed performance requirement, quantified wherever practicable, applicable to the organization or parts thereof, that arise from the environmental objectives and that needs to be set and met in order to achieve these objectives.

In contemplating the setting of goals, one should examine these distinctions very carefully. Broad, policy-like statements or goals are fundamentally the exposition of a vision of a desirable future. They set a direction, but not specific actions to be taken to move in that direction. They are *declarative*[13] in nature and their power to create action rests on the authority of the person or body issuing the goal. There is no promise in such goals, although they have strong normative power. The actualization is left to members and functions of the organization. Targets represent promises to achieve a determinable end-state by some certain time. Promises fall into a different category of speech acts, *commissives*,[14] which commit the speakers, individually or as a group, to meet some explicit set of conditions by some time. The completion of a promise can be observed by examining the degree to which the target is satisfied at the time set. Fulfillment of promises depends on the resources available to the actors, their competence, and the strength of their intention to do what they have said.

Goals and targets may be set by individual firms, by industries as a whole, or by external agents (i.e., regulators set targets for environmental and safety performance, and government consortia set goals for economic or trade performance of certain industries). The key feature of a corporate target is that it is a (public) statement that reflects a promise by an agent (the firm or industry) to take action to deliver a result to an external party (typically shareholders or the general public). Goals and targets, particularly those articulated in general terms, are not always met and it is often difficult to judge when a target has been achieved because there may be no metric by which to make this judgment. Some goals, like sustainability, that create a broad vision of the future are not intended to be met in a specific way, but to serve as symbols directed to the many actors and stakeholders that new meanings and values should replace or be added to the old set of cultural drivers.

If a target is not met, the organization or actor loses legitimacy in the eyes of the party to whom the promise was made. The action of setting explicit targets thus in some way commits the agent to a course of action, validates this action, and punishes the agent for not acting consistently with the target, providing a weak enforcement mechanism. Confusion over goals and targets by both the acting agents and the listening publics has frequently led to loss of trust and legitimacy. The public tends to hear declarations of broad goals as targets and looks for specific accomplishments when none were promised. Whatever processes are to be adopted in setting goals and targets, firms need to be very clear as to their distinctive nature in public disclosures in order to avoid negative consequences later.

Even with the potential down-side of setting goals and targets, there is still considerable value in the activity of goal-setting per se, regardless of outcome, because most players with environmental concerns believe that the setting of goals encourages a type of behavior that is favorable. In other words, working towards a goal may be as valuable as achieving the goal. This feature will come up later in the discussion of what we shall call "process" goals—i.e., goals set to encourage a way of doing things rather than a particular end-state. For the purposes of this paper, then, corporate goal-setting means the commitment of a firm or group of firms to a certain course of action or end-state, which may be tightly or loosely defined.

Environmental Goals and Stages of Corporate Environmental Management

Few, if any, *environmental* goals can be identified in the early, problem-solving stage of corporate environmental management. Companies at this stage interpret environmental problems as fitting into a more traditional area of concern. For example, when one company was faced with an explosion of some waste drums that had washed up on a public beach and caught fire (obviously that had been mishandled by their waste management contractor), it gave the job to the PR department because this was perceived as a threat to its image as an innovative consumer product company.[15] It would appear that the company saw this incident as belonging to a different category from environment. The same firm, today, has a highly elaborated set of environmental functions and would give this problem to them, as well as to the PR department. The compliance stage in the overall evolution has been characterized by indirect environmental goals, manifest primarily in commitments to comply with regulations that, by and large, set the path a company needs to follow as mandated by the "command-and-control" regulatory system. This was a period in which environmental concerns were considered to be externalities that, in terms of the dominant economic theory, were not incorporated into the firm's cost structures and, thereby, were out of its strategic ambit. Explicit technology goals were infrequently established during the legislative and regulatory process, but generally not by the industry (except in opposition to the proposed mandates). Environmental goals were set by the public bodies. In terms of goal-setting, industry response was often to argue its case during the legislative debate and regulatory proceedings. Its environmental goals might be cynically stated to cut the best deal it could.

In the compliance mode, industry ultimately accepts the goals set by the government and promises to comply with whatever implementing regulations follow. By far the bulk of past industrial efforts have been expended to keep such promises. There is an important set of secondary goals that accompany these formal promises of compliance. They are to comply as cheaply as is possible, taking account of normal capital and operating costs and also the costs of non-compliance. Penalties for non-compliance have been severe in terms of civil and criminal penalties and loss of public image. In this regard, industry reacts no differently to environmental rules than to any other set of public mandates.

As we shall see, corporate environmental goal-setting becomes much richer in the proactive phase of environmental management. Cost-cutting and other rationalizing actions remain a strong driver, but environmental goals begin to reflect what are underlying broad social environmental concerns more explicitly. This is the period in which environment emerges as an explicit area of concern in corporate policies and public communications.

In the last stage, managing for the environment, companies broaden goals to deal with problems such as global warming, ozone depletion, excessive resource depletion, and loss of productivity. With the publication of the Brundtland report in 1987, the overarching paradigm has become sustainability, even while no consensus on its operational meaning has been reached. The change towards aggregate goals such as the prevention of ozone depletion is problematic in terms of establishing discrete industry or firm goals as so many sectors and firms are causal agents. (See, however, the discussion of the Dutch target group approach, below.) At this stage, sectoral and collective approaches to goal-setting become more important.

The following sections discuss trends in and examples of corporate environmental goal-setting in the final three stages of environmental management. Much of the information presented below has been obtained from phone interviews and literature provided by industry associations and firms, corporate annual reports, and a number of recent books on the subject of corporate environmentalism. While it is by no means a comprehensive study, it is intended to provide an outline of some of the trends in environmental goal-setting and some conjectures for the future.

GOAL-SETTING IN THE STAGE OF ENVIRONMENTAL COMPLIANCE

The natural environment has been inextricably linked to human society since the earliest stages of human development. Much of what we call civilization consists of the technological artifacts that humans use to gain both sustenance and protection from the natural world. For much of human history, the environment was just there, to be treated as a regenerative resource for human use. Social consciousness about the environment in the United States became organized late in the last century when a few men (the gender-specificity is intentional) began to notice the loss of wilderness. Many of the themes could be found in early individualistic writings such as Thoreau's. Their concerns showed up in the form of conservation or preservation—two notions that were quite opposed. Preservationists saw the need to keep natural settings as they were while conservationists saw the possibilities of using the land at the same time as maintaining its productivity and other values. Early "environmental" organizations grew up around these two notions. Important goals were set in those days and came in the form of the acquisition or protection of large tracts of land as National Parks and National Forests. But many environmental concerns as we know them today were largely absent.

"Environment" entered the social scene in the 1960s with the publication of Rachel Carson's *Silent Spring*, in which she raises specific concerns over the integrity of the environment which she sees as threatened by the impact of certain industrial products and processes. Others picked up this theme and further articulated what they saw as an attack on the environment by the technological forces of modern society. The concerns of these sentinels grew into calls for public action and led to the large body of public policy instruments that have been promulgated since the 1970s. A few environmental statutes were passed in the 1960s, but none were very strong. It was not until 1970 that the Congress, through the passage of one environmental statute after another, declared that the protection of the environment was a national goal together with human health and welfare. These acts set the first targets and boundaries for industrial environmental performance. The federal Environmental Protection Agency (EPA), through such statutory instruments as the Clean Air Act of 1970, the Clean Water Act of 1977, and the Comprehensive Environmental Response, Compensation, and Liability (CERCLA, or "Superfund") Act of 1980, translates societal expectations of appropriate treatment of the environment into requirements for industry. Environmental statutes exist at all levels of government—federal, state, and local—and represent a decision made long ago "not to entrust environmental problems and disputes solely to markets, to the courts, or to mediation services" in the U.S.[16]

The EPA and other regulatory agencies set environmental goals that reflect a broad range of desired outcomes. Paul Portney notes that while many statutes reflect a zero-risk or threshold philosophy (for example, the Clean Air Act requires that ambient standards for common air pollutants provide "an adequate margin of safety"), a number of them establish a technical standard to be met (for example, "best available technology" is commonly espoused in laws that elsewhere embrace a zero-risk goal). Still others (e.g., the Toxic Substances Control Act and the Federal Insecticide, Fungicide and Rodenticide Act) require a balancing of economic and environmental costs and benefits.[17] Despite a confusing and sometimes conflicting prescriptive for desired outcomes, federal environmental policies are consistent in one manner, the means through which environmental goals should be pursued. The command-and-control approach has tended to dominate, with little attention paid until recently to incentive-based approaches.

Both the Clean Air Act of 1970 and the Clean Water Act of 1972 had explicit goals, with the Water Act taking the bold step (not to be realized) of setting a target date by which the goals were to be achieved. The Clean Air Act directed the EPA to establish air quality levels to serve as goals to be met through the use, primarily, of control devices on sources of six air pollutants. The regulations that followed to implement the standards relied mainly on some form of available technology. Thus the polluters had little or no incentive to do anything but purchase whatever types of control devices were being marketed by specialty firms making these devices. The regulatory system was predicated on the assumption that the basic production technologies would not change and that environmental quality targets would be met by the superposition of control devices. This gave virtually no incentives to companies to modify their production technologies to be less polluting in the first place. New technologies for control did emerge as pollution control companies attempted to develop and market cheaper or more efficient devices. The tack taken in the Clean Water Act was somewhat different with significant implications for the technologies that were to be implemented. Rather than set some form of ambient or water quality standards, which would then have to met by technological means, the Act directed the EPA toward technology-based standards. These were, in essence, performance standards that were established by examining the technological systems that were being used, and, in some cases, represented not-yet fully developed commercial prototypes.

There was one important exception, the technology-forcing requirement to reduce auto exhaust emissions to levels beyond those attainable according to the technologies available in 1970 when the law was passed. The law required industry to develop cars that reduced tailpipe carbon monoxide and hydrocarbon emissions by 90% (relative to 1970 levels) by the 1975 model year and a similar goal for nitrogen oxides with all the dates slipped one year.[18] This was the first instance of an environmental technology goal for industry, but it was one that was clearly established by the government. It was not set, however, without much intense lobbying by the U.S. automobile industry, who fought the deadline as unrealistic. It continued to fight and was permitted an extension by William Ruckelshaus, the first EPA administrator, but eventually produced cars meeting the standard by 1977. The technologies included the exhaust catalyst, exhaust gas recirculation, and changes in the combustion system. It is interesting to note that Honda developed an engine that could meet the standards without an exhaust catalytic converter. The U.S. firms ignored this approach, preferring to preserve the basic internal combustion systems they were comfortable with. Many of the technological advantages of later Japanese automobiles that fueled their competitive onslaught in the American market were spawned by these earlier engineering approaches taken to meet environmental standards.

Several problems arise when we consider the environmental goals set by government regulatory bodies and their impact on a firm's behavior. First, in cases where specific targets have been written into the laws, they have often been grossly overly ambitious to the point that partial compliance by industry is the best that can be hoped for. The Clean Air and Clean Water Acts call for

> the establishment of literally tens of thousands of discharge standards, mandate the creation of comprehensive monitoring networks . . . yet the laws allocated just 180 days for completion of many of these responsibilities. Today, more than seventeen years after the passage of the laws, many of those assignments have yet to be carried out.[19]

Second, federal environmental statutes often embody absolute goals (waters are to be "fishable and swimable") and provide little guidance on appropriate metrics to be used in assessing progress towards the goal. The result is that the inevitable balancing of environmental and economic costs is performed implicitly, by firms acting on their own or in concert with regulatory bodies. Third, environmental goals established by government inevitably embody broader societal goals, such as redistribution of wealth or maintenance of certain productive sectors in the face of adverse environmental outcomes.

GOAL-SETTING IN THE STAGE OF PROACTIVE ENVIRONMENTAL MANAGEMENT

Several highly visible events, which occurred in the 1980s—the Bhopal toxic chemical release; the discovery of a hole in the ozone layer; and the Exxon *Valdez* oil spill—inextricably linked environmental degradation with the actions of industry. While the public sought to blame the chemical, petroleum, and other industries for their environmental wrongs, firms were simultaneously driven to accept responsibility for the environment. As a result, corporate environmental rhetoric is vastly different now as opposed to twenty, and even ten, years ago. A growing number of firms are participating in voluntary initiatives to reduce waste and restrict emissions, and industry leaders across a broad range of industries are changing their processes and products to make them more environmentally friendly. For some firms, environmental issues have been taken on as explicit business goals as the strategic importance of reducing the environmental burden of operations, and of communicating these commitments to the public, is recognized. In the words of one company, S.C. Johnson Wax,

> The establishment of these goals was a recognition that environmental responsibility has become as integral a strategic element of the business as product performance and cost effectiveness.[20]

The problems associated with environmental goal-setting by government have prompted repeated calls from many industry sectors for more flexible and "sensible" regulation. Industry associations for manufacturers see lobbying to restrict the purview of environmental regulation as a key part of their mandate. Of the 22 activities reported on by the Society of the Plastics Industry in its quarterly issue activity report published in May of 1995, eleven dealt directly with environmental regulatory issues, and five more dealt with broader health and safety issues.[21] Parallel to these efforts, the last ten years have seen the emergence of a new form of positive corporate environmental goal-setting. Rather than focusing only on negotiating guidelines established by regulators, some firms and industry associations are beginning to see environmental goals as an element of overall strategic goal-setting. (See the discussion of ISO standards, below.) This approach may fundamentally supplant regulatory goals as an organization seeks to define the outcomes and methods that reflect desirable environmental and business conditions for itself. Strategic environmental goals set by firms rather than regulators more closely fit the definition given above of a goal as a promise by the agent to act in a certain way or achieve a certain end-state. Under this definition, the agent's promise is most credible if it reflects beliefs held and articulated by the agent itself. In other words, a "goal" of complying with regulation defined by government is not a strategic corporate goal in any robust sense. As former Chevron CEO, George Keller, noted,

> . . . as long as our environmental philosophy is framed by the concept of compliance, we won't get much credit for our positive actions. Compliance means that the moral initiative lies elsewhere—outside of industry.[22]

Thus, environmental goal-setting by firms and industry groupings can be expected to reflect fundamentally different aims from those of government. The strategic reasons for corporate environmental goal-setting may include any or all of the following:

- protection of right-to-operate by defusing public mistrust;
- reduction in operating or waste management costs;
- preemption of command-and-control regulation;
- obtaining "first-mover" advantage through marketing a new, cleaner product or service;
- creating a market for an environmentally sound substitute to an entrenched product (e.g., CFCs);
- being seen as a good corporate citizen and establishing a reputation as a "caring" firm; and
- reflecting an internal culture and values that attract higher quality employees, suppliers, and distributors.

These factors have contributed to the establishment of environmental goals by leading firms in a number of industries (e.g., Monsanto, Dow Chemical, 3M, AT&T, Hewlett Packard, Noranda, Xerox, ICI, Arco, to name a few)[23] and also by industry groups. In certain industries, notably chemicals and petroleum, the environmental issues have been considered sufficiently critical that nothing short of coordinated action by the industry as a whole could secure and ensure individual firms' right-to-operate. In these cases, the industry association has established a formal program that sets environmental goals and establishes principles and management practices.

Several other groups have added their voices to those of government and firms in the articulation of environmental goals for industry. These include national and local public interest groups, who may work either independently or in conjunction with industry. For example, the CERES (Coalition for Environmentally Responsible Economies) principles and public reporting requirements were established by environmental advocates and a socially responsible investment firm. Other cross-industry corporate groupings have formed, as well as industry-government coalitions. These include the Global Environmental Management Initiative (GEMI), a group of 27 multinationals from a variety of industries, which acts as a forum for benchmarking environmental management practices and stimulating new strategies; the International Chamber of Commerce's Business Charter for Sustainable Development, which outlines 16 principles for environmental management; and the Industry Cooperative for Ozone Layer Protection (ICOLP), formed to facilitate diffusion of technologies for eliminating CFC use by manufacturers in developing countries.[24] In each case, coordinated action was a response to the perception that no individual firm would or could adequately address the environmental questions at stake. Action by a group may minimize the potential costs of individual action. In the case of ICOLP, North American electronics manufacturers saw the early elimination of CFC-113 solvents from their processes as a potential cost disadvantage relative to manufacturers in countries that were not required to reduce greenhouse gas emissions on such an aggressive schedule. Group action also formalizes a vehicle for the exercise of peer pressure in the adoption of voluntary environmental practices.

A final set of actors in environmental goal-setting is the growing number of collaborative industry-government groups. Recently, various federal and state regulatory bodies have established voluntary programs with individual firms and industry groups as a replacement to traditional command-and-control regulation. These voluntary programs typically arise less in response to positive environmental initiatives on the part of a given industry and more through recognition that the control of certain substances cannot be efficiently achieved through inflexible, "one-size-fits-all" regulation. As a result, these voluntary programs consist of agreements between firms and the regulatory body to seek technological solutions or product replacements to achieve a reduction in the use or release of the substance of concern. The largest of these programs is the 33/50 program created by the EPA in 1991, which asked companies to voluntarily commit to a 33% reduction in environmental releases and off-site disposal of 17 toxic chemicals by 1992 and a 50% reduction by 1995. Companies may see their participation in the 33/50 program as an important step in avoiding additional regulation of the 17 named chemicals. We now turn to an anecdotal "roadmap" of the recent environmental goal-setting activities of three sets of actors: individual firms, industry groups, and industry-government coalitions.

Goal-Setting by Individual Firms

The Pollution Prevention Pays (3P) program introduced by 3M in 1975 is perhaps the earliest example of a U.S. firm taking explicit action in environmental management. The reasons cited for the 3P program were strategic; pollution represented waste, an inefficient use of resources, so eliminating pollution should improve efficiency. The 3P program provided incentives for employees to seek innovative ways to eliminate waste at the source, rather than merely recycling or recovery. It was aimed primarily at 3M's technical staff in its laboratories, manufacturing, and engineering divisions worldwide. Between 1975 and 1992, the 3P initiative involved more than 3,000 projects, prevented more than 1 billion pounds of emissions, and saved 3M more than $500 million.[25] These results led to emulation; in the mid-1980s, Dow launched its Waste Reduction Always Pays (WRAP) program and Chevron started an effort to "Save Money and Reduce Toxics (SMART)."

While 3M saw numerous benefits from the 3P program's results—lower operating and manufacturing costs, reduced regulatory compliance paperwork, fewer potential liabilities, improved competitive position, improved company reputation, and lower waste disposal costs—no specific goals had been articulated at the outset. The only explicit aims articulated by 3M and Dow were that their employees act according to a hierarchy of pollution prevention priorities that put source reduction at the top of the list, followed by recovery and recycling, waste treatment, and finally, disposal. Dow President and CEO, Frank Popoff, clearly had more than just cost reduction in mind when he spoke of the reasons for the WRAP program:

> . . . the public is skeptical of industry's efforts at environmental protection. If we fail to take the initiative, the result will be a regulatory crunch that costs us—and the public—dearly without achieving significant benefits. Through pollution prevention, industry can be viewed as part of the solution, not as part of the problem.[26]

Several of the strategic goals discussed earlier—cost reduction, restoration of public trust, and avoidance of inflexible regulation—are clearly evident in the actions taken by 3M, Dow, and Chevron in these early initiatives. However, these goals were never clearly articulated as commitments to the public or other stakeholders, and no specific targets for performance were set. Since the mid-1980s, corporate environmental goals have taken on both of these attributes.

The Monsanto Pledge, first articulated by CEO Richard Mahoney in 1990 in response to his company's Toxic Release Inventory data for 1987, committed Monsanto to several specific actions. First, it aimed to reduce toxic air emissions by 90% by the end of 1992, with an ultimate goal of zero emissions of all toxic and hazardous substances. Six other elements of the Pledge include the following: to ensure that no Monsanto operation poses any undue risk to employees or communities, to achieve sustainable agriculture through new technology and practices, to ensure groundwater safety, to keep plants open to communities and involve them in plant operations, to manage all corporate real estate with the benefit of nature as a serious operating factor, and to search worldwide for technology to reduce and eliminate waste from operations with the top priority being not to make it in the first place.[27] While few of the goals are directly measurable, Monsanto saw its Pledge as setting forth a "public target against which to measure the progress of our environmental programs."[28]

Mahoney notes that, prior to making the pledge, Monsanto had worldwide environmental guidelines in place for more than a decade, and waste elimination programs with specific targets were already established. But, in his words, the Pledge was necessary because it ". . . directs our attention to the future. It sets our course for becoming a corporate environmental leader. . . . It elevates these (environment, health, and safety) commitments above our waste elimination programs. . . . And it moves us far closer to public expectations."[29] These elements—to align corporate environmental goals with public expectations, and to set a course for environmental leadership—were not an explicit part of the earlier pollution prevention programs launched by 3M and others.

A number of other companies, while not displaying the depth and range of commitment exhibited by the Monsanto Pledge, began publicly to articulate quantifiable environmental goals with target dates. In response to growing concern about degradation of the ozone layer, both Northern Telecom and AT&T committed to reducing and eventually eliminating use of CFC-113 solvents in their manufacturing processes. Northern Telecom set a "Free in Three" goal in 1988 to eliminate the solvent from its operations worldwide by 1991.[30] Similarly, AT&T committed in 1989 to halve CFC-113 use by the end of 1991 and to eliminate the solvents entirely by the end of 1994.[31] While the Montreal Protocol of 1987 establishing a deadline for the elimination of CFCs by the year 2000 was clearly the impetus for their actions, both AT&T and Northern Telecom took the initiative to define their own business goals and implement internal programs (R&D on substitutes and design of new manufacturing processes) to achieve the goals.

The shift from internal incentive programs to encourage consideration of environmental issues in manufacturing, to the explicit articulation of specific targets for environmental performance is perhaps best seen within 3M itself. In 1989, 3M announced an enhancement of its 3P program, called the 3P+ program. The 3P+ program set 3M's first formal goals for pollution prevention and sought to bolster the voluntary nature of the 3P program with a more systematic approach. An overarching goal to reduce all hazardous and non-hazardous releases to air, water, and land by 90% and to reduce generation of all waste by 50% by the year 2000 was set. Beyond the year 2000, the goal is to approach a level as close to zero emissions as is technically possible. Shorter-term goals have been set to enable stepwise implementation; for example, air emissions were to be cut by 70% by 1993. In addition to emission reductions, the 3P+ program sets goals for resource recovery, energy reduction, phasing out of ozone-depleting substances, and removal of PCBs.[32]

Beyond its setting of measurable goals, the 3P+ program marks two other significant shifts in corporate environmental management. It replaces the award system used in the 3P program with a practice of using adherence to environmental goals to evaluate employee performance and set compensation levels. Furthermore, it explicitly incorporates environmental goal-setting into other goal-setting activities. Challenge '95, 3M's five-year corporate productivity program, includes, amongst its overall productivity goals, the aims of cutting waste generation by 35% and energy use per unit of production by 20% by 1995.[33]

One clear trend in corporate environmental goal-setting is that most quantifiable targets set by firms deal exclusively with waste reduction or elimination of hazardous substance release. From Veryfine, a family-owned producer of bottled fruit juices based in Massachusetts, to Noranda, a large public company operating in multiple commodity markets including lumber and mining, firms tend to put numbers only to a limited range of environmental commitments.

Veryfine set a goal for waste reduction in its operations and within ten years achieved a 40-fold reduction in waste sent to landfill sites. Noranda, who stressed in its 1990 annual report that "the environment is the most complex, challenging, and urgent issue we face as a company" and elevated environmental compatibility to the level of two other key operating strategies, competitiveness and financial strength, nonetheless has chosen to limit its explicit goal-setting statements to those dealing with waste reduction. The only goal-related statement made in its 1994 annual report was in reference to a program to reduce waste by 20% by 1996. The environmental management discussion section reported exclusively on spending on environmental research, site restoration, and new equipment to meet environmental standards, and referred to Noranda's participation in industry-government debates on global environmental issues.[34]

Companies differ widely in the approach they take to setting even this narrow range of measurable waste reduction goals. Some firms set targets that elicit greater attention to manufacturing processes, but do not fundamentally change the design of processes or the selection of products for manufacture. Certain companies, however, see waste reduction goals as an opportunity to force change within the organization that may lead to new thinking about environmentally friendly product and process design. Such companies set "stretch" goals—targets that are not technically feasible when set and are often initially met by skepticism or anxiety on the part of the organization. The Monsanto Pledge was such a goal. When CEO Richard Mahoney announced his pledge to cut Monsanto's toxic air emissions by 90% by 1992 and work towards the ultimate goal of zero emissions, the head of Monsanto's agricultural products division had a typical reaction: "You want us to do *what*? By *when*? *How*?"[35] But once the shock wore off, Monsanto employees set to the task of scrutinizing products and processes for waste reduction opportunities. Monsanto reports that the 90% program marked a turning point in the company's environmental culture, which has enabled them to achieve and exceed the original target. In its 1994 Annual Environmental Review, Monsanto reports on research efforts in process change and the development of new products from waste materials. For example, a waste stream that had been previously burnt for fuel has been upgraded to a non-chlorinated paint stripper product, eliminating 500,000 pounds of air emissions and 800,000 pounds of waste formerly disposed of through underground injection.[36] Mahoney sees the zero-emissions goal as "the only goal which will keep us stretching for ever greater improvement."[37] His rationale for setting a stretch goal for environmental performance parallels that for setting other stretch goals:

> Although many eyebrows were raised, zero is the only standard we should be judging ourselves against. Technically, it may be impossible, but in safety we strive for zero injuries and in quality we strive for zero defects. Our intention is to continue to earn the right to operate; therefore, zero effect is the only acceptable standard to strive toward.[38]

Xerox Corporation chooses to set its environmental goals in a manner consistent with total quality management, which sets goals for continuous improvement rather than stretch goals. By identifying the best practices of industry leaders and benchmarking one's own processes against these practices, goals are set to bring internal operations in line with best practices. Once such goals have been achieved, the target is raised. Xerox first set goals for site recycling in 1990 and, based on industry benchmarking, aimed for a 50% reduction in waste by the end of 1992. In early 1991, many manufacturing sites had exceeded the 50% goal, so a revised goal for quarter-upon-quarter continuous improvement was set.[39] Setting goals for continuous improvement is only meaningful if performance is continually measured and evaluated and methods for meeting goals are implemented and updated. Xerox cites its adoption of product life-cycle analysis as a key methodology that will enable continuous improvement in environmental performance.

AT&T, while not committed to continuous improvement in the sense Xerox has articulated it, has taken the role of process very seriously in setting its environmental goals. It has not only set extremely aggressive environmental goals, but has put in place a number of practices and tools to monitor and measure performance against environmental goals. AT&T is one of a growing number of firms who now publish annual environmental reports that state achievements, publish performance data, and articulate goals for the future. In its 1994 Environment and Safety Annual Report, AT&T reported on progress towards goals set in 1990 by CEO Robert Allen. CFC emissions were eliminated 19 months ahead of the target date of 1994; air emission were cut by 96% relative to 1987, exceeding the target of a 95% reduction by year end 1995. Manufacturing process waste disposal was decreased 66% from 1987, exceeding the goal of a 25% reduction by year end 1994. Finally, 65% of waste paper was recycled by year end 1994, exceeding a goal for 60% reduction.[40] While achievement of these goals is impressive, what really distinguishes AT&T's goal-setting activities is its commitment to the process and tools used, rather than merely the end-state achieved.

AT&T cites two methodologies as its "primary allies" in evolving environmental management: quality and Design for Environment (DFE). New practices and information systems are necessary in order to support product life-cycle analysis, develop benchmarks against other firms, and identify and implement plans for closing gaps. AT&T is developing several tools, including a Green Index software tool, which "scores" products and processes on their environmental impact. The performance of environmental self-audits is an important element in AT&T's plan to make its environmental goals visible and measurable. The importance of process in AT&T's environmental strategy is demonstrated by the fact that two of the five goals for the year 2000 published in the 1994 Environment and Safety Annual Report explicitly deal with methodologies. AT&T commits to (i) put in place internationally recognized environment and safety management systems for at least 95% of AT&T's products, services, operations, and facilities, and (ii) develop and apply Design for Environment (DFE) criteria that provide competitive, environmentally preferable products and services.[41] Two of the remaining three goals deal with reduction of greenhouse gas emissions and recycling of wastepaper, and the final goal addresses safety issues.

By setting aggressive, process-related (as opposed to end-state) environmental goals, AT&T seems to be recognizing that compliance with government regulation and the setting of waste reduction goals do not go far enough in transforming an organization. In the words of Brad Allenby, research vice president in AT&T's Technology and Environment Group:

> We did the easy stuff by making the proper environmental adjustments within the existing model. Now its time for the hard stuff—heavy duty, fundamental changes in the company—and we're going to have to break that old model and evolve to a new one in the process.[42]

The institutionalization of on-going efforts to minimize environmental impact is a driving force behind AT&T's goals for the year 2000. With this institutionalization will come recognition of AT&T as a leader in environmental management and corporate citizenship. Furthermore, the company sees incorporation of new environmental management processes as a strategic move to distinguish its products and services from those of competitors. AT&T's Environmental Vision is

> . . . to be recognized by customers, employees, shareholders, and communities worldwide as a responsible company which fully integrates life cycle environmental consequences into each of our business decisions and activities. Designing for Environment is a key in distinguishing our processes, products and services.[43]

AT&T's environmental goals hint at a number of possible future trends in corporate environmental goal-setting. First, they combine specific, measurable targets for desired "end-state" conditions with more general process-oriented goals. Second, they move beyond environmental goal-setting as merely an exercise in establishing targets for pollution prevention and waste reduction. Finally, they are "stretch" goals, which require the reconceptualization of environmental issues by a number of employees, beyond those involved in compliance activities. While this approach to environmental goal-setting is far from widespread, its mere existence marks an evolution from the early Pollution Prevention Pays programs initiated by 3M and others.

Goal-Setting by Industry Groups

In this section, we examine cases of coordinated action, which seem to be motivated by several key factors, notably,

- the need to establish industry-wide legitimacy,
- the desire to promote the industry's product as an environmentally attractive alternative,
- the demonstration of leadership to preempt negative public opinion, and
- the aim to restrict regulatory obligations.

Although specific programs are placed into one of these categories in the discussion that follows, many, if not all, serve more than one of the objectives in the list.

Coordinated Action to Establish Legitimacy

Two recent initiatives, by the chemical and petroleum industry associations, seek to guide the environmental activities of their members by articulating principles and practices for environmental management. These programs both arose following a major incident that threatened the industry's right to operate and seriously undermined public confidence. The chemical industry responded to Union Carbide's toxic release at Bhopal, India, by creating the Responsible Care program, and the petroleum industry responded to the Exxon *Valdez* spill in Alaska by creating a similar STEP (Strategies for Today's Environmental Partnership) program. Such voluntary programs establish guiding principles and management practices, create mechanisms for self-reporting and the public disclosure of information, but rely only on peer pressure and association membership to enforce compliance.

The U.S. Chemical Manufacturing Association's Responsible Care program was initiated in 1988 and now consists of ten guiding principles and six management codes to govern the health, safety, and environmental practices of the industry. Participation in the Responsible Care program, which involves committing to implement the management practices and reporting on progress against them, is a condition of membership in the Chemical Manufacturing Association (CMA). Although the program was announced seven years ago, the final two codes (Product Stewardship and Employee Health and Safety) were defined only in 1992. At this stage, opinion is split over whether the Responsible Care program has really changed the way firms operate. Self-evaluation is required on the part of each participating firm and the CMA publishes annual aggregate performance statistics, but critics argue that third-party verification should be adopted as an enforcement mechanism.[44]

The six codes cover a comprehensive set of environmental and safety practices, several of which go far beyond the waste reduction and pollution prevention goals established by many individual firms. In particular, the most rigorous code is the Product Stewardship code, which aims to "make health, safety, and environmental protection an integral part of designing, manufacturing, marketing, distributing, using, recycling and disposing of . . . products."[45] The Product Stewardship code provides guidance on the means by which companies should measure improvement in product stewardship, in addition to goals. A component of product stewardship is the selection of contract manufacturers and suppliers who use "appropriate practices for health, safety, and environmental protection."[46] A separate code deals specifically with the distribution of chemicals, with the goal to "reduce the risk of harm posed by the distribution of chemicals to the general public, to carrier, distributor, contractor and chemical industry employees; and to the environment."[47] The four other codes for management practices include: Pollution Prevention, Employee Health and Safety, Process Safety, and Community Awareness and Emergency Response.

While Dow Chairman and CEO Frank Popoff observes that "it's fair to say this self-generated, bold industry initiative has changed our operations and behavior,"[48] it is not clear that the public has witnessed this change. For an initiative that has as its first principle to "recognize and respond to community concerns about chemicals and our operations,"[49] the fact that less than 10% of the public is even aware of the program is troubling. Despite a $10 million per year print and television advertising campaign conducted by the CMA to communicate the Responsible Care principles, public perception of the chemical industry remains low—it ranks ninth out of ten key industries rated, above only the tobacco industry.[50]

The American Petroleum Industry's STEP program is modeled closely after the Responsible Care program. It has eleven guiding principles and seven management practice codes (pollution prevention, operating and process safety, community awareness, crisis readiness, product stewardship, proactive government interaction, and resource conservation). It differs from the Responsible Care program in that commitment to implementation of the codes is not a condition of membership in the industry association. However, the eleven guiding principles of STEP are written into the by-laws of the API. The STEP program was established in 1990, and the API started publishing annual reports on aggregate industry environmental performance in 1992.

The Deputy Director of Health and Environmental Affairs for the API, Eldon

Rucker, spoke of the motivation for firms' voluntary participation in the STEP program. He felt that public image was the key issue because the industry "hit rock bottom with the Valdez issue."[51] He believes peer pressure has also played a secondary role in getting firms to participate as participation has been growing over the few years the program has been in place. Mr. Rucker felt that public opinion has been less negative towards the industry recently, but attributes this shift to an overall improved level of environmental performance by the industry rather than a result of the STEP codes alone.

The International Organization for Standardization (ISO) will soon introduce a new voluntary code for environmental management, ISO 14000, which can be used by multiple manufacturing and service industries. The key to the ISO system is the words, *management standard*. The standards are *not* a set of environmental performance criteria, targets, or goals. In the words of the latest draft of ISO 14001, Environmental Management Systems—Specifications with Guidance for Use:[52]

> International environmental management standards are intended to provide organizations with the elements of an effective environmental management system which can be integrated with other management requirements, to assist organizations to achieve environmental and economic goals. [It further goes on to define an] environmental management system [as] that part of the overall management system which includes organizational structure, planning activities, responsibilities, practices, procedures, processes and resources for developing, achieving, reviewing and maintaining the environmental policy.

The part of the standard that is relevant to this paper comes as a further elaboration of what planning means. Section 4.2.3 on objectives and targets (see above for their definitions) specifies:

> The organization shall establish and maintain documented environmental objectives and targets, at each relevant function and level within the organization. When establishing and reviewing its objectives, an organization shall consider the relevant legal and other requirements, its significant environmental aspects, its technological options and its financial, operational, and business requirements and the views of the interested parties.

The standard goes on to add a section on checking and corrective action that specifies that the organization shall regularly check to see how it is performing relative to its objectives and targets, and take whatever corrective action is appropriate.

There is nothing in the ISO standard that, in any way, is determinative of the objectives and targets that a firm chooses, but it requires that such goals be set and that the firms' actions produce reasonable progress towards their achievement. The main parts of the ISO standard have been thoroughly vetted by the technical committees and are close to adoption which is expected in the next year. Other parts of the overall standard dealing with guidance for eco-labeling, life-cycle assessments, and environmental performance evaluation are in much early phases of the adoption process. One of the draft guidelines on Environmental Performance Evaluation (EPE) (to become ISO 14031) notes that EPE "is a process and tool which provides management with information linked to the achievement of an organization's targets and objectives in the environmental management system."[53]

It is much too soon to predict how the adoption of ISO 14000 will affect the way firms set objectives and targets, but it will certainly make whatever process they use more visible and subject to scrutiny. The standard does not require the firms to make public their progress to these goals, but many will choose to do so in their efforts to gain more legitimacy and public trust, as discussed earlier. And, since many expect that installing ISO 14000 will rapidly become a prerequisite for doing business in the same way that ISO 9000 has, many firms that have not yet become explicit about their intentions with respect to environment will have to first determine what they are going to do and then act to get there. Most U.S. firms ignored the potential power of ISO 9000 as a ticket of admission; this time around, with ISO 14000, they are determined not to make the same mistake and have taken a very active role in the drafting process.

Coordinated Action to Promote an Environmentally Attractive Product

Several industry associations have taken action to promote the products or processes of their member firms as environmentally attractive alternatives. For example, the Aluminum Association (AA) and the Steel Manufacturer's Association are stressing that the ability to recycle certain metals makes their industry less dependent on non-renewable ore resources, and therefore imposes a lower burden on the environment through reduced extraction activities and energy use.

Promoting recycling of aluminum has long been a focus of the AA. Barry Myers of the AA observes that the emphasis has shifted recently from merely recycling beverage cans, to working with auto manufacturers to develop recyclable car parts. He adds that, in 1994, the share of aluminum used in the auto industry surpassed the share used in packaging for the first time.

The plastics industry also sees scope for promoting recyclable plastics as an alternative to heavier materials typically used in durable products. A spokesperson for the American Plastics Council noted that there has been a shift in focus from four years ago when the greatest concern for the industry was reducing and recycling packaging material. There is now a greater emphasis on recycling durable plastics (which are expected to last three years or more) as opposed to non-durables (e.g., soda bottles). While people held the belief that plastic bottles were filling up landfill sites, durable plastics actually contribute two to three times the volume of non-durables in landfills. The APC spokesperson also noted a trend towards increasing the use of plastics in durable items like automobiles and appliances.

Coordinated Action to Signal Environmental Leadership

Several industry groups have taken action to preempt regulation or negative public reaction by initiating research efforts aimed to provide solutions to emerging environmental concerns. The aim of such programs is often to promote leadership amongst the industry on certain issues, rather than set formal targets and principles industry members must operate under. The semiconductor industry and forest products industry have been active in pursuing common research agendas and promoting technological solutions.

The American Forest and Paper Association (AFPA) published a report in November 1994 on its technology vision and research agenda for America's forest, wood, and paper industry. It states, "Recognizing the inability of humans to accurately predict the future, the focus is on direction and broad, general goals rather than specific endpoints and solutions."[54] The only numerical goal mentioned is an intention to increase the rate of paper recovery from recycling from its current 40% to 50% by the year 2000. However, the document outlines six research priorities for the industry as a whole, four of which deal directly with environmental issues. These include research into sustainable forest management, environmental performance, improved capital effectiveness, recycling, and sensors and control. The AFPA sees these actions as necessary to ensure the long-term success of the industry, which, along with other pressures on performance and competitiveness, faces "more demanding environmental requirements" as a "major burden . . . over the next decade and beyond."[55] While it believes that specific product research and development should be left to individual companies, the AFPA sees a need for the industry as a whole to partner with government, suppliers, national laboratories, and universities to leverage all available resources for its long-range research agenda.

Despite the many positive goal-setting activities described above, perhaps the greatest effort expended by industry groups is in lobbying to restrict or control environmental regulation. However, lobbying to manipulate regulatory boundaries seems to be increasingly aimed at finding flexible, voluntary solutions that may be more effective environmentally and more efficient economically than "one-size-fits-all" regulation.

The semiconductor and electronics industries also have taken proactive stances on developing research agendas for environmental and safety issues. The Industry Cooperative for Ozone Layer Protection (ICOLP) is an initiative of leading firms within the semiconductor industry working in cooperation with the U.S. EPA. ICOLP was founded by Northern Telecom (NT) and AT&T in response to the Montreal Protocol ban on CFCs and other ozone-depleting substances. A certain solvent, CFC-113, used in the electronic industry, lacked any suitable alternative at the time the ban was announced. Both AT&T and NT developed their own technological solutions and substitutes, and eliminated the solvent from their operations in the early 1990s, well ahead of schedule. ICOLP was seen as a mechanism to transfer this technology to manufacturers in developing countries where compliance with the ban was expected on a more relaxed schedule. By providing technical alternatives, the North American manufacturers hoped to dispel any competitive disadvantage they might otherwise have encountered as a result of making their own process cleaner.

The list of such initiatives is rapidly growing as industries recognize the importance of establishing some sort of collective image of leadership and commitment. The "Encouraging Environmental Excellence" initiative of the American Textile Manufacturer's Institute (ATMI) calls for companies to adopt a 10-point plan to improve the environment, including the establishment of a corporate policy and goals.

Coordinated Action to Seek Regulatory Relief

Perhaps the greatest consistency is seen between industry groups in their activities to restrict or control environmental regulation. While not a positive goal-setting exercise in itself, this form of action serves to define the regulatory boundaries within which firms have to operate. Lobbying to manipulate those boundaries has traditionally been the extent of industry associations' involvement in environmental issues and remains the key focus of several groups. However, these efforts seem increasingly to be aimed at finding flexible, voluntary solutions that may be more effective environmentally and more efficient economically than "one-size-fits-all" regulation.

The Society of the Plastics Industry (SPI), in its 1995 Issue Activity Report identifies a number of environmental issues it is addressing through lobbying. It strongly supports the risk assessment bill passed by the House of Representatives in February of 1995 and is a founding member of the Alliance for Reasonable Regulation, an industry group that led the lobbying effort. SPI has worked to obtain more flexible agreements with the EPA in several areas of environmental regulation. For example, SPI negotiated an effort to allow the industry to pursue a product stewardship program in lieu of certain toxicology testing for an epoxy resin compound. This is the first time the EPA has negotiated such an effort, and a second negotiation is forthcoming for a different epoxy chemical. The SPI estimates that the plastics industry will save more than $10 million with this approach.[56]

The SPI is also working with the EPA on the Sustainable Industry Project (SIP) which encourages hazardous waste minimization and pollution prevention. In particular, SPI is lobbying to allow polymerization to be considered as an acceptable control technology to handle hazardous wastes. Elsewhere in its Issues Activity Report, the SPI notes that it is actively representing the plastics industry on several other environmental issues including: hazardous waste, global warming/climate change, Clean Air Act implementation, clean water/Great Lakes initiative, and chlorine.

In discussing environmental policies in its 1995 Public Policy Agenda, the Steel Manufacturers Association reports on its lobbying efforts and expresses the industry's stance on various particular EPA regulations.[57] While it advocates regulatory relief so that voluntary environmental audit reports will not make a firm liable to litigation and argues for the harmonization of federal reporting requirements, the industry does not appear to be taking a proactive stance to define its own environmental goals and reporting requirements.

Goal-Setting by Government-Industry Coalitions

In this section, we discuss several programs in which government regulators are cooperating with individual firms or industries to jointly define goals. Earlier, we addressed the emergence of voluntary programs at the federal, state, and local level, and briefly described the EPA's 33/50 program, which consequently is not addressed in this section.

An agreement between the Big 3 (Chrysler, GM, and Ford) auto makers and Michigan's Department of Natural Resources is an example of a voluntary initiative that involves collaboration with a state regulatory body. Don Edmunds, Pollution Reduction Manager with the American Automobile Manufacturers Association (AAMA), cites this as one example of what he sees as a growing trend towards voluntary programs involving government bodies at the federal, state and local levels.[58] The Automotive Pollution Prevention Project (Auto Project) was established in 1991, and the parties committed to reduce the generation and release of persistent toxic substances in the Great Lakes basin. While it sets no numerical reduction goals or time limits and allows each auto company to establish its "own priorities and mechanisms for reducing the generation and release of GLPT (Great Lakes Persistent Toxic) substances," releases of GLPTs were reduced by 28.9% (adjusted for production volume), or by 20.2 % (unadjusted) in the first year of the program.[59]

The Progress Report issued by the Auto Project in 1994 sees this voluntary initiative as an aid to auto companies in establishing their individual environmental improvement priorities. It notes:

> The Auto Project represents a new way of doing business. It provides an example of how a flexible, voluntary, and cooperative government/industry environmental initiative can achieve and reconcile our mutual environmental and economic needs in a globally competitive marketplace.[60]

Similarly, the Aluminum Association has mediated negotiations to establish a voluntary agreement between individual aluminum companies and the federal EPA. The program is known as the Voluntary Aluminum Industry Partnership (VAIP) and was signed in January of 1995. Each firm commits to reducing levels of perfluorocarbons (PFCs), identified by the EPA as greenhouse gases, released during the aluminum smelting process. While the agreement acknowledges the Climate Change Action Plan of October 1993, which targets a reduction in PFC emissions by the aluminum industry of 30-60% by the year 2000, the VAIP itself does not set goals for reduction levels or time frames. Instead, it requires that the EPA and each firm work together to seek better methods for measurement of the gases, a better understanding of the relationship between gas generation and process design and control, and technically feasible and cost-effective solutions that are suited to each facility's unique characteristics. This type of agreement accommodates those firms who had made significant strides in reducing PFCs prior to adopting the code and allows solutions to be tailored to individual plants. While this may reduce the burden of compliance and thus improve chances for successful reduction in PFC levels, it is unclear to what extent this type of voluntary agreement reflects a desire on the part of industry to take responsibility for environmental consequences. Bob Strieter of the Aluminum Association noted that aluminum is seen as an environmentally friendly metal because of its ability to be easily recycled, so the industry wanted to preempt any negative public reaction that may result from inaction on reducing PFC emissions.[61]

Like the aluminum industry, the Semiconductor Industry Association (SIA) is pursuing voluntary reduction of PFCs. A Memorandum of Understanding (MOU) has been signed with the EPA that defines what the industry believes it can accomplish in terms of PFC reduction and commits it (through Sematech, a research consortium) to identify replacement chemicals. The MOU is only several months old, and formal mechanisms to address the issue do not yet seem to be in place. Lee Neal of the SIA adds that, while voluntary, the industry takes this initiative seriously because public credibility is important.[62]

GOAL-SETTING IN THE STAGE OF MANAGING FOR THE ENVIRONMENT

The last section contains initiatives that suggest the beginning of a transition towards sustainable development, a notion that has been gathering head since the publication of the so-called Brundtland report in 1987. The report, called *Our Common Future,* centered on a simple-sounding, but extraordinarily difficult-to-operationalize, notion of sustainable progress or development as that "which meets the needs of the present without compromising the ability of future generations to meet their own needs."[63] The process of change needed to move towards this new, broad goal has been subtle, although the UN (Rio) Conference on Environment and Development in 1992 signaled a sharper delineation and the start of many new initiatives with sustainability or sustainable development as the target.[64] Industry has begun to respond to a new set of demands created by increasing cognizance of the ecosystem nature of the "environmental" problematique. One of the earliest responses was the formation of the Business Council for Sustainable Development, a group of about 50 of the world's largest corporations, led by a Swiss industrialist, Stephan Schmidheiny. Under his guidance and authorship, the BCSD published a book designed to change the course of industrial practice towards sustainable development.[65] The book sets forth a vision of a different kind of company and of new practices, growing out of the kinds of proactive examples described in the prior section of this paper. The author points to the practices of the leading firms as creating a new target for industrial environmental performance—eco-efficiency. Eco-efficiency

> is [not to be] achieved by technological change alone. It is achieved only by profound changes in the goals and assumptions that drive corporate activities, and change in the daily practices and tools used to reach them. This means a break with the business-as-usual mentality and conventional wisdom that sidelines environmental and human concerns.[66]

Other sets of sustainability principles and goals can be found in the Business Charter for Sustainable Development. This set of 16 principles for environmental management was developed by the International Chamber of Commerce and has been adopted by many of their members. In some ways, it is a meta-goals setting initiative, exhorting firms

> to recognize environmental management as among the highest corporate priorities and as a key determinant to sustainable development; to establish policies, programmes and practices for conducting operations in an environmentally sound manner[67]

while avoiding any explicit vision of sustainability. The preamble to the ICC (BCSD) principles does, however, refer to the UNCED (Brundtland) notion of sustainable development. It is interesting to note that the ICC group who was responsible for these principles and the Schmidheiny group have joined forces to form the World Business Council for Sustainable Development. The ICC set of principles has been used by GEMI (see above) as the basis of its environmental self-assessment program (ESAP). This program can be used by firms to assess progress in the implementation of management systems and policies and in the achievement of explicit goals and targets.

CERES (Coalition for Environmentally Responsible Economies) has issued a set of principles that includes broad goals aimed at sustainability. The CERES strategy originally had companies sign its set of principles. This strategy has been modified to one of cross-endorsement wherein the firm and CERES endorse each other's principles as mutually reinforcing and as a commitment to action. Six of the 10 CERES principles are goal statements addressed to a specific broad environmental concern. For example, Principle 2 reads,

> We will make sustainable use of renewable natural resources such as water, soils, and forests. We will conserve nonrenewable resources through efficient use and careful planning.

One of the primary drivers for endorsing the CERES principles has been concerns over the image (legitimacy and trust) of the company.[68] The strong statements and commitments embodied in the CERES (formerly called the *Valdez*) principles have been a deterrent for many companies. Only a handful of large companies have endorsed these principles to date (GM, Sun Oil, H.B. Fuller). By far, the majority of signatories or endorsees is made up of small firms with a broad involvement in the arena of socially responsible enterprises.

The BCSD and other notions of sustainability are very broad statements of a vision of the future, but are difficult to translate into operational concepts and concrete goals. This section concludes with two approaches that offer some concrete guidance on goal-setting procedures for this emergent stage of evolution.

Industrial Ecology and Design for the Environment[69]

Design for environment and other sustainable development schemes guide decisions based on some set of normative objectives related to prevention of toxic impacts, resource conservation, avoidance of dysfunction in the natural system, and so on. There is no consensus on what this set of norms should be, nor is it likely that such a set will appear soon. This absence suggests that one of the key parts of any DFE procedure must be the selection of the environmental norms to guide the design. Our group at MIT and others have been working with one such set called "industrial ecology." Industrial ecology is a "holistic framework for guiding the transformation of the industrial system."[70] It springs from an interest in building models of societal structures and behavior that more fully integrate environmental and economic systems in such a holistic manner.[71] The term industrial ecology was first used in about 1971 in Japan by a research group developing industrial policy for the Ministry of International Trade and Industry (MITI).[72] Allenby, a U.S. pioneer in the field of industrial ecology, writes:[73]

> Somewhat teleologically, "industrial ecology" may be defined as the means by which a state of sustainable development is approached and maintained. It consists of a systems view of human economic activity and its interrelationship with fundamental biological, chemical, and physical systems with the goal of establishing and maintaining the human species at levels that can be sustained indefinitely—given continued economic, cultural, and technological evolution.

The first textbook on this subject moves towards a more practical context and defines industrial ecology as:[74]

> [T]he means by which humanity can deliberately and rationally approach and maintain a desirable carrying capacity, given continued economic, cultural, and technological evolution. The concept requires that an industrial system be viewed not in isolation from its surrounding systems, but in concert with them. It is a systems view in which one seeks to optimize the total materials cycle from virgin material, to finished material, to product, to waste product, and to ultimate disposal. Factors to be optimized include resources, energy, and capital.

A group of academics meeting in Colorado during the 1992 summer collectively characterized industrial ecology as "the study of how we humans can continue rearranging Earth, but in such a way as to protect our own health, the health of our natural ecosystems, and the health of future generations of plants and animals and humans."[75]

The Netherlands' planning approach to environmental policy development used a related set of concepts in the first National Environmental Policy Plan.[76] Among other objectives, the plan sought to

1. close substance (materials) cycles,
2. conserve energy and utilize sustainable sources of energy, and
3. promote the quality of products with a goal of extending the product life and maintaining the stock of materials in the economy.

The fundamental normative premise behind industrial ecology is that industrial systems should perform like natural ecosystems, which frequently (but not always) exhibit dynamic stability over a wide range of perturbing forces. One key characteristic of natural ecosystems is that there is no such thing as "waste." Natural ecosystems do not produce materials, in any significant amounts, that are not used by some organism within the system. Materials flow through the system in closed loop patterns. In this way, natural ecosystems behave in a "sustainable" fashion.

Fundamental to industrial ecology is that environmental impacts across the entire life cycle of products, from cradle to grave, be considered in decision-making. Industrial ecology provides principles by which to guide actions that affect the entire product life cycle. For example, the principles can be organized as follows:[77]

- Loop closing—circulating material flows within the system
- Dematerialization—reducing the material intensity in products that produce equivalent functions to those they replace
- Protecting the natural metabolism—restricting the flow of substances that are harmful to natural systems and living organisms
- Systematizing energy use—conserving energy and extracting as much of the available energy content from that which is used or, conversely, minimizing the flow of waste heat back to the environment.

Many of these principles are embodied in current environmental practices. They show up as more familiar terms:

- Pollution control—the treatment of pollutants in order to reduce their environmental impact upon emission.
- Pollution prevention—the reduction of pollutants across the entire life cycle. This may include the transfer of "pollutants" as inputs to other product life cycles.

- Toxic use reduction—the reduction of toxic materials throughout all flows, especially those which enter the environment or may impact humans.
- Waste reduction or minimization—the reduction of materials used in products so to minimize "waste" during disposal.
- Energy conservation—the reduction of energy required during all phases of the life cycle.
- Product use extension—the reduction of the temporal flow of materials from extraction to disposal by extending the user life of products.
- Recycling, remanufacturing, and reuse—at the end of consumer use, the recycling, remanufacturing, and reuse of materials and components in an attempt to close the product life-cycle loop.
- Environmentally conscious manufacturing—the reduction of environmental impacts during the manufacturing/production phase of products. This embodies many of the notions above such as pollution prevention and energy conservation as applied to the manufacturing phase.
- Industrial symbiosis—the "interaction among companies so that the residual of one becomes the feedstock of another."[78] Converting pollutant, disposal, and energy flows into inputs to other product life cycles.

The following quote is taken from a recent Monsanto report and indicates considerable interest in industrial ecology, even in its early stages, as an important factor for industry to consider in setting out its long-range goals.

> A new buzzword has been insinuating itself into the language of business and the environment—"industrial ecology." Some might call it an oxymoron—a contradiction in terms. Others see it as a useful tool or framework to encourage industry in its manufacturing processes to follow the patterns of natural ecosystems.
>
> That a debate about the concept is even occurring is evidence that the ground rules for business and the environment, which changed profoundly in the late 1980s, are changing once again.

That this statement was made by Richard Mahoney about industrial ecology is of special interest.[79] As noted above, Mahoney was the CEO at the time Monsanto had to submit its first TRI report as required by Title III of the Superfund Amendments and Recovery Act (SARA) of 1986. He is responsible for creating the Monsanto Pledge. Mahoney has continued as an active corporate leader in the environmental arena. When he touts industrial ecology as a new driver for corporate change, it is a signal to be taken seriously.

Design for environment (DFE) is emerging as a systematic approach to addressing the entire system of environmental impacts across the whole product life cycle. To this point, the term "design for environment" has been used in reference to a variety of practices and concepts. It has been defined both narrowly and broadly. In order to provide a foundation for discussion it is important that to establish a more concise definition of DFE. As stated by the U.S. Environmental

Protection Agency, "if used in a broad and simplistic way, DFE can be used to justify almost any efforts, even those that are not in the best interests of environmental preservation."[80] Conversely, if used too narrowly, DFE can fail to fully capture environmental improvement opportunities. The following definition attempts to bridge these extremes:

> *Design for environment is the systematic process by which firms design products and processes in an environmentally conscious way based on industrial ecology principles across the entire product life cycle.*

There are three critical points in this definition of design for environment:

- environmental impacts across the entire product life cycle are considered;
- impacts are addressed during the product development cycle; and
- decision-making is guided by a set of principles, based on industrial ecology or some set of system-configured, integrative principles.

Further, environmentally conscious means that the design process and all who participate in it consider potential environmental impacts explicitly alongside and in the same manner as all other factors that influence the design process. Such consciousness should reveal environmental impacts and problems to be addressed over the entire life cycle of the product, but it does not mean or refer to any specific set of design criteria that must be met.

Design for the environment considers the entire product life cycle. While other practices address particular flows or phases of the product life cycle, design for environment in its most robust form considers all environmental impacts. Figure 2 provides a schema summarizing the areas of potential environmental impact across the entire product life cycle. Each of the arrows in the figure represent a material flow, be it a gas, liquid, or solid. Not depicted in the figure is the energy use during each phase and flow. The production of by-products and the use of secondary inputs within each phase are also included with the "product" streams. These materials may come directly from the environment or go through various production sequences and represent other product life cycles, demonstrating the interconnectedness of economic activity. Pollution represents any output that is not used as an input to another product life cycle. The pollutant materials may be treated or released directly into the environment.

The product life cycle is a representation of the flows of materials and energy that accompany a product from the primary production of materials used in its construction to its ultimate end-of-life disposal, including its potential reincarnation as recovered parts or materials. Life-cycle analysis is a systematic framework to identify and account for (inventory) all of these material and energy flows and their embodied environmental impact. In current practice, the relation of this set of inventory and impact assessments is only tenuously related to the design process. One of the critical elements in any DFE or ECM system is how well the LCA base is tied to and informs the designers.

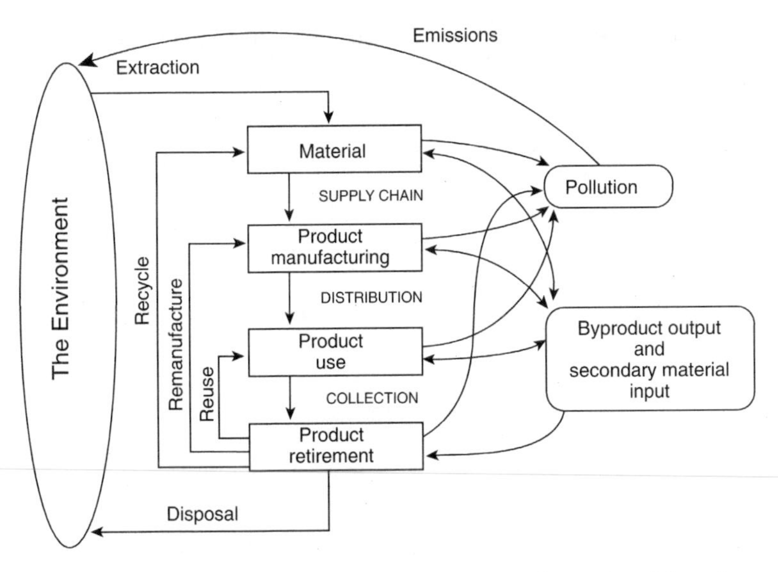

FIGURE 2 Potential environmental impacts across the entire product life cycle.

In subtle ways, LCA incorporates goals that may be invisible to the users. One standard system for performing life-cycle assessments comes from the work of an international professional society, The Society of Environmental Toxicology and Chemistry (SETAC). The SETAC LCA methodology involves three steps—inventory analysis, impact assessment, and improvement assessment.[81] The first step is basically a careful accounting of all physical, tangible inputs and outputs to the product life cycle that are environmentally consequential. The second step requires the user of the LCA or the designer of a formal assessment tool, such as a software package, to choose a specific set of impacts to be assessed. This set of environmental impacts or "burdens" forms the framework by which the inventory is mapped into quantities that represent human health risk, ecological productivity, global warming, and so forth. Since these impacts are then linked to the third and final step, improvement assessment, wherein the actors deliberately introduce features into a design or policy that should positively alter the impact, they implicitly set forth goals for the designer/policy analyst/strategist. These goals can be and are often very general in that they do not tell the actors how much improvement they should consider. Workers in the LCA field have used a variety of impacts in the past, as shown in Table 1.[82]

The large number and categorically dispersed nature of these impact objectives suggests that translating broad goals such as sustainable development is a very complex and problematic exercise: Which ones should the actors use and how should they aggregate them or trade them off in practice? Given the complexity of the phenomena and the need for large amounts of information, it ap-

TABLE 1 Generally Recognized Environmental Problems

Depletion	Pollution	Disturbances
Abiotic resources	Ozone depletion	Desertification
Biotic resources	Global warming	Ecosystem degradation
	Smog	Landscape degradation
	Acidification	Direct human impacts
	Human toxicity	
	Ecotoxicity	
	Eutrophication	
	Thermal impacts	
	Noise	
	Odor	
	Worker health and safety	

pears likely that any near-term practical system for identifying and acting to embed environmental goals into design and policy will fix on the immediate, tangible physical inventory of controllable inputs and outputs without much sense of how these actions will actually impact the world.

National Goals for Sustainability: The Dutch Covenants

The Dutch have taken a, some would say, radically different tack towards environmental goal-setting for industry, away from the firm-by-firm model of LCA, ISO, and some of the other approaches discussed above. After the Brundtland report was issued, the Dutch government began to look seriously at the implications of sustainable development on its own economy.

Following the publication of a national assessment of the state of the environment,[83] the government issued a sweeping National Environmental Policy Plan (NEPP).[84] The Plan recognized that the kind of future that was desired could not be achieved through the normal policy route (regulation) nor through business-as-usual practices (end-of-pipe). Deep structural transformation of the economy was critical if the necessary reductions in loadings of 70-90% were to be achieved in the 20-25 year time frame for the Plan. It was an explicit target that such change was to occur in a single generation. The plan sets forth concrete measures for the initial period (1990-1994) and broader goals for the next several decades. Following a change of government, which was to a considerable degree influenced by debate over the Plan and its significance, the initial Plan, which focused on specific areas of environmental concern, was replaced by a new one (NEPP Plus).[85] This plan adopted sustainability as one of the basic pillars of government policy. Closing cycles (integrated chain management), reducing energy consumption by 50% by 2000, and improving product quality to double the time materials stay in use were the primary routes by which the Plan's targets were to be realized.

Ambitious overall reduction targets have been set, for example,

- 90% reduction of SO_2 and NO_X (relative to 1985) by 2010,
- 3-5% reduction (relative to 1989) of CO_2 by 2000,
- 60% reduction (relative to 1985) of VOCs by 2000, and
- 80-90% reduction (relative to 1985) of heavy metal emissions by 2010.

This broad strategic framework has been segmented into target groups representing selected economic spheres of activity. Eight were selected initially:

- Agriculture
- Traffic and transportation
- Industry
- Energy
- Oil refining
- Construction
- Public waste utilities
- Consumers and retail trade

These categories may be broken down further into specific industrial sectors; for example, the chemical industry and the primary metals industry, have moved to develop targets for action. The process of setting targets—that is, explicit objectives for reductions of environmental impacts—involves a long series of consultations among the key stakeholders—industry, government, and NGOs. These public dialogues were begun shortly after the Plan was announced. In the 1992-1993 period, sectoral targets were agreed to by associations representing the chemical, printing, and primary metals industries. Their agreements have been converted to "covenants" or promises to abide by the agreement, signed by both governmental and industry authorities. Most of the 150 or so Dutch chemical companies have signed their sectoral covenant. The covenants are voluntary, but the potential for regulation always looms in the background.

Each company must then prepare a plan showing how it will contribute to the overall sectoral goal and submit its plan to the reviewing authorities who will determine whether or not the aggregate actions will achieve the objective for the entire target group.

The covenant approach, which rests heavily on a broad and extensive open planning process, is designed to provide certainty to the target sectors over a long time frame so that each company can make wise choices without expecting a change in the targets or means by which they are to be achieved. Each business is substantially free to choose its own target and approach, so long as the overall target is met. There is ample room for negotiation all along the way. The establishment of sectoral goals should minimize the risk of strategic competitive actions by individual firms.

The Dutch approach addresses one of the more problematic aspects of setting environmental goals, particularly for the "global" issues, that is, how to translate aggregated targets into goals for individual actors like firms or consumers. How does each firm, for example, decide how much it needs to reduce carbon dioxide emissions as its contribution towards some agreed upon international or regional target? It is too early to assess the actual results from the process.

CONCLUSION

The concluding section discusses, very briefly, the issue of measuring performance against corporate goals and finishes with a short analysis of a set of questions that must be addressed in looking at goal-setting in the current scientific, economic, and political context.

Measuring Performance Against Goals

Several mechanisms for monitoring a company's performance against its environmental goals have been mentioned in previous sections. Individual firms, like Monsanto and AT&T, publish annual environmental reports that make public their achievements against certain goals. These reports contain factual information on emissions, releases, and transfers, sometimes listed by site, chemical, or medium. Some of the data they contain is identical to that required by U.S. government reporting requirements, namely the toxic release inventory (TRI) required by the EPA. Industry associations also publish environmental performance reports of a similar nature, but they generally aggregate the data at the level of the industry, rather than report on performance by individual firms. The American Petroleum Institute started publishing annual reports on aggregate industry environmental performance around 1992. The most recent of these, published in May of 1995, provides data on chemical releases, refinery residuals, oil spills in U.S. waters, underground storage tanks, used oil, gasoline vapor controls, workplace safety, and environmental expenditures. Chemical release data for the industry were obtained from TRIs, and other data were collected through surveys of voluntarily participating API members.

As noted earlier, key publicly available metrics for tracking the environmental performance of firms are TRI data submitted by industrial companies as required by SARA. While they track the releases and off-site transfer of more than 300 named chemicals, it gives a narrow view of overall corporate environmental performance. First, they are limited to toxic chemicals and say nothing about material recycling efforts, product and process redesign, or a number of other "broader" environmental issues. Second, until recently, TRI data have not been normalized for production volume and thus cannot accurately track waste reduction per unit. Despite this, there are several interesting trends identified by INFORM, a non-profit environmental research organization, in its study of TRI and BRS (Biennial Reporting System) data submitted to the EPA.[86]

INFORM found that there was no net reduction in aggregate TRI waste between 1991 and 1992 (the last year for which a comprehensive data set is available). While several facilities, companies, and industries achieved significant reductions, the gains were more than offset by increased releases from other industries. Only 5 of 22 industry groups showed a net production-related decrease in TRI waste between 1991 and 1992. Projected data for 1992 through 1994 show that 16 of 22 industry groups reported lower TRI levels, but increases in the Primary Metals, Chemicals, and Stone/Clay/Glass groups offset these. INFORM expects that aggregate TRI levels between 1992 and 1994 will have remained fairly constant.

The chemical industry dominated all production-related waste reported in 1991 and 1992 TRIs. It was responsible for 53% of all TRI waste in 1992 and 69% of all carcinogens in TRI waste. Five industry groups were responsible for 85% of TRI waste: Chemicals, Primary Metals, Petroleum, Paper, and "Multiple Codes in 20-39 (SIC code) Range." TRI waste is also highly concentrated at a relatively small number of facilities within each industry. The five facilities reporting the largest production-related waste within each industry group contributed 36% of all TRI waste. Two firms were responsible for one-third of the chemical industry's TRI waste and 18% of total production-related toxic waste.

What does this mean for the setting and achieving of environmental goals? First, it emphasizes why waste reduction and pollution prevention goals have been so important to certain firms and industries, but does not support the conclusion that all firms should focus on waste reduction as their first environmental priority. The dominance of a few firms and industries in the creation of toxic waste and the inability of efforts of the past few years to make significant reductions in aggregate levels point to a need to reexamine goals and actions against them in certain industries. The TRI data presented above, because they aggregate release of some 300+ chemicals, do not indicate where progress has been made in eliminating the most toxic substances. Progress reported by firms participating in the EPA's voluntary 33/50 program suggests that many industries have met the goal of 50% reduction in the release of 17 toxic chemicals.

For example, the American Forest and Paper Association reports that the pulp and paper industry's participation in the EPA's 33/50 program resulted in its achieving a 50% reduction in the use of the listed chemicals by 1993, two years ahead of the EPA schedule.[87] INFORM's report indicates that this level of achieved reduction may not be isolated to this industry. It states that "the progress reported by the 33/50 participating companies towards the Program's goals has been substantial, but suggests that these goals were not as ambitious as was thought at the outset."[88] According to TRI reports, the 33% reduction had been met by 1991 (one year early) in an aggregate measure of all industries, and by 1992 total releases exceeded the 1995 goal by only about 20%.[89]

In the future, the ISO Standards are likely to add a significant new onus to the overall environmental management functions of an individual firm both in the setting of its own objectives and targets and in measuring its progress towards meeting them. Many hope that all of the several monitoring and reporting requirements set by internal and outside institutions can be integrated. The current status in ISO that would not require public disclosure of the results of progress assessments tends to make such a move problematic. On the other hand, some of the regional standards that will live side by side with ISO (the EU EMAS system) do require such public disclosure.

Outstanding Questions (and a Few Guesses as to the Answers)

This survey of corporate environmental goal-setting raises a number of questions that must be addressed before the effectiveness of goal-setting by industry can be speculated upon. First, the firms that have taken proactive steps to define environmental goals are typically large industry leaders. To what degree have the firms' size, resources, and public visibility impacted their desire and ability to set environmental goals? If these factors are important in motivating environmental goal-setting, what can be expected from firms or industries who have fewer resources or are less visible in the public eye? Second, the benefits of products of some industries are more clearly valued by the general public, and this differential perception necessarily impacts public response to the environmental costs of the industry. For example, the petroleum industry has "bounced back" from severe environmental accidents (spills, pipeline leaks), and the automobile industry resists most efforts to limit air pollution due to the perceived balance between public benefits and costs. An industry like the chemical industry manufactures products that are largely not directly visible in consumer goods, and consequently the associated benefits may not be appreciated by the public.

If we speculate that environmental goal-setting by industry actors represents a new trend in environmental management, we must consider its prominence relative to other corporate goal-setting activities. It seems that even the most proactive company in environmental goal-setting still implicitly places environment on a lower level than other strategic goals. AT&T, despite its ambitious environmental goals and commitment to adoption of green processes, its publication of an annual environmental and safety report, and the presence at its most senior levels of management of professionals with responsibility for environmental issues, mentions virtually nothing about these commitments in its 1994 annual report to shareholders. The focus of CEO Robert Allen's address in this document is on traditional strategic business objectives—financial performance, alliances and partnerships with related businesses, R&D leadership in the development of new products and services, and market share.[90] Allen sees the major challenges facing AT&T as the selection of product offerings and strategic alliances to cope with new competitive pressures in the information market. He does not mention environmental goals or performance in his letter, although he does refer to other initiatives (the corporate diversity program and a commitment to quality and continuous improvement).

Environmental performance comes up on the last page (page 20) of the text section of AT&T's 1994 annual report and shares the page entitled "We Keep Our Word" with other goals that "better position (AT&T) to serve customers and the communities in which we live and work." Of the five achievements listed, two report on environmental performance (air emissions reduction and waste paper recycling) and the other three report on the hiring of women and minorities, and the support of non-profit and charitable groups. When a leading company chooses to relegate its substantial commitments to environmental performance to such a position in a report to shareholders, it begs the question whether environmental responsibility really will "become as integral a strategic element of the business as product performance and cost effectiveness."[91]

Besides these questions that bear on "social responsibility" and concomitant commitments of companies to meet environmental norms, setting goals has many complex technical aspects. First, it is difficult to set norms. The ecological connections and consequences of economic activities are poorly understood at present, particularly for phenomena of large scale. But even at the molecular level, scientists agree on the effects in humans and other species for only a handful of the 80,000 or so chemicals in commerce. Setting goals for the use of chlorine-containing or chlorine-related chemicals, for example, has precipitated a global controversy, with some parties calling for a wholesale ban on chlorine use while others argue that insufficient data now exist with which to make rational choices about setting targets and limits.

One way out of the analysis/paralysis dilemma is to set broad goals that do not rely on large quantities of information and a traditionally "rational process." The notion of precaution, for example, is embodied in the ICC principles and permeates the calls for bans and other targets that do not need data to set intermediate numerical levels. Another approach is to drive innovation and change by a set of "process" principles such as those described above in the section on industrial ecology. Pushing change that embodies the principles of dematerialization or loop closing, for example, should, in most cases, move the continuous innovation-investment-implementation-consequences-assessment cycle towards sustainability or equivalent societal environmental goals. Continuous improvement is the cornerstone of TQM and is similarly applicable to meeting environmental goals, with some inherent difficulties. The environmental equivalent to product quality is, as already noted, much more loosely connected to the technical characteristics of products and processes, reflecting the imperfect state of knowledge today. Environmental quality targets *must* be set through some sort of "scientific" assessment as designers and analysts cannot ask the environment the key question about quality, Are you satisfied with what we are doing?

And even if there appears to be sufficient evidence, the myriad of parties who claim an interest in setting societal goals regarding the environment rarely agree on specifics. Setting environmental goals inevitably creates some kind of spillover to other arenas, requiring changes in patterns of production and consumption. This same problem is a major hindrance in private actions as well. Setting a company's environmental goals always impacts other norms held by the firm. Environment is often the new kid on the block, and any commitments in this arena that would limit a firm's activities in the more traditional and familiar domains tend to be viewed with suspicion, thereby setting up barriers. These barriers should continue to fall, as they have been doing for quite some time, while environment becomes more and more familiar and gains acceptance alongside the traditional elements, such as the bottom line, in the basic set of corporate values. The combination of great uncertainty as to the connection of human actions and their environmental consequences and the diversity of values as to what is the right environmental goal suggests, if not demands, that whatever goal-setting process becomes established, that process should be continuous.

Last, the problem of parsing out targets to individual firms so that the aggregate results achieve a broad societal goal is technically and politically very complicated. The rules for doing this need to be established. Should the shares be meted out proportionate to some measure of output? Or to the resources that a firm can bring to the problem? Or should there be some consideration of past use of environmental capital? The ongoing arguments about how to allocate whatever reductions in carbon dioxide are to be made to avoid global warming reflect these three options among others. The Dutch target group approach is extremely interesting in this regard as it is a conscious and deliberate attempt at both goal-setting and implementation. The open process involving all interested parties accepts the inherent differential values of these parties and the inability to set goals through purely analytic means. But the Dutch society is relatively small and less diverse than ours in the United States. Open planning, with strong governmental leadership, has been long used in many social domains besides environment. Nevertheless, the fact that a society can come together and set remarkably tough goals—goals that would fundamentally change the way the economy and the system works—should serve as a target in and of itself for the United States. These efforts are reminiscent of the observation made on the first moon landing—a small step for man, a giant leap for mankind. The path to ecological and social sustainability is certainly as daunting a task as reaching the moon ever was.

NOTES

1. Christopher B. Hunt and Ellen R. Auster, "Proactive Environmental Management: Avoiding the Toxic Trap," *Sloan Management Review*, Vol. 31, No. 2, 1990, pp. 7-18.

2. Paul Shrivastava, "Strategic Responses to Environmentalism," *Business Strategy and the Environment*, Autumn, 1992, pp. 9-21.

3. John R. Ehrenfeld and Andrew Hoffman, "The Importance of Culture in the Greening Process," Designing the Sustainable Enterprise Conference, Boston, MA, November, 1993.

4. Edward Prewitt, "Allied-Signal: Managing the Hazardous Waste Liability Risk," Case Study #N9-793-044, Harvard Business School, Cambridge, MA, 1992.

5. U.S. Congress, Office of Technology Assessment, *Serious Reduction of Hazardous Waste*, Washington DC, Government Printing Office, 1986.

6. Catherine Morrison, "Managing Environmental Affairs: Corporate Practices in the U.S., Canada and Europe," The Conference Board, New York, NY, 1991.

7. Hilary de Boerr, "Green Jobs at the Top," *The Financial Times*, June 17, 1992, p. 12.

8. David Sarokin et al., *Cutting Chemical Wastes*, Inform, New York, NY, 1985; Stephan Schmidheiny, *Changing Course*, MIT Press, Cambridge, MA, 1992.

9. Mary Melody, "Boatmaker Finds Solvent Substitute, Cuts Emissions, Costs," *Hazmat World*, February 1992, pp. 36-39.

10. George W. McKinney, "Environmental Technology for Competitiveness: A Call for a Cooperative Pollution Prevention Initiative," presented at the National Technology Initiative Conference, Massachusetts Institute of Technology, Cambridge, MA, February 12, 1992.

11. Hewlett Packard 1994 Annual Report, CEO's letter written by Lew Platt.

12. Unofficial ISO Draft International Standard 14001 (ISO/DIS 14001), June 26, 1995.

13. See Chapter 1, "A Taxonomy of Illocutionary Acts," in John Searle, *Expression and Meaning*, Cambridge University Press, 1979.

14. Ibid.

15. Yiorgos Mylonadis, "The Green Challenge to the Industrial Enterprise Mindset: Survival Threat or Strategic Opportunity?" Doctoral Thesis, MIT, 1993.

16. Paul Portney, ed., *Public Policies for Environmental Protection*, Resources for the Future, Washington, DC, 1990, p. 20.

17. Ibid., p. 20-21.

18. CAA §202(b); 42 U.S.C. § 1857f-1(b), 1970.

19. Ibid., p. 22.

20. Bruce Smart, *Beyond Compliance: A New Industry View of the Environment*, World Resources Institute, Washington, DC, 1992, p. 121.

21. Issue Activity Report of the Society of the Plastics Industry Management Staff, May, 1995, available from Lew Freeman, Vice President of Government Affairs.

22. Smart, op. cit., p. 102.

23. Several recent books provide case-studies and discussion of the environmental goals and actions of these and other major firms. See Smart, op. cit.; Bruce Piasecki, *Corporate Environmental Strategy: The Avalanche of Change Since Bhopal*, John Wiley & Sons, New York, 1995; Schmidheiny, op. cit.; and J-O. Willums and Ulrike Goluke, *From Ideas to Action: Business and Sustainable Development*, International Chamber of Commerce, Oslo, Norway, 1992.

24. For a discussion of the CERES principles, see Jennifer Nash and John Ehrenfeld "Private Codes of Management Practice: Assessing Their Role in Environmental Protection," Working Paper, MIT Technology, Business and Environment Program, May, 1995. ICOLP, GEMI, and the ICC initiative are discussed in Smart, op. cit., and Piasecki, op. cit.

25. Schmidheiny, op. cit., p. 190.

26. Ibid., p. 265.

27. Richard Mahoney, "People, Commitment, Results: The Monsanto Pledge," remarks made at the Annual Meeting of Monsanto Shareholders, April 27, 1990.

28. Smart, op. cit., p. 98.

29. Mahoney, op. cit., p. 3.

30. Elizabeth Rose and Arthur FitzGerald, "Free in Three: How Northern Telecom Eliminated CFC-113 Solvents from Its Global Operations," *Pollution Prevention*, Summer 1992, p. 297.

31. "AT&T to Eliminate Ozone-Depleters from Product Manufacturing," in *E&S Report*, a publication of AT&T E&S Engineering, Vol. 5, No. 2, 2nd Quarter, 1993, p. 1.

32. Detail on the goals and features of the 3P+ program are given in case studies in both Smart, op. cit., and Schmidheiny, op. cit.

33. Smart, op. cit., p. 16.

34. Noranda Annual Report, 1993, Management Discussion section, A. Powis, Chairman, and D. Kerr, President.

35. Smart, op. cit., p. 123.

36. Monsanto Company, Environmental Annual Review, 1994, p. 7.

37. Schmidheiny, op. cit., p. 116.

38. Smart, op. cit., p. 123.

39. Ibid., p. 129.

40. AT&T, 1994 Environment and Safety Annual Report, p. 4-7.

41. Ibid., p. 16.

42. Ibid., p. 12.

43. Ibid., p. 15.

44. See Lois Ember, "Responsible Care: Chemical Makers Still Counting on It to Improve Image," *Chemical and Engineering News*, May 29, 1995, for a full discussion on the pros and cons of third-party verification.

45. "Responsible Care Product Stewardship Code of Management Practices," Chemical Manufacturers Association.

46. Ibid.

47. "Responsible Care Distribution Code of Management Practices," Chemical Manufacturers Association.

48. Ember, op. cit., p. 18.

49. "Responsible Care Guiding Principles," Chemical Manufacturers Association, April, 1991.

50. Ember, op. cit., p. 10.

51. Phone interview with J. Eldon Rucker, Deputy Director, Health and Environmental Affairs, American Petroleum Institute, June 12, 1995.

52. "Environmental Management Systems - Specifications with Guidance for Use," ISO/DIS 14001, June 26, 1995, Unofficial version.

53. "Guidelines on Environmental Performance Evaluation," ISO 14031, Draft proposal from Swedish working group experts, June 19, 1995.

54. "Agenda 2020: A Technology Vision and Research Agenda for America's Forest, Wood, and Paper Industry," Prepared by the American Forest & Paper Association, November, 1994, p. 2.

55. Ibid., p. 3.

56. The Society of the Plastics Industry, 1995 Issue Accomplishments, May 1995, p. 3.

57. "1995 Public Policy Agenda," Steel Manufacturers Association.

58. Telephone interview with Donald Edmunds, Manager, Pollution Reduction, Facility Environment Department, American Automobile Manufacturers Association, June 12, 1995.

59. Automotive Pollution Prevention Project, Progress Report, February, 1994, available through the American Automobile Manufacturers Association.

60. Ibid., p. 1.

61. Telephone interview with Bob Strieter, Aluminum Association, June 7, 1995

62. Lee Neal, op. cit.

63. Anonymous, *Our Common Future*, New York, Oxford University Press, 1987.

64. The Rio Conference also added a set of social dimensions, centered on equity and justice, to those more directly related to the global ecosystem. The discussion in this paper has touched on the latter set only.

65. Schmidheiny, op. cit.

66. Ibid. p. 10.

67. Willums and Goluke, op. cit., p. 322.

68. Anne Gelfand, "The CERES Principles: Does Adopting a Voluntary Code of Management

Produce Corporate Accountability?" M. S. Thesis, MIT, June, 1995; Andrew Hoffman, "The Environmental Transformation of American Industry: An Environmental Account of Organizational Evolution in the Chemical and Petroleum Industries (1960-1993)," Ph. D. Dissertation, MIT, February, 1995.

69. This section is adapted from Michael Lenox and John Ehrenfeld, "Design for the Environment: A New Framework for Making Decisions," *Total Quality Environmental Management*, Vol. 4, No. 4, 1995, p. 37.

70. Ernest Lowe, "Industrial Ecology: A Context for Design and Decision," in Joseph Fiksel, ed., *Design for Environment: Principles and Practices, to be published by McGraw-Hill.*

71. L. W. Jelinski, T. E. Graedel, et al., "Industrial Ecology: Concepts and Approaches," *Proceedings of the National Academy of Sciences,* Vol. 89, February, 1992, pp. 793-797.

72. Chihiro Watanabe, "Japan's Approach to Energy Issues," *Industrial Ecology: U.S.-Japan Perspectives.* National Academy Press, Washington D.C., 1994.

73. Braden R. Allenby, "Achieving Sustainable Development Through Industrial Ecology," *International Environmental Affairs*, Vol. 4, No. 1, 1992, pp. 56-68.

74. Thomas E. Graedel and Braden R. Allenby, *Industrial Ecology*, Prentice-Hall, New York, 1995.

75. Bette Hileman, "Industrial Ecology Route to Slow Global Change Proposed," *Chemical and Engineering News*, August 14, 1992

76. Anonymous, *To Choose or to Lose, The National Environmental Policy Plan (NEPP)*, Netherlands Ministry of Housing, Physical Planning and Environment, 1989.

77. John R. Ehrenfeld, "Industrial Ecology: A Strategic Framework for Product Policy and Other Sustainable Practices," in *Green Goods*, E. Ryden and J. Strahl, eds., Ecocycle Delegation (Kretsloppsdelegationen), Stockholm, Sweden, 1995.

78. Lowe, op. cit.

79. Introduction to the 1994 Monsanto Corporation Environmental Report.

80. Jean Parker and Beverly Boyd. "An Introduction to EPA's Design for the Environment Program," *IEEE Symposium on Electronics and the Environment*, Washington, D.C., May, 1993.

81. Frank Consoli et al., eds., *Guidelines for Life-Cycle Assessment: A "Code of Practice,"* SETAC, Pensacola, Fla., 1993.

82. Jeroen B. Guinée et al., "Quantitative Life Cycle Assessment of Products, 2. Classification, Valuation and Improvement Analysis," *J. Cleaner Prod.*, Vol. 1, No. 2, 1993, p. 81.

83. Anonymous, "Concern for Tomorrow," National Institute of Environmental Protection, The Hague, 1988.

84. NEPP, op. cit.

85. Anonymous, *The National Environmental Policy Plan-Plus (NEPPPLUS)*, Netherlands Ministry of Housing, Physical Planning and Environment, The Hague, 1990.

86. INFORM's comprehensive study and analysis is entitled "Toxics Watch 1995" and is available through INFORM, Inc., 120 Wall Street, New York, NY 10005.

87. American Forest and Paper Association, "Pulp and Paper Industry Commitments to Improve Environmental Quality."

88. INFORM, Inc., op. cit., p. 504.

89. Ibid.

90. CEO Robert Allen's letter, AT & T 1994 Annual Report.

91. Smart, op. cit., p. 121

Status of Ecological Knowledge Related to Policy Decision-Making Needs in the Area of Biodiversity and Ecosystems in the United States

WALTER V. REID
World Resources Institute

CONTENTS

Scientific understanding of the structure and function of ecological systems has advanced tremendously over the past three decades. During this time, ecology and evolutionary biology have changed from descriptive disciplines into experimental and theoretical sciences with an ever-growing capability for prediction. As a result, ecological knowledge has become more and more useful to policy-makers. The utility of ecological information has been further enhanced by advances in research tools and technologies—particularly the development of sophisticated mathematical models of ecological systems and the growing use of geographical information systems—that help to provide information to decision-makers in forms tailored to the decisions that they face.

Still, the limits of scientific understanding are obvious. Even where general trends, such as wildlife population declines or changing stream quality, are clear, scientists are often unable to determine the impact of a specific action on those trends with any precision (or even whether the trends are a consequence of previous human actions or are natural). Problems of cumulative effects, lack of site-specific ecological knowledge, and the natural variability of ecological systems conspire to add substantial uncertainty to almost all uses of scientific knowledge in environmental decision-making. As a consequence, we must place as much emphasis today on techniques and policies for coping with uncertainty as we do on efforts to reduce that uncertainty.

In this paper, I assess the influence of recent advances in ecological knowledge on environmental policy decision-making, the current status of policy-relevant ecological knowledge, and key opportunities where advances in knowledge (or technologies) would improve decision-making.

Ecological knowledge influences three general aspects of environmental policy. First, choices about basic societal goals are made based on knowledge—or assumptions—about how the world works. Consequently the overarching goals we set for environmental management are based, in part, on our knowledge of ecological systems. For example, it was long assumed that biological communities were highly co-evolved equilibrium systems of organisms. Based on that assumption, a common goal of resource management has been to maintain certain systems in their "natural" state. But research that has found that biological communities are neither highly co-evolved nor regulated around equilibria calls into question the very meaning of a natural state. We can't objectively define a natural state in the absence of equilibrium states because we have no way of knowing what the structure and function of a system would have been if humans had never intervened. While we might set a goal of minimizing human intervention in certain systems, our current knowledge of ecology suggests that we should not set a goal of maintaining a system in a "natural state."

Second, ecological knowledge undergirds our ability to predict the ecological—and economic and social—consequences of human impacts, such as filling a wetland, changing stream flows, or introducing chemicals into the environment. Until recently, virtually all environmental management has been reactive rather than anticipatory, simply because we did not have sound information that would enable prediction of consequences of our actions. In effect, our management of resources has been a series of large scale—and often irreversible—experiments from which we often learned little because of the lack of information on baseline conditions or monitoring of trends. Now, with more sophisticated experimental research and modeling tools and information, we are better able to base decisions on anticipatory planning.

Third, ecological knowledge helps to define the types of policy tools and resource management practices that can be used to achieve various social and ecological objectives. For example, between the 1950s and 1970s, fisheries management relied on deterministic fisheries population models that were used to calculate harvest levels based on the maximum sustained yield of the population. As knowledge of fisheries population dynamics has grown, managers now realize that the non-linear nature of the population dynamics of many species, combined with the substantial uncertainty surrounding factors influencing recruitment and survival, requires different approaches to fisheries management—specifically, reliance on adaptive management techniques and highly conservative quotas.

ESTABLISHING GOALS FOR ENVIRONMENTAL MANAGEMENT

Do we know enough about ecological systems to be confident of the environmental management goals that we set? As our understanding of ecological systems has increased, we have refined environmental management goals and, in some cases, substantially changed them. For example, many of the regulations that were initially issued under the Clean Air Act and Clean Water Act in the early 1970s were based on limited knowledge of dose-response curves, environmental thresholds, and relative risks. With further research and experience with the consequences of various pollutants, those goals have been modified. More generally, though, how firm is the science on which we base our choice of environmental goals today?

Biodiversity conservation. Advances in scientific understanding of the importance of biodiversity and the threat that it faces is contributing to significant rethinking of biodiversity conservation objectives and goals. When the Endangered Species Act (ESA) was enacted in 1973, the goal of biodiversity conservation was straightforward—to save endangered species from extinction. The goal become substantially more complex, however, as scientists began to recognize the value of protecting all levels of biological diversity—genes, species, and biological communities. It became still more complex as the full magnitude of the species extinction problem in the U.S. was revealed. Where it may have seemed conceivable to protect all of the nation's species from extinction based on the knowledge available in the late 1960s, it is clear now that this would be extremely difficult and costly—not a surprising conclusion given the extensive habitat transformation that has taken place in the U.S.

Yet although current knowledge leads us to question aspects of the biodiversity policy goals implicit in legislation like the ESA, it is not yet clear what a new set of goals should be. For example, there is still no clarity on what the objective of conservation of biodiversity above the level of species should be. Is our goal simply to protect representative samples of biological communities in their own right, to protect samples of communities that also protect the widest diversity of species, or to protect communities and ecosystems that provide valuable ecosystem services? The problem is even more confusing when we consider that the composition and structure of communities will change through time as climate changes. Should we seek to protect the "arena" in which different distinctive communities exist, rather than attempting to protect "vignettes" of existing communities? If so, how would we choose those arenas? Similarly, although complete protection of all biodiversity is clearly an unrealistic goal, we are unable to determine what is realistic or desirable because we can't accurately estimate the social, economic, and ecological consequences of its loss.

Ecosystem services. We have long recognized the important role that ecosystems play in providing such "free" services as flood control (e.g., wetlands), water purification, maintenance of soil fertility, nutrient cycling, regulation of the micro-climate and regional climates, protection of coastal zones, dispersal and breakdown of wastes and cycling of nutrients, control of crop pests and disease vectors, and pollination of crops. And, to some extent, we have attempted to protect some of these services through legislative and regulatory actions. For example, Section 404 of the Clean Water Act, and the no-net-loss-of-wetlands policies of recent administrations, both seek to protect wetlands and the associated services that they provide. Similarly, requirements to maintain buffer strips along stream and river corridors help protect streams from sedimentation and changes in temperature and water flow.

More generally, however, we have established relatively few goals related to the protection of ecosystem services. This is due, in part, to the fact that with a few exceptions like wetlands and riparian habitat, our understanding of the relationship between various human pressures or changes in the structure or composition of ecological systems and the services they provide is very limited. Thus, even when the maintenance of the service is of obvious value, we are unable to define a management goal that will ensure the protection of the service. In some cases, where the maintenance of the service is particularly important we may set overly conservative goals. For example, protection of stream flows is often seen as essential for city or agricultural water supplies, and strict protection of the watershed is often the preferred means of ensuring those stream flows. In reality, changes could be made in an ecosystem (e.g., the addition or removal of species, harvesting of certain species, some transformation in the community structure) without any consequence for stream flows. But because in all but the most intensively studied systems, ecologists would have difficulty identifying what specific changes could or could not be made without disrupting the hydrology, we tend to set conservative goals when protection of the stream flow is a priority.

Without better understanding of how various ecological changes affect ecosystem services, we will be hard pressed to set goals for the protection of those services. One attempt to circumvent this problem is to define goals based on the need to protect the "integrity" or "health" of ecosystems (Noss 1990, Karr 1991). Although intuitively appealing, these goals face even greater problems than defining goals based on the maintenance of specific ecosystem services. Although it is entirely appropriate for political or educational reasons to seek to maintain a "healthy ecosystem," just as we seek a "strong economy," the former provides no more guidance to the biological manager than the latter provides to the economist. Establishing when an ecosystem is "sick" or disintegrating is entirely subjective. Is the eastern deciduous forest more healthy today than it was before climate change led to the migration of oak and hemlock into the community 10,000 years ago? Did the loss of the chestnut decrease the biological integrity of the system? Like "ecosystem services," concepts like health or integrity are fundamentally anthropocentric—we define the health of an ecosystem based on the characteristics and services that we want the ecosystem to provide.

Resource management goals. The development of the field of population biology in the 1950s fostered the emergence of the fields of wildlife, forestry, and fisheries management as professional disciplines. Population models enabled the determination of quantitative goals for resource harvesting. Concepts like "maximum sustained yield" and "optimum sustained yield" became the guiding principles of resource management. Clearly, though, no single harvested species exists in isolation from other species in the community, and virtually all ecosystems provide far more products and services than just the few harvested species. As a result, harvesting at the optimum sustained yield for one species may prove suboptimal for the benefits we obtain from the ecosystem if the management practice depletes other resources. For example, there was considerable concern in the 1980s that the Southern Ocean krill fishery would have negative impacts on other populations dependent on that krill, such as whales, seals, and marine birds. Thus, today's management goals must often be multi-species or ecosystem-wide goals.

Yet how such an ecosystem-wide management goal should be defined is not clear. We know that we should consider values other than just those of the harvested species—such as the value of ecosystem services and various values associated with biodiversity (e.g., ethical, existence, option, and use values), but we typically don't have good estimates of those values for any given ecosystem, we don't know how changes in the population of a harvested species might affect those values, and we don't know what relative weight to give the various types of values. Indeed, ecosystem management goals must necessarily have political as well as scientific dimensions. For example, while scientists and economists could, in theory, determine the economic consequences of continued forest harvest in the Pacific Northwest as a result of impacts on tourism, fisheries, and so forth, they are not in a position to weigh the ethical values of conserving a species like the Northern Spotted Owl against various economic values associated with continued loss of its habitat.

ASSESSING ENVIRONMENTAL IMPACTS

One of the few generalizations that can be derived from the past two decades of ecological research is that few generalizations can be made about the dynamics of ecological systems. This conclusion represents a significant change from the beliefs of researchers in the 1950s and 1960s, when studies suggested that ecological systems functioned in a state of dynamic equilibrium and that communities were tightly co-evolved assemblages of species. Considerable effort was devoted to the elaboration of underlying rules that would transform ecology into a predictive science. Today, there are relatively few general theories that researchers would be willing to apply to communities or to ecological interactions in a predictive manner. This change has come about as scientists came to realize that most communities appeared to be non-equilibrium systems with non-linear dynamics (Botkin 1990, Buzas and Culver 1994).

With few predictive theories at hand, considerable emphasis must be placed on empirical studies of specific systems to develop the ability to assess the consequences of change in those systems. Our ability to evaluate the impacts of any given perturbation depends strongly on the availability of long-term data on the particular system and in particular on the availability of experimental information bearing on the change being assessed. Even where scientists can predict the direct consequences of a specific perturbation on certain species (for example, the impact of a pesticide on a particular species), we are unable to predict the secondary effects of those changes on the structure and function of the ecosystem without much more detailed information.

The status of our knowledge about the impacts of human-caused changes differs depending on the type of perturbation involved:

Pollution and toxic chemicals. Our ability to predict the direct effect of a number of different classes of toxic chemicals on various wildlife populations is quite good. For many families of pesticides, for example, we have a good understanding of the mechanism by which the chemical works, and can predict which species might be affected by related chemicals and how they might be affected. Moreover, laboratory studies can often be used to identify particularly threatening chemicals relatively quickly and cheaply.

More worrisome, however, are chemicals that do not exhibit direct toxic effects on plants and animals but may have chronic effects. For example, a growing body of evidence suggests that a broad array of "estrogenic" chemicals may be influencing the reproductive physiology of many wildlife species (Colborn and Clement 1992, Kelce et al. 1995). Even where evidence of effects exists, there are substantial difficulties in distinguishing these effects from natural variation due to the lack of baseline information. Given the rapid growth in synthesized chemicals being released into the environment, and the potentially profound impacts on ecological and human social systems if some of these materials were responsible for chronic effects that changed basic life history characteristics of the species involved, there is a substantial need for further research on these potential risks.

The effects of many other widespread air and water pollutants—such as nitrogen in water or sulfur dioxide in air—on ecological systems has received considerable study over the past two decades. Although site-specific predictions of the impacts of such pollutants are often uncertain, the general effects of those pollutants in different types of ecosystems can often be predicted with reasonable accuracy. As the experience with the Exxon *Valdez* oil spill in Prince William Sound in Alaska demonstrates, however, considerable uncertainty and inaccuracy can surround the evaluation of impacts of even widespread pollutants like oil. In this particular case, the widespread initial predictions of an ecological disaster proved far too pessimistic. While the spill may still have substantial long-term impacts on the ecological systems, it is unlikely that it will be possible to distinguish those impacts from natural fluctuations since the baseline of data in the region is not extensive.

Habitat loss and fragmentation. One ecological generalization that has withstood the test of time is that the number of species found in a given habitat varies with the size of the habitat (see Figure 1). Consequently, when habitat is lost the number of species present will be reduced. While this species-area effect does allow a general assessment of the likely consequences of habitat change, it cannot provide detailed assessments of impacts on biodiversity. The actual impact of habitat conversion depends strongly on which areas are lost. Because species with restricted ranges are sometimes clustered in "hot spots" of diversity, protection of those hot spots could greatly reduce the effect of habitat loss on species diversity.

For a given species for which detailed demographic and life history information is available, we are able to make projections of the probability of survival of the species when its habitat is lost by use of Minimum Viable Population methodologies (Shaffer 1981). Insufficient demographic information is available for the vast majority of species, however, to allow such calculations.

The study of the impacts of fragmentation on ecological systems is still a relatively new field. Fragmentation can have significant effects both due to the increased ratio of edge habitat to core habitat (edge habitats have a different micro-climate and different species composition), and due to the access edge habitat may provide for certain species into core regions that would be otherwise inaccessible, as well as the barriers it may create for the dispersal and migration of

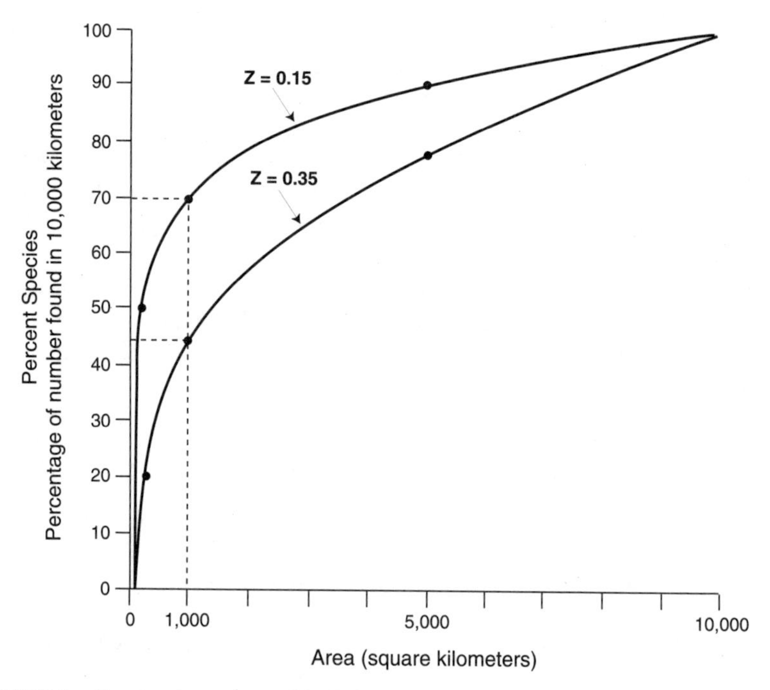

FIGURE 1 Species-area curve. SOURCE: M.L. Shaffer. 1981. "Minimum Population Sizes for Species Conservation." *BioScience* 31.

other species. Fragmented habitats are clearly far less desirable for conservation purposes than continuous habitats, but the specific impacts of fragmentation on a given species can only be determined through careful study of that species.

Species addition or removal. The loss of keystone species from ecological communities can have profound effects on the system because these species play important functional or ecological roles. Keystone species can include (i) predators, herbivores, parasites, and pathogens that maintain diversity among competing organisms by reducing the abundance of dominant competitors and preventing competitive exclusion; (ii) mutualists that link the fate of many species, such as specialized pollinators; and (iii) species that provide resources critical to the survival of dependent populations during periods of low resource availability. It is not possible, however, to determine which species serve "keystone" roles in a biological community without detailed experiments. In particular, neither trophic level, abundance, nor body size of a species is a good predictor of keystone status.

Introductions of certain species can also profoundly change ecosystem services. For example, the introduction in southwestern U.S. of the deep-rooted *Tamarix* tree, which has the ability to draw water from unsaturated alluvial soil—an ability missing in native trees in the region—has altered ecosystem services related to groundwater levels and stream flows. While it is extremely difficult to predict the consequences of an introduction, certain types of species are clearly higher risk than others. Weedy ("r"-selected) species that have large numbers of offspring, rapid growth, and high dispersal abilities, are more likely to become established in new habitats than "K"-selected species with lower population growth rates.

While the removal of a keystone species will, by definition, have profound impacts on a community and the services it provides, the removal or addition of some species in communities appears to have little demonstrable effect either on other species or on an ecosystem process. In these situations, other species compensate for the absence of the target species, at least over the short term. (It is not known for sure if all functions of the species in question are compensated for; in fact, it is rarely understood what the full range of functions is for each species.) As is the case with keystone species, it is not possible to predict based on general information about the biology of a species whether its loss would have little effect on the ecosystem. However, the one group of species whose loss would tend to have relatively little impact contains species that are currently in danger of extinction, simply because the primary impact of their loss would already have been felt when the species was originally reduced in population size to the point of endangerment.

Thresholds may sometimes exist in the response of ecosystem services to changes in abundance or distribution of species. In many of the world's biomes, ecosystems can exist in two or more alternative states that differ widely in rates of productivity or some other ecosystem services. Semiarid grassland can remain productive as long as grazing density is modest, but when density rises above a threshold the ecosystem shifts rapidly to a state of low productivity. Often, large and sustained reductions in exploitation are necessary to return ecosystems to high-productivity states, and in some cases, recovery from the undesirable state is not possible. At present, scientific knowledge is insufficient to be able to predict either the presence or the level of thresholds in ecosystem behavior.

Climate change. We currently have insufficient knowledge to make accurate assessments of the potential consequences of human-induced climate change on ecological systems. A few general predictions can be made: species found only high on mountains will be threatened with loss; protected areas with significant elevational or latitudinal gradients will be more secure than those without; species restricted to coastal regions will face significant threats from sea level rise, etc. By virtue of more complete knowledge of the requirements of crop plants, our ability to make predictions of impacts is somewhat better for agro-ecosystems than for natural systems. But our predictive ability is seriously constrained due to the high regional uncertainty of predictions from climate models and the lack of detailed autecological information (information on the interaction of a species and its physical environment) on many important species.

The predictive accuracy of climate models is quite low at a regional scale—the very scale where climatic factors become ecologically relevant. Until more precise regional estimates of changes in rainfall, temperature, seasonality, and weather disturbance frequencies are available, only general impact assessments are possible. But even when more detailed regional information becomes available, assessments will be hindered by the lack of information on the response of individual species to changes in climate. Because communities are not deterministic equilibrium assemblages of species, any climatic changes will have different consequences for different species. For tree species, which are some of the most structurally significant species in many North American communities, we do have relatively good paleoecological information on their rates of dispersal in response to past climate changes. Dynamic forest models also allow some prediction of how changes in climatic factors will alter forest community composition (Botkin 1990). For most other species, however, changes in response to climate shifts cannot be accurately predicted.

Moreover, the effect of changing distributions of species on various ecosystem services is even less certain. For example, changes in species distribution as a result of climatic changes will alter patterns of disease incidence. Such an effect can be seen in evidence that links the outbreak of cholera in South America in 1991 to El Niño (Epstein et al. 1993, Stone 1995). The warming of the waters off the coast of South America may have stimulated growth of a plankton harboring the cholera bacterium. While the increased frequency of El Niño in recent years cannot be conclusively tied to human-caused changes in climate, the example demonstrates how ocean current changes that are likely to occur in the event of global warming could have substantial effects on human health. Current models of changes in the distribution of disease vectors under likely future climates suggest that developing countries will see an increase in malaria, schistosomiasis, sleeping sickness, dengue, and yellow fever. The outbreak of the hantavirus in the southwestern U.S. has also been linked to weather conditions, possibly also associated with El Niño.

Introduction of genetically engineered organisms. Introductions of new cultivated varieties of crops (developed either through traditional breeding methods or through genetic engineering) into natural or agroecosystems can pose significant risks, such as the introgression of genes into wild populations, weediness, and pathogenicity. Genetic material from introduced plants, animals, and microorganism can be transferred into wild populations of related species through the formation of fertile hybrids. The new genetic material can then potentially alter the ecological interactions of that wild relative. For example, a disease or frost resistance gene transferred into a wild weedy relative of the crop could extend the range of that wild relative. To date, most gene flow in agricultural systems has taken place between crops and their weedy relatives, but in aquatic systems, substantial gene flow occurs between hatchery-reared fish and wild populations.

Current ecological knowledge can help minimize risks from the introduction of transgenic species. For example, risks are clearly higher in regions where wild relatives of the species are present.

These five examples demonstrate that substantial knowledge gaps exist that prevent a clear assessment of the ecological effects of any particular pressure or perturbation on a system. For certain types of well-studied perturbations, such as the application of the pesticide DDT, scientists can make detailed predictions of consequences. For certain well-studied species, such as deer, we can also make precise assessments of the impact of any change in the population of that species. But, by and large, unless a particular system has been the subject of intense research and monitoring, the current status of ecological knowledge only enables us to identify a list of potential consequences of a given action and to identify the set of data or experiments that would enable that uncertainty to be reduced.

MANAGING RESOURCES

Ecological knowledge helps to define the types of management practices and policy tools available for resource management. Current management practices have been influenced by advances in ecological knowledge, particularly increased understanding of (i) the natural variability of ecological systems; (ii) the non-linear nature of many ecological interactions; and (iii) the site-specific nature of many ecological interactions and processes (Ludwig et al. 1993).

For decades, wildlife and fisheries managers relied on demographic models that assumed far less natural variability in populations and more linear population responses than is now recognized to generally be the case. The knowledge available at the time led to the design of management techniques like calculating the maximum sustainable yield for a population and then setting harvest levels based on those estimates. Now, we recognize that the non-linear aspects of the population dynamics of most species place insurmountable limits to the precision of population projections. As a consequence, managers increasingly set goals for population management based on a safe minimum standard rather than a prediction of the optimum sustained yield.

Natural variability of ecological systems also makes it exceedingly difficult to evaluate the consequences of any particular management intervention. Adaptive management practices have been developed to reduce uncertainty and improve resource management through time. Adaptive management is a technique for managing biological systems so as to simultaneously reduce uncertainty about the functioning of the systems and respond to the changing social, biological, and physical environment. The principal elements of adaptive management are (i) management interventions are made in an experimental manner so that the outcome of the intervention can be used to reduce uncertainty about the system; (ii) sufficient monitoring prior to and during the intervention enables detection of the results of the management intervention and thereby allows managers to learn from past experience; and (iii) based on the feedback to managers, communities, and other constituencies, management interventions are then refined.

Recognition of the inherent variability in natural systems has also led to greater emphasis being placed on the need for long-term monitoring. The recent detection of a dramatic drop in phytoplankton abundance off the coast of Southern California was made possible only due to a monitoring program initiated in 1951 by the California Cooperative Fisheries Investigations (CalCOFI) (Roemmich and McGowan 1995). In contrast, the causes of coral bleaching episodes in the Caribbean remain controversial, in part because of the lack of long-term monitoring of water temperatures in the region. Without long-term records, it is difficult to distinguish human-caused changes from natural changes in most systems. Even with long-term records, however, conclusive evidence of human-caused influences typically also requires some form of experimental verification.

Finally, because of the predictive limits implicit in the fact that ecological communities are not tightly co-evolved assemblages of species existing in an equilibrium state, a premium is placed on site-specific information about the identity and biology of the species that occur in a given area. Resource managers are typically constrained by the absence of inventories of species in most regions of the United States.

To some extent, the need for inventories could be diminished if we had better knowledge of the complementarity of patterns of distribution of various groups of species. It has long been assumed, for example, that the pattern of diversity of plants would be a good predictor of the pattern of diversity in other groups of terrestrial organisms, because of the importance of the physical structure provided by plants for the diversity of other taxa in the community. We know that the distribution of plant diversity does not completely mirror other groups (Prendergast et al. 1993), but there are relatively few studies that seek to determine just how coincident patterns actually are.

WHERE CAN IMPROVED ECOLOGICAL KNOWLEDGE BEST AID DECISION-MAKING?

Setting Goals

The greatest need for improved ecological knowledge currently relates to its role in setting environmental goals. We are in the midst of a fundamental rethinking of environmental goals, but while our science is good enough to call into question some of our past goals, we cannot yet be certain of their replacements. In the past we set narrowly defined goals for species protection, wetlands conservation, or sustainable harvests. Yet progress toward these narrow goals did not always achieve appropriate balance among the many benefits we received from living systems and were often in conflict. Worse still, many of our environmental management goals were designed to be reactive to threats to the environment and, as a consequence, did not serve well as means to anticipate and prevent environmental threats. For example, we sought to prevent endangered species from going extinct, but paid less attention to the need to prevent species from becoming endangered.

It is fashionable now to speak of ecosystem management as the new environmental management paradigm. This may be true, but it is not a clear environmental management goal. Ecosystem management is a natural outgrowth of the recognition that the products and services we obtain from living systems are interdependent. It also is in keeping with the need to develop anticipatory policies to replace our current reactive regulations. But until we develop specific goals for "ecosystem management," it will be all things to all people. Advances in knowledge bearing on the following questions would significantly aid in the development of sound environmental management goals:

1. *What is the distribution of biodiversity in the U.S.? How much complementarity exists between regions of high conservation value for genetic, species, and community diversity? How sensitive is the protection of biological diversity to the area protected? How much opportunity is there to protect biodiversity in disturbed landscapes?*

One of the most difficult challenges for sustainable resource management in the U.S. is the protection of biological diversity. Our current focus on protecting endangered species from extinction needs to be complemented by efforts to prevent species from becoming endangered. This can be done most efficiently by protecting critical ecosystems from transformation. We still do not know the extent to which the protection of certain hot spots of biological diversity would protect the nation's biodiversity or the extent to which biodiversity can be protected in disturbed landscapes. We also do not know how effectively the protection of hot spots of species diversity also serves to protect community and genetic diversity. Finally, for many lesser-known taxa such as many invertebrates and fungi, we do not know enough about their status or distribution to know whether they are being adequately protected under current conservation policies. The answers to these questions bear on both issues related to management strategies (how can we most efficiently protect biodiversity) and to the goals themselves (what and how much are we trying to protect).

2. *What is the economic value of various services we obtain from ecosystems and what is the economic value of biological diversity? How should we weigh instrumental with intrinsic values of diversity?*

Decisions about the protection of many ecosystem services are fundamentally economic. The loss of the flood protection service of wetlands, for example, can be partially offset by the construction of dams and levees. However, because we do not have good estimates of the costs incurred by many environmental changes, they are not factored into the cost-benefit calculus of decision-making.

Particularly given the political climate that stresses greater and greater attention to such cost-benefit calculations, it is imperative that the ecological values be well represented. Consider three examples. The commitment to the goal of no-net-loss of wetlands is admirable. However, given the substantial benefits that wetlands provide in the form of flood control it may well be the case that an economically justified goal should be net gain in wetlands. Similarly, in the Pacific Northwest, the amenity value of relatively undisturbed ecosystems has a high economic value, demonstrated by the rapid growth in "footloose" industries in the Northwest. A more accurate calculus of the economic costs and benefits of resource conservation in the Northwest would probably give greater weight to the goal of biodiversity conservation. Finally, although the protection of habitat in rapidly growing urban areas is often criticized as being economically unsound, the maintenance of greenbelts and open space in urban areas is often found to have significant economic benefits. Again, better calculus of the economic values of ecosystem services and natural systems could often lead to the incorporation of those values as one goal of management. This would aid in defining goals not only for the protection of ecosystem services but also for multi-species resource management.

3. *How can planning tools be made to be more interactive and accessible to a broader cross-section of the public?*

Clearly, the goals of resource management cannot and should not be based only on economic considerations. Many values of biodiversity, for example, involve ethical or moral considerations and thus cannot be incorporated into cost-benefit calculus. Moreover, the protection—or use—of particular resources may have profound cultural significance that would be difficult to appreciate in studies of valuation. (How else to explain the surprisingly strong support for grazing-land management policies that benefit a tiny fraction of people at great cost to taxpayers and to the environment?)

Geographic information systems provide a tool that with sufficient development could be used by a broader array of civil society to learn about both the values of the resources in their region and the consequences of various management options. These tools would enable people to better understand the consequences of various options for both the short-term and the long-term economic and ecological outlook in a region. Without such tools, neither the civil society nor its representatives in government are able to adequately weigh the pros and cons of different resource management options.

Assessing Impacts and Managing Resources

The obstacles to enhanced ability to predict the consequences of human impacts on specific systems and the ability to successfully manage resources are similar. Most significantly, they relate to the need for significantly improved site-specific inventory and research and to the need for long-term monitoring (ESA 1991, NRC 1993).

4. *What species and communities occur in specific ecosystems and regions? What are their ecological requirements? How does the diversity of species influence various ecosystem services? What species play particularly important—keystone—roles in the system? How does fragmentation influence the key species? What methods can be used to successfully restore degraded habitats or ecosystem services?*

Site-specific information is of fundamental importance for successfully managing biological systems and for predicting the consequences of changes in those systems. This does not mean that we need to know everything about all systems. But it does mean that unless we have certain basic knowledge we will have limited ability to manage systems or to prevent unwanted changes. Perhaps the most fundamental need is to improve understanding of what species and resources are present in the system (NRC 1993).

5. *What are the long-term trends in the structure and function of given biological communities and what is the variability in various measures?*

There are only two ways of obtaining information about human impacts on systems. First, the perturbation can be made in a controlled experiment. Second, and much less conclusive, the effect of a perturbation can be compared to the long-term behavior of the system. While experimental ecology will continue to provide the primary means of teasing apart the workings of ecological systems, we must turn to long-term monitoring as the only practical means of identifying unexpected impacts and enabling adaptive management in situations where the replication of an experiment is impossible. However, the history of ecological research and funding, driven as it is by proposals to test specific hypotheses, is fundamentally in opposition to the need for monitoring and long-term ecological research.

6. *What are the chronic effects of chemicals being released into the environment on plant and animal populations?*

Sufficient evidence of endocrine disruption of wildlife populations caused by chemicals released into the environment now exists to warrant a substantial research effort to pin down the causality and to determine what classes of chemicals represent the greatest threats. (The potential impacts on human reproduction also warrant study.) The combination of the potentially dramatic effect of these chemicals on ecological systems and the rapid growth in the release of these chemicals makes this a high priority concern.

7. *How will species in various ecosystems respond to changes in temperature, precipitation, disturbance, and CO_2 levels as predicted under climate models?*

The ability of human societies to adapt to global changes will be linked in no small part to our success in ensuring that ecological systems adapt to change (NAS 1991). As predictions of changes in weather patterns become more refined, ecologists need to be in a position to model the responses of ecological systems. This will enable us to identify ecosystem services that might be threatened or species whose conservation will require which management interventions, and to evaluate potential economic costs or public health risks from ecological changes.

8. *What environmental indicators can be developed that bear on the achievement of environmental management goals?*

Indicators of the status of ecological systems in relation to some agreed goal or target, or of the value of various ecosystems or services, both provide an important means of communicating the status of ecological systems to decision-makers and the public, and present the opportunity for designing new conservation policies based on market incentives rather than regulatory approaches. With better indicators of the value of different types of resource management practices for the conservation of biodiversity, for example, it would be possible to create incentives (e.g., tradable permits, impact fees, tax breaks) that would encourage landowners to manage resources in a fashion that protects biodiversity.

CONCLUSIONS

Consider a classic case where ecological knowledge is called on to aid environmental decision-making: a proposal to build a housing development on natural habitat near an urban area. Should the development be allowed? What will its impact be on the environment? Are there alternative sites where the development could be placed with less impact? What information is needed to aid in the decision? What changes can be made to minimize the adverse impact?

Based on the analysis of this paper, the following insights on the role of scientific and technical information bear on the case. First, prior to assessing the impacts we need to clarify what the goals of environmental protection are for the region. The science is fairly clear on some aspects of those goals: the need to protect important services such as stream flow and flood control, the need to protect the diversity of species in the region. Some other goals, such as the need to protect representative samples of communities, may be less clear. Is the intent to actually protect the natural community that exists or simply to ensure that a representative sample of the bioregion is protected, recognizing that the species composition may change through time? More work is needed to clarify environmental goals based on recent scientific findings.

At the level of actual impact assessment, a qualitative assessment and evaluation of alternatives can be made with relatively little site-specific knowledge. Certain critical habitats, such as riparian communities, wetlands, steep slopes, and so forth, need to be protected to maintain ecological services in any region.

Relatively little site-specific information is needed to also allow a proposal to be modified to minimize its impact—minimizing fragmentation of habitats, avoiding migratory corridors, using native species in landscaping, avoiding riparian or wetland habitats, etc. But beyond these general statements, an assessment requires site-specific information on at least the identity, distribution, and range of the various species and community types. Ideally, sound planning also requires time-series information and knowledge of the ecological interactions in the ecosystem. Obviously, that ideal situation is rarely found. At best, impact assessments can only (i) identify critical resources (e.g., habitats of rare species) and (ii) weigh the relative risk of various alternatives. It is rare that any precise information can be given on the actual amount of impact on an ecosystem that would take place, much less the economic consequences of those impacts.

We do know enough, however, to design human interventions in ecosystems in a manner that will help us learn more about those ecosystems, through adaptive management. But for this approach to succeed, the financial resources need to be available for long-term studies and monitoring. Such long-term research, with its—by definition—delayed payoff, tends to be highly vulnerable to budget cuts.

The nation's biodiversity and ecological systems are resources that are growing in importance as they become increasingly imperiled and as society places ever greater values on the diversity of biological assets for tourism, recreation, ecological services, and quality of life, not just on their value as extractive resources. But our policies and actions still do not reflect this growing value—land is still "developed" only when we've altered the natural communities, even though the wisest economic development of much of our land now is to keep it in a natural or semi-natural state. Ecological knowledge can help improve our management decisions. But because we still undervalue ecological systems as resources, we also undervalue the need for the inventory, monitoring, and research that will enable us to anticipate and prevent their loss and degradation.

BIBLIOGRAPHY

Botkin, D.B. 1990. *Discordant Harmonies*. Oxford University Press, New York, N.Y.

Buzas, M.A., and S.J. Culver. 1994. Species pool and dynamics of marine paleocommunities. *Science* 264:1439-1441.

Colborn, T., and C. Clement, eds. 1992. *Chemically Induced Alterations in Sexual and Functional Development: The Wildlife/Human Connection*. Princeton Scientific Publishing, Princeton, N.J.

Culotta, E. 1995. Will plants profit from high CO_2? *Science* 268: 654-656.

Epstein, P.R., T.E. Ford, and R.R. Colwell. 1993. Marine ecosystems. *The Lancet* 342:1216-1219.

ESA [Ecological Society of America]. 1991. The sustainable biosphere initiative: an ecological research agenda. *Ecology* 72(2):371-412.

Karr, J.R. 1991. Biological integrity: a long-neglected aspect of water resource management. *Ecological Applications* 1:66-84.

Kelce, W.R., C.R. Stone, S.C. Laws, L.E. Gray, J.A. Kemppainen, and E.M. Wilson. 1995. Persistent DDT metabolite *p,p'*-DDE is a potent androgen receptor antagonist. *Nature* 375:581-585.

Ludwig, D., R. Hilborn, and C. Walters. 1993. Uncertainty, resource exploitation, and conservation: lessons from history. *Science* 260:17, 36.

NAS [National Academy of Sciences]. 1991. *Policy Implications of Global Warming*. National Academy Press, Washington, D.C.

Noss, R.F. 1990. Can we maintain biological and ecological integrity? *Conservation Biology* 4: 241-243.

NRC [National Research Council]. 1993. *A Biological Survey for the Nation*. National Academy Press, Washington, D.C.

Prendergast, J.R., R.M. Quinn, J.H. Lawton, B.C. Eversham, and D.W. Gibbons. 1993. Rare species, the coincidence of diversity hotspots and conservation strategies. *Nature* 365:410-432.

Roemmich, D., and J. McGowan. 1995. Climatic warnings and the decline of zooplankton in the California Current. *Science* 267:1324-1326.

Shaffer, M.L. 1981. Minimum population sizes for species conservation. *BioScience* 31:131-134.

Stone, R. 1995. If the mercury soars, so may health hazards. *Science* 267:957-958.

The Federal Budget and Environmental Priorities[1]

ALBERT H. TEICH
American Association for the Advancement of Science

CONTENTS

[1]This paper was prepared at the request of the National Research Council by the author writing as an individual. The interpretations and opinions contained herein are solely those of the author and should not be taken to represent positions of the American Association for the Advancement of Science, its Board, or its Council. The author wishes to thank Kei Koizumi for his assistance in the preparation of the paper.

INTRODUCTION

Focus

The federal budget is the central process of American government. Through the budget process, incommensurable quantities are compared, priorities are set, and the plans of federal agencies are laid out for Congress and the public to review, approve, modify, or reject. If one is to ask, therefore, what are the nation's priorities with regard to the environment, the federal budget is an obvious place to look for the answer. This paper is an attempt to take a rough first cut at examining national environmental priorities using federal budget data as its primary source. Specifically, it will attempt to respond to several questions:

- How are federal environmental priorities reflected in budgetary patterns and trends?
- How have these priorities shifted over the past several years?
- What part does environmental R&D play in overall federal funding for environmental programs?
- How does funding for environmental R&D compare to overall federal R&D funding?
- What can be said about the relation between funding for environmental programs and current statements of environmental goals?

Caveats Regarding the Analysis

Budget analysis is a complex and tedious undertaking. Time and resource constraints limit this paper to no more than a first approximation at responding to the above questions. In addition, several conceptual and practical issues that may limit the utility of the analysis should be noted.

First is the difficulty of defining environmental programs in an unambiguous manner. A huge variety of government activities impact the environment and could be classified as environmental programs. These range from energy production and conservation programs to agricultural efforts (e.g., promotion of integrated pest management) to outdoor recreation programs. At some level, it is necessary to make fairly arbitrary distinctions in order to limit the scope of "environmental." The umbrella covering environmental programs is discussed below. Differing definitions will, of course, yield different results.

Second is the difficulty of obtaining budget data at an appropriate level of aggregation. Extracting information about individual programs from federal budget documents is an arduous task when the subject is the current year's budget. The job is made infinitely more difficult in looking at trend data, as the budget documents tend to be ephemeral and organizational changes often make it difficult to track programs from year to year.

Third, and more fundamental, is the fact that budgetary data are an imperfect indicator of priority. The spending levels of different federal programs do indicate that society (through its government) has decided to allocate more money for one than the other, which can be construed as an indicator of preference, or a measure of priority. Yet some things cost more by nature. Launching a satellite to monitor conditions in the upper atmosphere may be considerably more expensive than a project that involves conducting national field studies of water quality. Their costs may differ by a factor of ten or even 100. This does not necessarily mean that one is 100 times more important than the other.

Fourth, in judging the relations between environmental budgets and statements of environmental goals, there is the problem of whose goals are "national goals." In our pluralistic society, there may be numerous conflicting views of what constitutes an appropriate set of national goals for the environment. Indeed, the very existence of the Forum for which this paper has been prepared, is evidence of the lack of consensus on such goals.

Finally, it should be remembered that the federal government is not the only source of funds for environmental programs. State and local governments as well as the private sector expend considerable sums on programs and activities related to the environment. None of these are included in the present analysis. The federal government may choose to spend more or less on a particular activity not because it is more or less important, but because others are already covering it adequately.

ENVIRONMENTAL PRIORITIES AND TRENDS IN THE FEDERAL BUDGET

EPA—A First Approximation

The Environmental Protection Agency (EPA) is the federal government's flagship agency for environmental regulation, for mitigation and remediation of problems caused by environmental pollution, for environmental monitoring, and for R&D related to these missions. It was formed in 1970 by bringing together units from several different federal departments and agencies, including components of the Department of the Interior and the then-Department of Health, Education, and Welfare. EPA's responsibilities include administration of programs under the Clean Air Act, the Clean Water Act, the Toxic Substances Control Act, the Safe Drinking Water Act, and several other major pieces of environmental legislation. Although most definitions of environmental programs include a considerable number of other federal activities, the level of funding for EPA and trends in the agency's budget are useful indicators of the priority of environmental protection at the federal level.

Overview of EPA's FY 1995 Budget

EPA's FY 1995 budget totals $5.731 billion. This figure includes $1.510 billion in rescissions enacted by the 104th Congress in the spring of 1995. Originally appropriated FY 1995 budget authority for EPA was $7.241 billion. The largest share of the rescission ($1.302 billion) was taken out of EPA's Water Infrastructure Financing program, which also represents the largest single element of the agency's budget. In the current budget, Water Infrastructure Financing represents 29 percent of EPA's total budget; prior to the rescission, it constituted 41 percent of EPA's budget. The upper portion of Table 1 shows EPA's FY 1995 budget by function, providing one view of its program priorities. (NOTE: All tables and charts are found at the end of this paper.)

As noted above, nearly a third of the agency's budget (down from two-fifths prior to the rescission) is devoted to the Water Infrastructure Financing program which provides support to state and local governments for construction and improvement projects to help meet water quality standards and ensure drinking water safety. The Clean Water State Revolving Fund awards grants to state programs that provide low cost financing to municipalities for sewage treatment projects. The Drinking Water State Revolving Fund offers loans to help governments improve their drinking water systems.

Second to Water Infrastructure in EPA's FY 1995 budget is the "Abatement, Control and Compliance" line under which EPA funds contracts, grants, and cooperative agreements for pollution abatement, control and compliance activities, as well administrative activities, including regulatory enforcement. These programs represent nearly one-fourth of EPA's current funding, a total of $1.405 billion.

Superfund, at $1.331 billion, is EPA's third major budget element in FY 1995, representing 23 percent or almost a quarter of total spending. This program is responsible for cleanup of hazardous waste sites and associated activities. Together, the three top program areas account for more than three-quarters of EPA's budget.

The "Research and Development" line, $335 million in FY 1995, represents less than 6 percent of EPA's budget. This is somewhat misleading, however. The R&D appropriations account finances mainly extramural research through grants, contracts, and cooperative agreements with industry, universities, nonprofits, and other federal agencies, as well as some in-house activities. The costs of most in-house R&D (including personnel and related costs) are funded through the "Program and Research Operations" account, and some is included in "Abatement, Control and Compliance." In addition, nearly $70 million of R&D is supported through Superfund and smaller amounts of R&D are supported under two other trust funds. Thus, EPA's total R&D in FY 1995 is estimated at $600 million, representing about 10.5 percent of the agency's budget. This is discussed in more detail below.

The "Program and Research Operations" account, it should be noted, also includes personnel costs and travel expenses associated with administering many other (non-R&D) EPA programs, excluding Superfund, the Water Infrastructure Financing program, and a few other areas.

A somewhat different view of EPA's budget priorities may be gained from the lower portion of Table 1, which displays FY 1995 funding by media. Again, Water Infrastructure Financing stands out, representing more than a quarter of the agency's budget when sliced in this manner. In fact, if one lumps this line together with the "Water Quality" and "Drinking Water" lines, it becomes apparent that water is EPA's dominant concern, at least in dollar terms ($2.148 billion or 37.5 percent of the total). Hazardous waste (combining Superfund and the "Hazardous Waste" line) comes a close second ($1.623 billion, 28.3 percent), followed by air at just under 10 percent and "multimedia" at 7.6 percent ($438 million). Toxics, pesticides, and radiation account for small fractions. It should be noted that these figures do not include more than $650 million in "Management and Support" costs. The latter should be allocated among the programs and would augment their percentage shares somewhat. Figures 1 and 2 display this information in graphical format.

Trends in EPA's Budget

The FY 1995 rescission, following cuts in FY 1994, has turned EPA's budget trend sharply downward. As it stands now, EPA's FY 1995 budget is actually 4 percent below its level of ten years ago in constant dollars. Table 2 shows this long term trend over the period FY 1985-1995, and Figures 3 and 4 display EPA's total budget in current and constant dollars over the decade. Table 2 also indicates that EPA's budget has been declining relative to other components of domestic discretionary spending in the federal budget. In FY 1985, EPA's budget represented 2.73 percent of total domestic discretionary spending. This figure peaked at 3.50 percent four years later in FY 1989 and has now declined to 2.67 percent. The table and charts also include the House-passed FY 1996 appropriation level. (As of this writing, the Senate has not yet considered EPA's appropriation.) As can be seen, EPA's budget would take an unprecedented cut under this legislation.

Environmental Programs in Other Federal Agencies

Defining "Environmental Programs"

While EPA is the major focus of environmental concern in the federal government, it does not come close to representing the full extent of federal efforts relating to the environment. Delimiting the federal role in the environment is not a straightforward task. In one sense, nearly everything the federal government

TABLE 1 Environmental Protection Agency Budget for Fiscal Year 1995 (budget authority in millions of dollars)

	FY 1995[a]	Percent of EPA Budget
By Function		
Program and Research Operations	922.0	16.1%
Research and Development	334.6	5.8%
Abatement, Control & Compliance	1,404.6	24.5%
Buildings and Facilities	–39.4	—
Oil Spill Response	20.0	0.3%
Asbestos Loan Program	0.0	0.0%
Hazardous Substance Superfund	1,331.3	23.2%
L.U.S.T Trust Fund[b]	70.0	1.2%
Water Infrastructure/State Revolving Funds	1,659.6	29.0%
Other	28.5	0.5%
Total, EPA Budget	5,731.2	
By Media		
Air	558.4	9.7%
Radiation	42.8	0.7%
Water Quality	516.7	9.0%
Drinking Water	163.8	2.9%
Water Infrastructure Financing	1,467.1	25.6%
Pesticides	94.2	1.6%
Toxic Substances	124.9	2.2%
Multimedia	438.1	7.6%
Hazardous Waste	291.6	5.1%
Hazardous Substance Superfund	1,331.3	23.2%
L.U.S.T.[b]	70.0	1.2%
Oil Spill Response	20.0	0.3%
Management and Support	651.7	11.4%
Buildings and Facilities	–39.4	—
Total EPA Budget	5,731.2	

SOURCE: EPA Budget Justification for FY 1996 and text of Public Law 104-19.

[a]Adjusted to reflect rescissions enacted in Public Law 104-19. These rescissions total $1,509.6 million, of which $1,302.2 million are from Water Infrastructure/SRF. These rescissions are of FY 1995 as well as prior-year funds, but are scored against FY 1995 budget authority. The original FY 1995 budget authority was $7,240.8 million.

[b]Leaking Underground Storage Tanks.

does, from conducting military operations to shaping national economic policy, has an impact on the environment. In addition, the federal government owns nearly one-third of the land in the United States, holds title to the resources of the outer continental shelf, and manages fisheries and marine mammal populations in waters within 200 miles of the coast (Gramp, Teich, and Nelson 1992, p. 7).

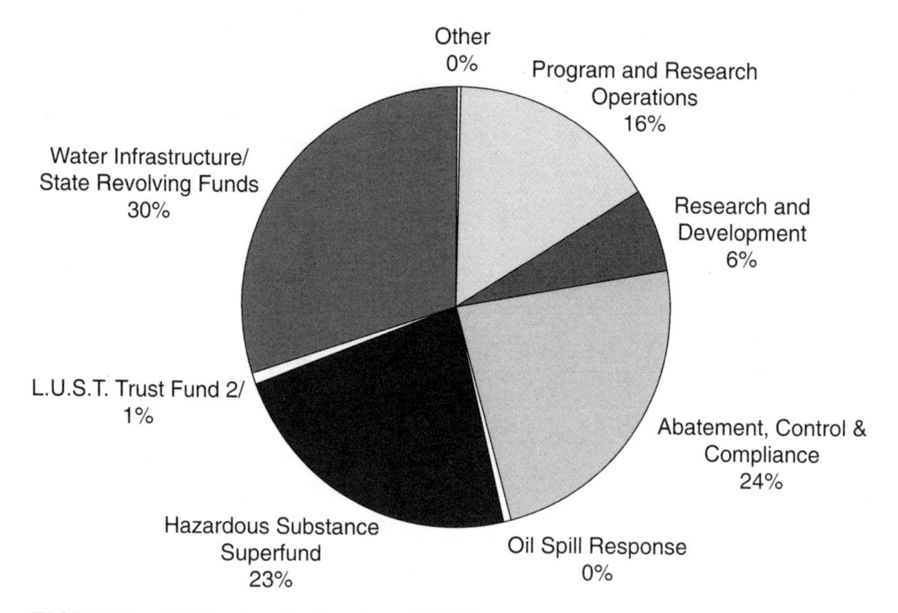

FIGURE 1 EPA budget by function, FY 1995.

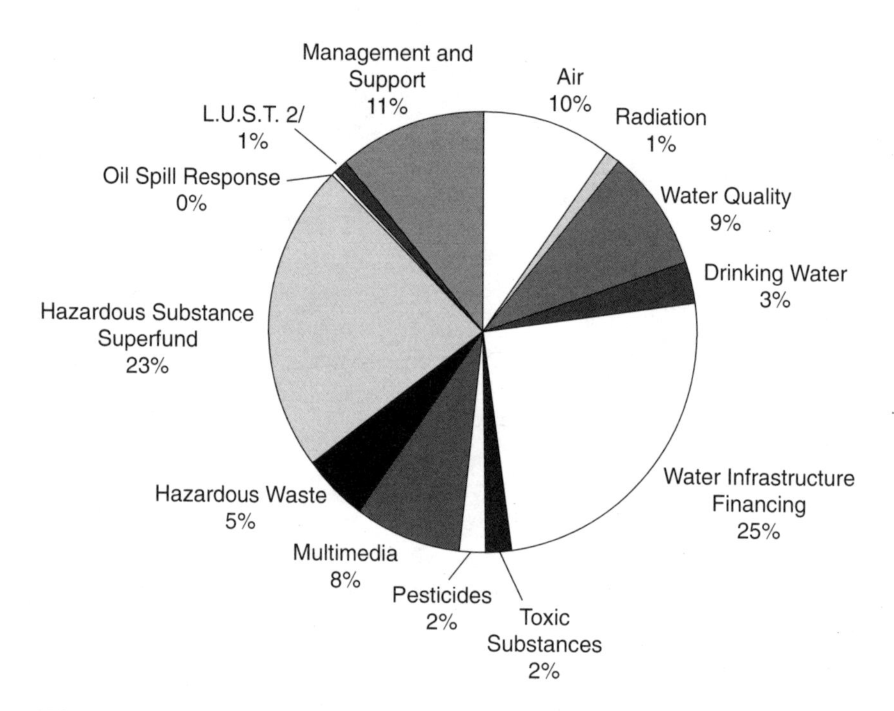

FIGURE 2 EPA budget by media, FY 1995.

TABLE 2 Environmental Protection Agency Budget (budget authority in billions of dollars, by fiscal year)

	FY 1985	FY 1989	FY 1992	FY 1993	FY 1994	FY 1995[a]	House FY 1996[b]
EPA budget	4.35	5.08	6.46	6.74	6.44	5.73	4.89
Total domestic discretionary spending	159.3	145.2	209.3	212.3	229.0	214.5	N/A
EPA as % of total domestic discretionary[c]	2.73%	3.50%	3.09%	3.17%	2.81%	2.67%	—
EPA budget (in constant FY 1987 dollars)[d]	4.61	4.69	5.33	5.43	5.09	4.41	3.66

SOURCE: Office of Management and Budget, *Budget of the United States Government Fiscal Year 1996* and OMB, *Mid-Session Review of the 1996 Budget.*

[a]Adjusted to reflect rescissions enacted in Public Law 104-19.
[b]Based on House-approved VA-HUD appropriations bill.
[c]Excludes discretionary spending in defense and international programs.
[d]Deflated using fiscal year GDP deflators from OMB.

Furthermore, the environment is integral to such federal functions as assisting the agricultural and energy sectors, managing the National Parks, predicting the weather, and assisting with natural disaster preparedness and recovery.

Rather than include practically everything the government does under the "environmental" rubric, we have taken a fairly narrow definition of environment, and included federal programs relating to pollution control and abatement, conservation and management of natural resources, and managing policy related to global climate change. This definition is essentially contiguous with the federal government's "Environment and Natural Resources" budget function, except that it *excludes* funding for reclamation projects, flood control, public lands acquisition, and other land management-oriented programs and it *includes* NASA environmental programs classified under "General Science and Space" budget function.

The scope includes environmental R&D in the various agencies, but goes well beyond R&D into operational programs. It encompasses activities in at least 12 federal civilian agencies plus the Department of Defense (DOD). Data are presented here for the eight largest among the civilian agencies. DOD (with the exception of the Army Corps of Engineers) is not included because data on its environmental programs were not available in a timely manner. Comments on DOD are included in a few places based on data obtained in the author's 1992 study of federal funding for environmental R&D (Gramp, Teich, and Nelson 1992).

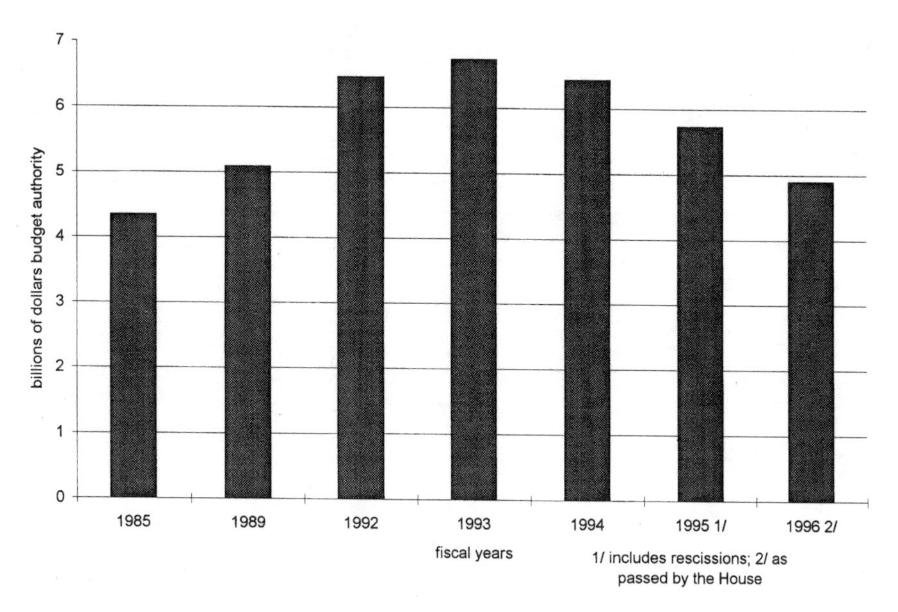

FIGURE 3 EPA budget, FYs 1985–1996 (House).

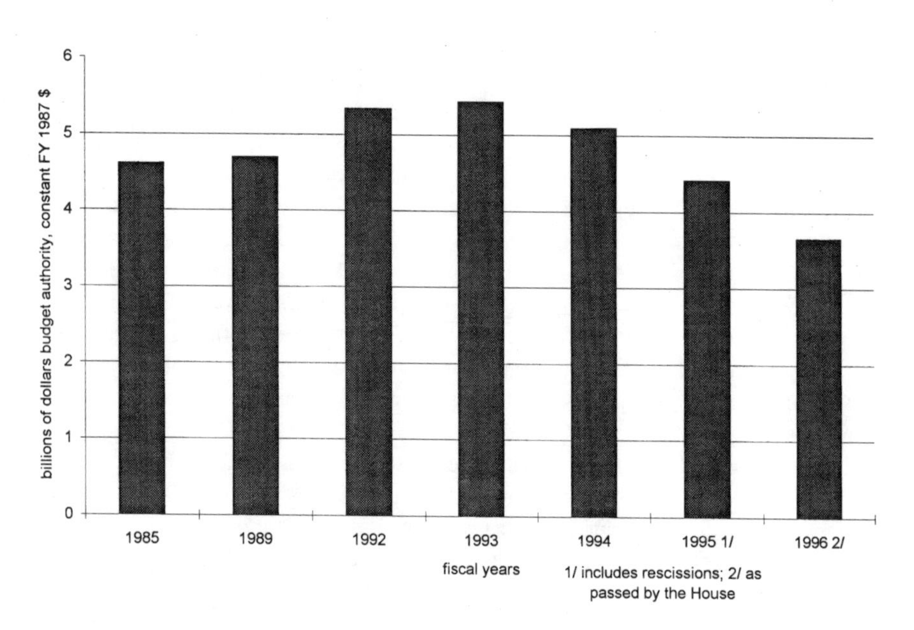

FIGURE 4 EPA budget in constant dollars, FYs 1985–1996 (House).

Budget Trends

Table 3 presents a summary of federal programs in environment and natural resources that fit the definition presented above. The information in the table is based on federal budget documents. The total shown on the table, $22.7 billion in FY 1995, is probably a low-end estimate of the government's total effort relating to the environment. Environmental R&D not included in the data set would add approximately $1 billion to this total (including DOD, the four civilian agencies omitted, and an estimate of environmental health research not covered in this analysis). The total shown represents a bit more than 10 percent of the federal government's total domestic discretionary spending.

Somewhat surprisingly, EPA shows up not at the top of the list but in second place among the agencies whose programs are identified in this table. Ranking first is the Department of Energy, mainly because of its Defense Environmental Restoration program (i.e., the nuclear materials and weapons facilities cleanup), which includes $4.855 billion in non-R&D activities and its nondefense Environmental Restoration and Waste Management program, funded at $723 million in FY 1995. Together, these efforts total $5.578 billion, nearly as much as the entire EPA budget. DOE also conducts a significant amount of environmental R&D, which, together with its other programs, including nuclear waste disposal and uranium enrichment decontamination, bring its total environmental portfolio to more than $7 billion.

Other major players among federal agencies include the Forest Service (within the Department of Agriculture); NASA, with life sciences research and Mission to Planet Earth; and NOAA, among whose programs and agencies are the National Weather Service, the National Marine Fisheries Service, the National Ocean Service, and a range of oceanic and atmospheric research programs. Also on the chart are the Departments of the Interior (with just over $1 billion in FY 1995), the National Science Foundation, and R&D conducted by the Army Corps of Engineers. The three-year trend in the bottom line of this table is essentially flat in current dollars, indicating a loss in constant dollars. And it does not take a clairvoyant to see that the trend over the next several years is likely to be even more negative.

FUNDING FOR ENVIRONMENTAL R&D

Environmental R&D in Relation to Other Environmental Programs

As discussed in the previous section, the agencies with the largest environmental programs are DOE and EPA. In both of these agencies, *environmental R&D* is a relatively small part of overall spending for environmental programs. EPA's R&D represents a total of $600 million out of the agency's $5.7 billion budget, about 10.5 percent. DOE's environmental R&D, estimated at $580 million, constitutes only 8 percent of its $7.1 billion in environmental programs,

TABLE 3 Estimate of Environmental Programs in the Federal Government (budget authority in millions of dollars, by fiscal year)

	Actual FY 1993	Est. FY 1994	Est. FY 1995[a]	Avg. Annual % Change FY 1993-95
Environmental Protection Agency	6,737	6,436	5,731	–8%
NASA	1,024	1,255	1,481	20%
Life Sciences (incl. non-terrestrial)	140	187	141	5%
Mission to Planet Earth	884	1,068	1,340	23%
NOAA	1,534	1,796	1,816	9%
National Ocean Service	168	202	180	5%
Nat'l Marine Fisheries Service	223	265	269	10%
Oceanic and Atmospheric Research	202	233	259	13%
National Weather Service	533	683	659	12%
NESDIS	347	351	387	6%
Marine Services	61	63	62	1%
National Science Foundation	594	655	685	7%
Environmental Programs R&D	530	591	622	8%
Non-R&D (logistical support)	63	64	63	–1%
Department of the Interior	945	1,021	1,069	6%
Department of Agriculture[b]	5,626	5,483	4,459	–11%
Natural Resources Conservation Service	2,021	1,375	837	–36%
Forest Service (non-R&D)	3,259	3,751	3,260	1%
Environmental R&D	345	357	363	2%
Department of Energy	6,612	7,173	7,126	4%
R&D Activities	589	782	580	3%
Environment, Safety and Health	138	105	127	–2%
Nuclear Safety Policy	24	15	17	–11%
Environmental Analysis	12	9	0	–62%
Env. Restor. and Waste Mngmt. (non-defense)	692	673	723	2%
Defense Env. Restoration (non-R&D)	4,580	4,922	4,855	3%
Uranium Enrichment Decontamination	201	286	301	24%
Defense Nuclear Waste Disposal	100	120	129	14%
Nuclear Waste Disposal Fund	275	260	393	23%
Corps of Engineers Environmental R&D	29	31	36	12%
National Institutes of Health	251	259	267	3%
Total Environmental Spending	23,351	24,108	22,670	–1%

SOURCE: Author's estimates based on agency budget justifications. Does not include funds for reclamation projects, flood control, public lands acquisition and other land management-oriented programs.

[a]FY 1995 figures reflect rescissions of $1.5 billion from EPA's budget and rescissions from NOAA enacted in Public Law 104-19, but do not reflect other rescissions enacted in Public Laws 104-6 and 104-19. Detailed information on other rescissions not available at this time.

[b]Includes environmental R&D and programs classified as "Natural Resources and Environment."

which are dominated by the huge costs of restoring environmental quality in DOE's defense facilities. Elsewhere in the government, however, R&D is a major (or in some cases *the* major) part of agency environmental programs.

Definition and Overview

An overview of environmental R&D in the federal government is contained in Table 4. Details on the environmental R&D activities of the major federal agencies are shown in Tables 5 through 17. As elsewhere in this paper, these tables do not include information on Department of Defense military activities. The tables show a three-year trend, from FY 1993 through FY 1995. The definition of environmental R&D is largely the same as in the author's previous work on federal funding for environmental R&D:

- environmental sciences, including (a) environmental life sciences, such as environmental biology, forestry, marine biology, and related fields, and (b) physical environmental sciences, such as oceanography, geology, and atmospheric sciences, excluding extraterrestrial research;
- engineering and other sciences related to the impacts of natural and anthropogenic activities on the environment, including prevention, control, amelioration, and regulation;
- social sciences related to the environment, such as environmental economics, and social science research on cultural and institutional factors affecting sustainable development, pollution prevention, adaptation to global change, etc.; and
- information and data sciences related to the environment, such as computer sciences and specialized information management R&D identified in agency budgets (Gramp, Teich, and Nelson 1992, p. 2).

There are, however, a few differences. The tables in this paper include the National Institute of Environmental Health Sciences, a unit of the National Institutes of Health, which was not included in the 1992 report because the working definition chosen for that report excluded environmental health. They exclude (for reasons of time and resources) several agencies with relatively small environmental R&D portfolios: the Agency for International Development (USAID), the Smithsonian Institution, the Tennessee Valley Authority (TVA), and the Department of Transportation (DOT).

The coverage of those agencies included is generally analogous to that in the 1992 report. However, because of organizational changes and changes in budget presentation, as well as possible inconsistencies in data collection techniques, the data from the 1992 tables, which cover FY 1990-1992, have not been integrated with the FY 1993-1995 data collected for this paper. To allow readers to gain a rough sense of the trends over six years, the tables from the 1992 report are

TABLE 4 Estimate of Total Environmental R&D (budget authority in millions of dollars, by fiscal year)

	Actual FY 1993	Est. FY 1994	Est. FY 1995[a]	Avg. Annual % Change FY 1993-95
National Aeronautics and Space Admin.	967	1,152	1,402	20%
Department of Energy	589	782	580	3%
National Science Foundation	530	591	622	8%
Department of the Interior	546	568	569	2%
USGS	353	365	357	0%
Other	193	204	212	5%
Department of Agriculture	345	357	363	2%
ARS	127	134	142	5%
CSRS	105	102	99	–3%
Forest Service	104	111	114	5%
ERS	10	9	9	–5%
Environmental Protection Agency[a]	497	588	600	10%
National Oceanic and Atmospheric Admin.[a]	357	443	447	13%
AID[b]	N/A	N/A	N/A	—
Smithsonian[c]	N/A	N/A	N/A	—
TVA[d]	N/A	N/A	N/A	—
Corps of Engineers	29	31	36	12%
Transportation[e]	N/A	N/A	N/A	—
National Institutes of Health	251	259	267	3%
Department of Defense[f]	N/A	N/A	N/A	—
Total Environmental R&D	4,111	4,771	4,885	9%

SOURCE: Author's estimate based on agency budget justification and other supporting documents.

[a]Does not reflect rescissions enacted in FY 1995 except for NOAA and EPA (see agency tables). Detailed information on rescissions in other agencies not available at this time.
[b]FY 1992 level for AID was $45 million.
[c]FY 1992 level for Smithsonian was $33 million.
[d]FY 1992 level for TVA was $31 million.
[e]FY 1992 level for DOT was $17 million.
[f]FY 1992 level for DOD was $577 million.

N/A= Data not available at this time.

included as Appendix A. The appendix tables also provide information on environmental research in DOD, not available for the more recent years. In FY 1992, DOD spent $577 million on environmental R&D, representing about 13 percent of the total federal investment in environmental R&D in that year. Adding this amount, about $100 million for the agencies with smaller environmental R&D programs, and about $400 million for environmental health not included suggests that data here understate overall federal spending on environmental R&D in FY 1995 by a bit over $1 billion.

Agency Highlights

The leading agency in environmental R&D—at least in terms of budget levels in FY 1995—is NASA, which is expected to spend $1.4 billion, the largest share of which will go to its Earth Observing System (EOS). Details are shown in Table 5. NASA's capabilities in space transportation provide it with the means for conducting important large-scale research on the earth's atmosphere and surface. The nature of these studies, and their context within a large and costly space program, make them the most expensive single element of federal environmental R&D. While some of the work relates to pollution control and abatement, the largest share is devoted to aspects of global change and other large-scale atmospheric, oceanic, and geologic processes. NASA's environmental R&D has grown at an average annual rate of 20 percent since FY 1993, although this is not likely to be sustained in coming years.

DOE (shown in Table 6) supports a wide range of environmental R&D, most of it associated with mitigation of the effects of energy production on the environment. Environmental restoration and waste management R&D, part of the much larger environmental restoration effort comprises the greatest share. Clean coal technology has also been a major focus of DOE's work, although it was sharply reduced in FY 1995. DOE's research also contributes to the national effort on global change through studies of carbon dioxide in the atmosphere, conducted mainly at national laboratories and universities. DOE's environmental R&D programs total $580 million in FY 1995, down about $200 million from FY 1994.

Basic research on the environment is the province of NSF, whose efforts total some $622 million in FY 1995 and have grown at a rate of 8 percent a year over the FY 1993-1995 period. (See Table 7.) Most of these efforts are in environmental sciences, including ecology and environmental biology, ocean sciences, atmospheric sciences, earth sciences, and polar programs. NSF also funds social science research related to global change as well as engineering research in a number of environmental areas, including earthquake hazard mitigation. NSF funds most of its environmental research through investigator-initiated grants to academic researchers and in FY 1992 NSF was responsible for nearly half of federal grant funding for environmental R&D (Gramp, Teich, and Nelson 1992, p. 26).

The Department of the Interior is another key source of support for environmental R&D, as shown in Tables 8 and 9. The U.S. Geological Survey, whose long-standing mandate is to classify and analyze the nation's water, mineral, energy, and other geologic resources, also conducts work on water quality, nuclear waste, energy development, natural hazards, and global change. Other agencies in the Interior Department that conduct environmental R&D include the National Biological Service, Bureau of Mines, Minerals Management Service, and National Park Service. Interior's environmental R&D programs total nearly $570 million.

Environmental research is supported by many units of the Department of

TABLE 5 Estimate of Environmental R&D at the National Aeronautics and Space Administration (budget authority in millions of dollars, by fiscal year)[a]

	Actual FY 1993	Actual FY 1994	Est. FY 1995	Avg. Annual % Change FY 1993-95
Environmental Sciences				
Earth Observing System (EOS)	264	393	591	50%
Earth Probes	99	96	82	–9%
Payload and Instrument Development	35	26	20	–26%
Mission Operations and Data Analysis	94	98	97	1%
Airborne Science and Applications	21	25	26	12%
Interdisciplinary Analysis	4	5	42	376%
Process Studies	119	126	118	–0%
Modeling and Data Analysis	43	44	42	–1%
Ocean Color Data	16	3	1	–73%
Life Sciences, Research and Analysis	53	55	51	–2%
Construction of Facilities, Global Change	0	12	17	—
Estimated Intramural Research	70	75	80	6%
Subtotal	818	958	1165	19%
Information and Data R&D				
EOS Data and Information Systems	131	188	231	33%
CIESIN[b]	18	5	6	–26%
Subtotal	149	193	237	26%
Total, NASA Environmental R&D	967	1,152	1,402	20%

SOURCE: Author's estimate based on agency budget justification and cross-cut by field of science.

[a]These estimates exclude funding for environmental sciences in NASA's planetary and commercial applications programs.

[b]Consortium for International Earth Science Information Networks.

Agriculture (USDA), as shown in Tables 10 through 13, which include the Agricultural Research Service (ARS), the Cooperative State Research, Education, and Extension Service (CSRESS), the Economic Research Service (ERS), and the Forest Service (FS). Together, these units bring the USDA total to $364 million. ARS, nearly all of whose work is done in house, represents almost 40 percent of this total. Its major foci include soil and soil-water relationships, as well as environmental biology and ecology. Other ARS studies relate to water conservation, watershed management, and pollution issues relating to agriculture. CSRESS provides nearly $100 million for grants on environmental issues associated with agriculture. These include both formula and competitive grants. Research related to management of publicly owned national forests and rangelands, including over 12 million acres of wetlands, is carried out by the Forest Service, whose expenditures on environmental R&D totaled $114 million in FY 1995.

TABLE 6 Estimate of Environmental R&D at the Department of Energy (budget authority in millions of dollars, by fiscal year)

	Actual FY 1993	Actual FY 1994	Est. FY 1995	Avg. Annual % Change FY 1993-95
Environmental Sciences				
Atmospheric Science	12	13	13	4%
Marine Transport	7	7	7	–1%
Terrestrial Transport	18	18	19	2%
Ecosystem Functioning and Response	7	4	6	–2%
CO_2, Core Program	19	16	22	9%
CO_2, CHAMMP[a]	11	10	10	–3%
CO_2, Atmospheric Radiation Measurement	31	34	30	–0%
CO_2, Oceans Research	5	5	4	–9%
CO_2, Nat'l Institute for Global Env. Change	11	11	11	2%
CO_2, Education	3	3	3	–7%
Basic Energy Sciences, Geosciences	20	19	19	–0%
Bonneville Power, Fish and Wildlife[b]	1	1	1	13%
Subtotal	145	142	146	0%
Social Sciences				
CO_2, Global Change Integrated Assessmt.	2	2	3	39%
Engineering and Related R&D				
Clean Coal Technology Program	0	222	37	—
Atomic Defense Env. Waste and Restoration	248	248	238	–2%
Coal Environmental Research	186	166	154	–9%
Oil Shale Research	5	0	0	—
Subtotal	440	635	429	6%
Information and Data R&D				
CO_2, Information and Integration	2	2	2	–8%
Total, DOE Environmental R&D	589	782	580	3%

SOURCE: Author's estimate based on agency budget justification.

[a]Computer Hardware, Advanced Math and Model Physics program.

[b]Represents obligational authority financed by Bonneville Power Administration ratepayers.

The distribution of EPA's R&D budget by media, adding up to $600 million in FY 1995, is shown in Table 14. In contrast to the agency's overall budget in which expenditures for water quality and drinking water dominate, multimedia R&D, at $329 million, is the largest area of R&D at EPA, followed by air quality and Superfund (hazardous waste). The shift of resources into multimedia R&D is an important and relatively recent development. Until about five years ago, the agency's research agenda was largely reactive, focusing on areas dictated by vari-

TABLE 7 Estimate of Environmental R&D at the National Science Foundation (budget authority in millions of dollars, by fiscal year)

	Actual FY 1993	Actual FY 1994	Est. FY 1995	Avg. Annual % Change FY 1993-95
Environmental Sciences				
BIO, Ecological Studies	24	27	28	8%
BIO, Systematics and Population Biology	22	22	24	5%
BIO, Long-Term Environmental Biology	24	25	26	4%
BIO, Center for Ecological Analysis & Synthesis	1	1	2	50%
GEO, Ocean Sciences Research Support	92	99	103	6%
GEO, Oceanographic Centers and Facilities	52	51	51	–1%
GEO, Ocean Drilling Program	36	39	40	5%
GEO, Atmospheric Sciences Research Support	69	82	85	11%
GEO, National Ctr. for Atmospheric Research	50	53	58	8%
GEO, Earth Sciences Project Support	52	53	54	3%
GEO, Instrumentation and Facilities	18	21	20	7%
GEO, Continental Dynamics	6	7	7	8%
MPS, Environment and Global Change	0	14	18	1319%
U.S. Polar Programs	50	57	56	6%
Subtotal	496	550	574	8%
Social Sciences				
SBE, Environment and Global Change	12	17	21	33%
Subtotal	12	17	21	33%
Engineering and Related R&D				
ENG, Environmental and Ocean Systems	6	6	7	10%
ENG, Natural and Tech. Hazard Mitigation	3	3	3	5%
ENG, Earthquake Hazard Mitigation	14	15	17	11%
Subtotal	22	24	27	10%
Total, NSF Environmental R&D	530	591	622	8%

SOURCE: Author's estimate based on agency budget data and cross-cut by field of science.

NOTE: NSF directorates include Biological Sciences (BIO), Geosciences (GEO), Social and Behavioral Sciences (SBE), and Engineering (ENG).

ous environmental statutes. As described in a recent AAAS report, EPA has sought to redirect its research program to "support both near-term studies needed by regulatory offices and the longer-term research needed to resolve scientific uncertainties about the interrelationships among environmental problems and their efforts on ecological and human communities (AAAS 1995, p. 123). A more complete description of EPA's R&D program is contained in Appendix B.

TABLE 8 Estimate of Environmental R&D at the Department of the Interior, U.S. Geological Survey (budget authority in millions of dollars, by fiscal year)

	Actual FY 1993	Actual FY 1994	Est. FY 1995	Avg. Annual % Change FY 1993-95
Environmental Sciences				
Global/Climate Change	11	11	10	–5%
Marine and Coastal Surveys	37	36	36	–1%
Earthquake Hazards	50	54	49	–0%
Volcano Hazards	20	20	20	–0%
Landslide Hazards	2	2	2	–0%
Water Resources Research Program[a]	60	75	79	15%
Water Research Grants	6	6	4	–11%
Mineral Resource Surveys	48	47	45	–4%
Energy Resource Surveys	27	26	25	–3%
Geomagnetism	2	2	2	–1%
Deep continental studies	3	3	3	0%
Subtotal	266	281	275	2%
Information and Data R&D				
National Geologic Mapping	22	23	22	–0%
Research and Technology[b]	12	11	12	0%
Advanced Cartography	7	6	4	–18%
Federal-State Water Program, Data/Analysis	24	24	24	–1%
Water Data Collection and Analysis	22	20	21	–4%
Subtotal	87	84	82	–3%
Total, USGS Environmental R&D	353	365	357	0%

SOURCE: Author's estimate based on agency R&D cross-cut.

[a]Incorporates programs in Regional Aquifer System Analysis, Core Hydrologic Research, Water Resources Assessment, Toxic Substances Hydrology, Acid Rain, Global Change Hydrology, Truckee-Carson Program, and National Water Quality Assessment Program.

[b]Incorporates programs in Cartographic and Geographic Research, Nat'l Cartographic Requirements Coordination and Standards, and Geographic and Spatial Information Analysis.

Research on climate and weather, on marine sciences, on coastal and ocean management, and on fisheries is conducted by NOAA, the National Oceanic and Atmospheric Administration, a unit of the Department of Commerce. NOAA's environmental research totals $447 million. The Office of Oceanic and Atmospheric Research (OAR) is responsible for much of this work, and its efforts include NOAA's contributions to the interagency Global Change Research Program as well as studies called for by the 1990 Clean Air Act Amendments. Other units involved are the National Ocean Service and the National Marine Fisheries Service. NOAA's R&D program is presented in Table 15.

TABLE 9 Estimate of Environmental R&D at the Department of the Interior, Other Agencies (budget authority in millions of dollars, by fiscal year)

	Actual FY 1993	Actual FY 1994	Est. FY 1995	Avg. Annual % Change FY 1993-95
Environmental Sciences				
NBS, Population Dynamics[a]	15	14	14	–3%
NBS, Ecosystems	47	47	49	2%
NBS, Cooperative Research Units	18	15	15	–9%
NBS, Research Center Maintenance	15	16	17	6%
National Park Service	20	24	19	1%
Bureau of Land Management	0	0	2	84%
Bureau of Reclamation, Global Change	1	1	0	–24%
Subtotal	117	118	116	–0%
Social Sciences				
Office of the Secretary	0	0	0	–55%
Engineering and Related R&D				
Bureau of Mines	17	21	32	37%
Minerals Management Service	6	20	22	115%
Bureau of Reclamation	8	9	6	–11%
Subtotal	31	50	59	39%
Information and Data R&D				
NBS, Inventory and Monitoring	31	22	22	–14%
NBS, Information Transfer	15	14	14	–0%
Subtotal	45	35	36	–9%
Total, DOI, Other Agencies' Env. R&D	193	204	212	5%

SOURCE: Author's estimates based on agency budget justification and R&D cross-cuts.

[a]National Biological Service. Some R&D activities of other agencies were transferred to NBS in FY 1994. FY 1993 totals have been adjusted for comparability.

The Army Corps of Engineers contributes a small, but important, component of the federal government's environmental R&D. Its programs, which are mainly in engineering and total $36 million in FY 1995, are shown in Table 16. They include modest efforts in coastal engineering, in flood control and navigation, in aquatic plant control, and a variety of other areas.

The budget for the National Institute of Environmental Health Sciences (NIEHS), which totals $267 million is shown in Table 17. In 1992, total environmental health R&D in the federal government was estimated at about $700 million. This figure included activities at NIEHS as well as environmental health R&D in the Food and Drug Administration, the Centers for Disease Control, the

TABLE 10 Estimate of Environmental R&D at the Department of Agriculture, Agricultural Research Service (budget authority in millions of dollars, by fiscal year)[a]

	Actual FY 1993	Est. FY 1994	Est. FY 1995	Avg. Annual % Change FY 1993-95
Environmental Sciences				
Appraisal of Soil Resources	2	2	2	
Soil, Plant Water Nutrient Relationships	33	35	37	
Management of Saline Soils	2	2	3	
Alternative Uses of Land	0	0	0	
Improvement of Range Resources	7	7	8	
Wildlife and Fish Ecology	5	5	5	
Environmental Biology	14	15	16	
Subtotal	63	67	70	5%
Engineering and Related R&D				
Conservation and Efficient Use of Water	13	14	14	
Efficient Drainage and Irrigation Systems	5	5	5	
Watershed Protection and Management	20	21	22	
Protection from Pollution	4	4	4	
Alleviation of Pollution	19	20	21	
Subtotal	60	63	67	5%
Information and Data R&D				
Remote Sensing	4	4	4	5%
Total, ARS Environmental R&D	127	134	142	5%

SOURCE: Authors' estimates.

[a]ARS budget authority has been projected based on 1993 data in the Current Research Information System for goals pertaining to the environment and for environmental sciences research supporting other goals.

Department of Energy, and EPA. Except for NIEHS and EPA, none of this R&D is included in this paper.

Environmental R&D in the Context of Overall Federal R&D

Environmental R&D is a relatively small component of total federal R&D. According to NSF figures, in FY 1985, R&D devoted to the "environment and natural resources" budget function represented 2.1 percent of total federal R&D and 6.5 percent of nondefense R&D. By FY 1994, with defense R&D shrinking and health and space research growing, environmental and natural resources R&D had risen to 2.7 percent of total R&D, but declined to 6.0 percent of nondefense

TABLE 11 Estimate of Environmental R&D at the Department of Agriculture, Cooperative State Research, Education, and Extension Service (budget authority in millions of dollars, by fiscal year)

	Actual FY 1993 [a]	Est. FY 1994 [a]	Est. FY 1995 [a]	Avg. Annual % Change FY 1993-95
Environmental Sciences				
Special Grants	13	11	9	–16%
Competitive Grants	21	22	22	3%
Hatch/Agricultural Experiment Stations	20	21	21	0%
Cooperative Forestry	3	3	3	6%
1890 Colleges and Tuskegee	3	3	3	1%
Animal Health	*	*	*	0%
Subtotal	59	59	57	–2%
Social Sciences				
Special Grants	1	0	0	–16%
Competitive Grants	2	2	2	3%
Hatch/Agricultural Experiment Stations	6	7	7	0%
Cooperative Forestry	5	6	6	6%
1890 Colleges and Tuskegee	*	*	*	1%
Subtotal	14	15	15	2%
Engineering and Related R&D				
Special Grants	16	13	11	–16%
Competitive Grants	5	5	5	3%
Hatch/Agricultural Experiment Stations	10	10	10	0%
Cooperative Forestry	0	0	0	6%
1890 Colleges and Tuskegee	*	*	*	1%
Animal Health	*	*	*	0%
Subtotal	31	29	27	–7%
Total, CSREES Environmental R&D	105	102	99	–3%

SOURCE: Author's estimates.

*Less than $500,000.

[a]CSRS budget authority has been projected based on 1993 data in the Current Research Information System for goals pertaining to the environment and for environmental sciences research supporting other goals.

R&D (NSB 1993, table 4-26). AAAS's federal budget data indicate that in the post-rescission FY 1995 budget, environment and natural resources constitutes 6.2 percent of nondefense R&D.

To gain a sense of the scale of environmental research compared to other areas of research in various agencies, one can compare the figures for environ-

TABLE 12 Estimate of Environmental R&D at the Department of Agriculture, Economic Research Service (budget authority in millions of dollars, by fiscal year)[a]

	Actual FY 1993	Est. FY 1994	Est. FY 1995	Avg. Annual % Change FY 1993-95
Social Sciences				
Natural Resource Management	10	9	9	–5%
Total, ERS Environmental R&D	10	9	9	–5%

SOURCE: Author's estimates.

[a]ERS budget authority has been projected based on 1993 data in the Current Research Information System for goals pertaining to the environment and for environmental sciences research supporting other goals.

TABLE 13 Estimate of Environmental R&D at the Department of Agriculture, Forest Service (budget authority in millions of dollars, by fiscal year)[a]

	Actual FY 1993	Est. FY 1994	Est. FY 1995	Avg. Annual % Change FY 1993-95
Environmental Sciences				
All Forest Service Env. Sciences	75	81	82	
Subtotal	75	81	82	5%
Social Sciences				
Economics of Timber Production	3	3	3	
Subtotal	3	3	3	5%
Engineering and Related R&D				
Protection from Pollution	8	9	9	
Alleviation of Pollution	3	4	4	
Watershed Protection and Management	14	15	16	
Subtotal	26	28	28	5%
Total, FS Environmental R&D	104	111	114	5%

SOURCE: Author's estimates.

[a]FS budget authority has been projected based on 1993 data in the Current Research Information System for goals pertaining to the environment and for environmental sciences research supporting other goals.

TABLE 14 Estimate of Environmental R&D at the Environmental Protection Agency (budget authority in millions of dollars, by fiscal year)

	Actual FY 1993	Est. FY 1994	Est. FY 1995[a]	Avg. Annual % Change FY 1993-95
By Media				
Multimedia	164	265	329	43%
Air Quality	87	86	79	–5%
Acid Deposition	16	10	2	–58%
Global Change	24	31	23	0%
Water Quality	30	27	23	–13%
Drinking Water	19	20	22	7%
Hazardous Waste	38	31	27	–17%
Pesticides	14	13	14	–2%
Toxic Substances	27	22	18	–17%
Management	6	6	7	11%
Buildings and Facilities	3	5	0	–5%
Rescission from FY 1995:	–15			
Trust Funds				
Superfund	65	70	67	2%
Leaking Underground Storage Tanks	0	0	0	2%
Oil Spill Response	2	2	2	–7%
Total EPA Environmental R&D	497	588	600	10%

SOURCE: Author's estimates based on agency budget justification, R&D cross-cuts, and other supporting documents. Includes R&D management support, environmental health R&D, and other environmental R&D.

[a]Adjusted to reflect rescissions enacted in Public Law 104-19.

mental R&D by agency in Table 4 to the overall R&D budgets of these agencies. These overall R&D budgets are shown in a table drawn from the most recent AAAS R&D Report and included in Appendix C. As might be expected, these figures vary widely. At one extreme is EPA, where 100 percent of the agency's R&D is (not surprisingly) classified as environmental. At the other is NIH, where the environmental research in the relatively small NIEHS comprises only 2.5 percent of NIH's total R&D. Others at the high end include the Department of the Interior (83 percent), NOAA (76 percent), and the Corps of Engineers (66 percent), while those at the low end include NASA (14 percent), DOE (9 percent), and NSF and USDA (each at 24 percent). Although FY 1995 data for DOD are not available, FY 1992 figures indicate that environmental research makes up something like 1 percent of DOD's R&D portfolio.

TABLE 15 Estimate of Environmental R&D at the National Oceanic and Atmospheric Administration (budget authority in millions of dollars, by fiscal year)[a]

	Actual FY 1993	Actual FY 1994	Est.[b] FY 1995	Avg. Annual % Change FY 1993-95
Environmental Sciences				
OAR, Climate and Global Change	43	64	57	19%
OAR, Weather Research	27	42	41	28%
OAR, Long-Term Climate and Air Quality	22	27	33	22%
OAR, Interannual/Seasonal	7	7	7	-3%
OAR, Marine Prediction	18	18	18	1%
OAR, Sea Grant / Env. Sciences	12	19	25	43%
OAR, Undersea Research	8	21	16	69%
NWS, National Weather Service	28	48	33	19%
NOS, Observation and Prediction	2	0	0	–26%
NOS, Estuarine and Coastal Assessment	0	0	0	—
NOS, Coastal Ocean Science	13	12	9	–14%
NOS, Coastal Management	0	2	2	72%
NOS, Ocean Management	0	0	0	9%
NMFS, Fisheries Resource Information	90	92	102	6%
NMFS, Fishery Information Analysis	21	22	26	11%
NMFS, Protected Species Management	11	11	11	-0%
NMFS, Marine Fisheries Grants to States	6	6	9	27%
NMFS, Fisheries Development	16	17	17	3%
Subtotal	326	409	408	13%
Social Sciences				
OAR, Sea Grant	1	1	1	0%
NMFS, Fisheries Industry Information	18	19	22	12%
Subtotal	19	20	23	11%
Engineering and Related R&D				
OAR, Sea Grant	2	2	2	0%
Information and Data R&D				
NOS, Geodesy	2	0	0	–32%
NOS, Mapping and Charting	0	4	4	435%
National Env. Data and Information Service	8	8	8	4%
Subtotal	10	12	13	12%
Total, NOAA Environmental R&D	357	443	447	13%

SOURCE: Author's estimate based on agency R&D cross-cut and budget justification.

[a]Agencies within NOAA include Oceanic and Atmospheric Research (OAR), National Weather Service (NWS), National Ocean Service (NOS), National Marine Fisheries Service (NMFS), and the National Environmental Data and Information Service.

[b]Adjusted to reflect rescissions enacted in Public Law 104-19.

TABLE 16 Estimate of Environmental R&D at the Corps of Engineers (budget authority in millions of dollars, by fiscal year)

	Actual FY 1993	Actual FY 1994	Est. FY 1995	Avg. Annual % Change FY 1993-95
Environmental Sciences				
Long-Term Env. Effects of Dredging	1	1	1	8%
Social Sciences				
Economic Impact of Global Warming	0	0	0	–42%
Risk Analysis	1	1	2	10%
Subtotal	2	1	2	2%
Engineering and Related R&D				
Coastal Engineering	6	4	6	9%
Flood Control and Navigation	4	4	5	19%
Environmental Quality	3	2	4	33%
Water Resource Planning	0	0	0	–2%
Aquatic Plant Control	3	6	7	47%
Wetlands Research	7	5	0	–62%
Coastal Inlet Research	0	3	4	—
River Confluence Ice Research	0	0	1	—
Zebra Mussel Control	1	2	2	42%
Oil Spill Research	0	0	0	—
Subtotal	24	28	31	13%
Information and Data R&D				
Surveying and Remote Sensing	2	1	2	24%
Total, Corps of Engineers Env. R&D	29	31	36	12%

SOURCE: Author's estimate based on agency budget justification and related data.

ENVIRONMENTAL FUNDING AND NATIONAL ENVIRONMENTAL GOALS: CONCLUSIONS AND OUTLOOK

A quick look at Tables 3 and 4 demonstrates the pitfalls of judging the priority of environmental programs simply by looking at budget numbers. In strict dollar terms, the leading agency is the Department of Energy at $7.1 billion (out of total of $22.7 billion), the largest share of which is for defense environmental restoration. In second place is EPA ($5.7 billion) whose budget, as noted above, is dominated by Water Infrastructure Financing and Superfund—i.e., sewage treatment grants and hazardous waste site cleanup. Following are Agriculture, Interior, NOAA, and NASA. The picture is a bit different in environmental R&D

TABLE 17 Estimate of Environmental R&D at the National Institutes of Health (budget authority in millions of dollars, by fiscal year)

	Actual FY 1993	Est. FY 1994	Est. FY 1995	Avg. Annual % Change FY 1993-95
Environmental Health Sciences				
National Institute of Environmental Health Sciences	251	259	267	3%
Subtotal	251	259	267	3%
Total NIH Environmental R&D	251	259	267	3%

SOURCE: Author's estimate based on agency budget justification.

(Table 4), where NASA dominates, spending more than a quarter of the federal government's resources for environmental R&D.

These are the areas in which the federal government is investing the largest shares of the money devoted to environmental programs. They are certainly important to the future of the nation and its environment. Few would argue, however, that these are the nation's top environmental priorities or that this distribution corresponds in any meaningful way to virtually anyone's statement of environmental goals.

To get out of this trap, one first of all needs much more fine-grained data on environmental programs so as to allow one to associate the programs with particular goals. Even more importantly, one needs to look beyond the raw numbers at such issues as the relative costs of achieving different goals, the role of the federal government relative to other actors for various goals, and the availability of technical means to achieve the goals. The issue then becomes not simply the level of spending for each of the various programs, but the *sufficiency* of the budgetary level relative to the estimated cost of achieving the goal and federal role in its achievement. Such an analysis, unfortunately, is well beyond the means available for this paper.

A glimpse at how one might begin such an analysis, however, might be gained from a look at the author's 1992 study (Gramp, Teich, and Nelson 1992). That study took a preliminary look at federal environmental R&D priorities by breaking down and reaggregating the data by scientific and engineering focus. Of the $4.5 billion in federal environmental R&D identified in that study, the bulk ($3.1 billion) was estimated to support R&D in the environmental sciences. About 70 percent of that amount ($2.2 billion) involved fields such as oceanography, geology, chemistry, and atmospheric sciences, while the balance ($0.9 billion) went to environmental life sciences, including environmental biology, forestry, biology, and marine biology.

Engineering and other R&D related to environmental impacts of anthropogenic and natural activities accounted for $1.2 billion; a large share of this went to mitigation efforts related to fossil-fuel consumption. The smallest component of environmental R&D identified in the 1992 study was social sciences, which came in at $41 million. Also modest in financial terms was the total for information and data systems, some $0.2 billion, much of it related to space-borne observing platforms and global change studies.

These data in themselves do not answer questions about goals and priorities, but they do point to a means whereby financial data can be related to arrays of environmental goals, such as those identified by the National Science and Technology Council's Committee on Environment and Natural Resources.

REFERENCES

American Association for the Advancement of Science, Intersociety Working Group, *AAAS Report XX: Research and Development, FY 1996* (Washington, D.C.: 1995).

Kathleen M. Gramp, Albert H. Teich, and Stephen D. Nelson, *Federal Funding for Environmental R&D: A Special Report* (Washington, D.C.: American Association for the Advancement of Science, 1992).

National Science Board, Science and Engineering Indicators—1993. (Washington, D.C.: U.S. Government Printing Office, 1993). (NSB-93-1).

APPENDIX A

From: Kathleen M. Gramp, Albert H. Teich, and Stephen D. Nelson, *Federal Funding for Environmental R&D: A Special Report* (Washington, D.C.: American Association for the Advancement of Science, 1992), pp. 20-26 and 55-68.

Overview Tables

Table I-1. Estimate of Federal Funding For Environmental R&D by Agency[1]
(budget authority in millions of dollars, by fiscal year)

	FY 1990	FY 1991	FY 1992	Avg. Annual % Change FY 1990-92
Agency for International Development	43	38	45	2%
USDA, Agricultural Research Service	129	144	162	12%
USDA, Cooperative State Research Service	87	107	119	17%
USDA, Economic Research Service	6	6	7	14%
USDA, Forest Service	89	106	115	13%
Corps of Engineers	19	22	27	20%
Department of Defense (military)	445	599	577	14%
Department of Energy	882	709	799	-5%
Environmental Protection Agency	307	307	347	6%
DOI, Fish and Wildlife Service	71	82	85	9%
DOI, Geological Survey	310	360	367	9%
DOI, Other Agencies	42	59	72	30%
National Aeronautics and Space Admin.	575	735	826	20%
National Oceanic and Atmospheric Admin.	250	290	319	13%
National Science Foundation	440	492	541	11%
Smithsonian Institution	27	31	33	11%
Tennessee Valley Authority	19	25	31	27%
Department of Transportation	9	10	17	40%
Total, Environmental R&D	3,748	4,121	4,489	9%
Defense[2]	632	768	801	13%
Nondefense	3,116	3,353	3,687	9%

Source: Authors' estimates.

[1]Excludes administrative overhead and R&D related to environmental health.

[2]Defense includes Department of Defense and the Department of Energy's Atomic Energy Defense Programs.

Table I-2. Environmental and Other Nondefense R&D by Character of Work, FY 1992
(budget authority in billions of dollars)

	Nondefense Env. R&D[1]	All Other Nondefense	Total Nondefense
Basic Research	1.4	10.7	12.1
Applied Research	1.3	7.6	8.9
Subtotal, Research	2.7	18.3	21.0
Development	0.9	6.4	7.3
R&D Facilities	0.1	2.2	2.2
Total, Nondefense R&D	3.7	26.9	30.6

Source: Authors' estimates based on *AAAS Report XXVII, Research and Development FY 1993* (AAAS, March 1992).

[1]Excludes administrative overhead and R&D related to environmental health.

Table I-3. Estimate of Environmental R&D by Agency Mission[1] (budget authority in millions of dollars, by fiscal year)

	FY 1990	FY 1991	FY 1992	Avg. Annual % Change FY 1990-92
RESEARCH AGENCIES				
National Aeronautics and Space Admin.	575	735	826	20%
National Science Foundation	440	492	541	11%
DOI, Geological Survey	310	360	367	9%
National Oceanic and Atmospheric Admin.[2]	136	166	184	16%
Smithsonian Institution	27	31	33	11%
Subtotal	1,487	1,784	1,951	15%
SECTOR-SPECIFIC AGENCIES				
Department of Energy	882	709	799	-5%
Department of Defense (military)	445	599	577	14%
USDA, Agricultural Research Service	129	144	162	12%
USDA, Cooperative State Research Service	87	107	119	17%
USDA, Economic Research Service	6	6	7	14%
Agency for International Development	43	38	45	2%
Tennessee Valley Authority	19	25	31	27%
Bureau of Mines	15	19	18	10%
Department of Transportation	9	10	17	40%
Subtotal	1,635	1,653	1,775	4%
MANAGEMENT AGENCIES				
Environmental Protection Agency	307	307	347	6%
National Oceanic and Atmospheric Admin.[3]	113	124	135	9%
USDA, Forest Service	89	106	115	13%
DOI, Fish and Wildlife Service	71	82	85	9%
DOI, Other Agencies	27	40	54	40%
Corps of Engineers	19	22	27	20%
Subtotal	627	679	763	10%
Total, Environmental R&D	3,748	4,121	4,489	9%

Source: Authors' estimates.

[1]Excludes administrative overhead and R&D related to environmental health.

[2]Includes Oceanic and Atmospheric Research, National Weather Service, and National Environmental Data and Information Service.

[3]Includes National Marine Fisheries Service and National Ocean Service.

Table I-4. Estimate of Distribution of Environmental R&D by Field and Agency, FY 1992[1] (budget authority in millions of dollars)

	Environmental Sciences	Engineering & Related R&D	Social Sciences	Information & Data	Total
AID	31	5	9	0	45
USDA	324	61	16	3	403
Corps	1	23	1	2	27
DOD (military)	432	146	0	0	577
DOE	129	667	1	2	799
EPA	190	149	1	8	347
DOI	406	31	*	86	524
NASA	718	0	0	108	826
NOAA	305	2	1	11	319
NSF	503	26	12	0	541
Smithsonian	33	0	*	0	33
TVA	0	31	0	0	31
DOT	0	17	0	0	17
Total, Environ. R&D	3,072	1,156	41	219	4,489
Defense[2]	432	370	0	0	801
Nondefense	2,641	786	41	219	3,687
Estimated Research[3]	2,533	388	39	147	3,107

Source: Authors' estimates.

* Less than $500,000.

[1] Excludes administrative overhead and R&D related to environmental health.

[2] Defense includes Department of Defense and the Department of Energy's Atomic Energy Defense Programs.

[3] Basic and applied research.

Table I-5. Estimate of Federal Funding for R&D in Physical Environmental Sciences (budget authority in millions of dollars, by fiscal year)

	FY 1990	FY 1991	FY 1992	Avg. Annual % Change FY 1990-92
Agency for International Development	5	3	4	-15%
USDA, Agricultural Research Service	13	15	17	16%
USDA, Forest Service	10	10	10	2%
Department of Defense (military)	411	448	429	2%
Department of Energy	90	105	117	14%
Environmental Protection Agency	81	74	105	14%
DOI, Geological Survey	239	273	281	8%
DOI, Bureau of Reclamation	0	2	3	--
National Aeronautics and Space Admin.	507	603	632	12%
National Oceanic and Atmospheric Admin.	117	146	163	18%
National Science Foundation	320	362	398	11%
Smithsonian Institution	0	1	1	--
Total, Physical Environmental Sciences	1,793	2,044	2,160	10%

Source: Authors' estimates.

Table I-6. Estimate of Federal Funding for Environmental Life Sciences R&D[1] (budget authority in millions of dollars, by fiscal year)

	FY 1990	FY 1991	FY 1992	Avg. Annual % Change FY 1990-92
Agency for International Development	28	26	27	-2%
USDA, Agricultural Research Service	87	95	104	9%
USDA, Cooperative State Research Service	71	90	99	18%
USDA, Forest Service	72	87	93	14%
Corps of Engineers	1	1	1	0%
Department of Defense (military)	2	2	3	20%
Department of Energy	14	12	12	-8%
Environmental Protection Agency	69	74	84	10%
DOI, Fish and Wildlife Service	71	82	85	9%
DOI, National Park Service	18	23	27	22%
DOI, Bureau of Land Management	3	5	11	89%
National Aeronautics and Space Admin.	60	88	86	20%
National Oceanic and Atmospheric Admin.	121	131	142	9%
National Science Foundation	90	98	106	8%
Smithsonian Institution	27	30	32	10%
Total, Environmental Life Sciences	734	844	912	11%

Source: Author's estimates.

[1]Includes all life sciences in environmental programs, including biology, environmental biology, marine biology, forestry, etc.

Table I-7. Estimate of Federal Funding for Engineering and Other R&D Related to Environmental Impacts (budget authority in millions of dollars, by fiscal year)

	FY 1990	FY 1991	FY 1992	Avg. Annual % Change FY 1990-92
Agency for International Development	2	2	5	56%
USDA, Agricultural Research Service	29	33	38	15%
USDA, Cooperative State Research Service	13	15	17	11%
USDA, Forest Service	6	6	6	2%
Corps of Engineers	17	19	23	17%
Department of Defense (military)	32	149	146	114%
Department of Energy	774	590	667	-7%
Environmental Protection Agency	157	154	149	-2%
DOI, Other Agencies	22	28	31	20%
National Oceanic and Atmospheric Admin.	2	2	2	3%
National Science Foundation	24	24	26	3%
Tennessee Valley Authority	19	25	31	27%
Department of Transportation	9	10	17	40%
Total, Engineering and Related R&D	1,103	1,055	1,156	2%

Source: Authors' estimates.

Table I-8. Estimate of Federal Funding for Social Sciences R&D Related to the Environment (budget authority in millions of dollars, by fiscal year)

	FY 1990	FY 1991	FY 1992	Avg. Annual % Change FY 1990-92
Agency for International Development	7.0	7.1	8.7	11%
USDA, Cooperative State Research Service	2.7	2.9	3.2	8%
USDA, Economic Research Service	5.7	6.1	7.4	14%
USDA, Forest Service	1.9	2.0	5.1	62%
Corps of Engineers	0.0	0.0	1.4	--
Department of Energy	2.0	0.0	0.9	-33%
Environmental Protection Agency	0.0	0.5	0.5	--
DOI, Other Agencies	0.0	0.1	0.2	--
National Oceanic and Atmospheric Admin.	0.9	0.9	1.5	26%
National Science Foundation	4.8	8.0	12.2	60%
Smithsonian Institution	0.0	0.1	0.3	--
Total, Social Sciences R&D	25.1	27.8	41.4	28%

Source: Authors' estimates.

Table I-9. Estimate of Federal Funding for Information and Data R&D Related to the Environment (budget authority in millions of dollars, by fiscal year)

	FY 1990	FY 1991	FY 1992	Avg. Annual % Change FY 1990-92
USDA, Agricultural Research Service	1	1	2	46%
USDA, Forest Service	0	1	1	--
Corps of Engineers	1	2	2	22%
Department of Energy	2	2	2	0%
Environmental Protection Agency	0	4	8	--
DOI, Geological Survey	71	86	86	11%
National Aeronautics and Space Admin.	8	44	108	267%
National Oceanic and Atmospheric Admin.	10	11	11	5%
Total, Information and Data R&D	92	150	219	54%

Source: Authors' estimates.

Table I-10. Estimate of Environmental R&D by Agency and Character of Work, FY 1992[1] (budget authority in millions of dollars)

	Basic	Applied	Development	Facilities	Total
AID	1	39	4	0	45
USDA, ARS	82	60	10	10	162
USDA, CSRS	55	64	0	0	119
USDA, ERS	1	6	0	0	7
USDA, FS	44	68	3	0	115
Corps	3	10	14	0	27
DOD (military)	167	171	240	0	577
DOE	133	118	513	34	799
EPA	73	196	79	0	347
DOI, FWS	6	66	13	0	85
DOI, USGS	180	165	18	4	367
DOI, Other	7	58	5	1	72
NASA	224	244	309	49	826
NOAA	0	286	33	0	319
NSF	516	18	0	8	541
Smithsonian	33	0	0	0	33
TVA	3	5	23	0	31
DOT	0	6	11	0	17
Total, Environmental R&D	1,528	1,580	1,275	106	4,489
Defense[2]	167	256	344	34	801
Nondefense	1,360	1,324	931	72	3,687

Source: Authors' estimates.

[1]Excludes administrative overhead and R&D related to environmental health.

[2]Defense includes Department of Defense and the Department of Energy's Atomic Energy Defense Programs.

Table I-11. Estimate of Environmental R&D by Agency and Performer, FY 1992[1]
(budget authority in millions of dollars)

	Federal Agencies	Colleges & Universities	Other Extramural	Total
Agency for International Development	2	12	30	45
USDA, Agricultural Research Service	160	0	2	162
USDA, Cooperative State Research Service	0	115	4	119
USDA, Economic Research Service	7	*	0	7
USDA, Forest Service	104	9	2	115
Corps of Engineers	13	2	11	27
Department of Defense (military)	173	130	275	577
Department of Energy	26	60	713	799
Environmental Protection Agency	84	53	211	347
DOI, Fish and Wildlife Service	77	8	0	85
DOI, Geological Survey	325	26	16	367
DOI, Other Agencies	50	12	10	72
National Aeronautics and Space Admin.	150	81	595	826
National Oceanic and Atmospheric Admin.	250	55	15	319
National Science Foundation	4	405	132	541
Smithsonian Institution	33	0	0	33
Tennessee Valley Authority	17	*	14	31
Department of Transportation	7	2	8	17
Total, Environmental R&D	1,481	971	2,037	4,489
Defense[2]	189	130	482	801
Nondefense	1,292	841	1,554	3,687

Source: Authors' estimates derived by applying agency trends to environmental programs. These figures represent general trends, and should not be viewed as exeact amounts.

*Less than $500,000.

[1]Excludes administrative overhead and R&D related to environmental health.

[2]Defense includes Department of Defense and the Department of Energy's Atomic Energy Defense Programs.

Table I-12. Estimate of Federal Grant Funding for Environmental R&D, FY 1992[1]
(budget authority in millions of dollars)

	Estimated Grants
Agency for International Development	38
USDA, Cooperative State Research Service	117
Department of Defense (military)	100
Department of Energy	57
Environmental Protection Agency	89
DOI, Geological Survey	26
National Aeronautics and Space Admin.	67
National Oceanic and Atmospheric Admin.	90
National Science Foundation	497
All Other	0
Total, Environmental R&D Grant Funding	1,080

Source: Authors' estimates derived from agency object classification reports, except for DOD and NASA, which are projected based on agencies' grant management data.

[1]Excludes administrative overhead and R&D related to environmental health.

Agency Tables

Table II-1. Estimate of Environmental R&D at the National Aeronautics and Space Administration[1] (budget authority in millions of dollars, by fiscal year)

	FY 1990	FY 1991	FY 1992	Avg. Annual % Change FY 1990-92
ENVIRONMENTAL SCIENCES				
Earth Observing System (EOS)	74	151	184	58%
Earth Probes	14	52	93	162%
Topex	85	80	60	-16%
Upper Atmosphere Research Satellite	55	62	0	-100%
Payload Instrument Development	76	49	40	-28%
Mission Operations & Data Analysis	24	31	59	57%
Airborne Research	19	20	20	2%
Process Studies	114	116	124	4%
Modeling & Data Analysis	39	44	49	13%
Interdisiplinary Analysis	9	12	3	-46%
Construction of Facilities, Global Change	0	9	17	--
Life Sciences, Research & Analysis	3	4	5	23%
Estimated Intramural Research	56	60	65	8%
Subtotal	567	691	718	13%
INFORMATION AND DATA R&D				
EOS Data and Information Systems	0	36	83	--
CIESIN[2]	8	8	25	77%
Subtotal	8	44	108	267%
Total, NASA Environmental R&D	575	735	826	20%

Source: Authors' estimates based on agency budget justification and cross-cut by field of science.

[1]These estimates exclude funding for environmental sciences in NASA's planetary and commercial applications programs, which are estimated to total $305 million in FY 1990, $328 million in FY 1991, and $322 million in FY 1992.

[2]Consortium for International Earth Science Information Networks.

Table II-2. Estimate of Environmental R&D at the Department of Energy
(budget authority in millions of dollars, by fiscal year)

	FY 1990	FY 1991	FY 1992	Avg. Annual % Change FY 1990-92
ENVIRONMENTAL SCIENCES				
CO_2, Core Program	15	18	17	5%
CO_2, Quantitative Links (ARM)	16	22	26	28%
CO_2, CHAMMP[1]	6	7	10	33%
CO_2, Nat'l Inst. for Global Env. Change	6	9	11	37%
CO_2, Oceans Research	2	4	5	47%
CO_2, Education	0	2	2	--
Terrestrial Transport	14	15	14	1%
Atmospheric Science	12	10	10	-7%
Ecosystem Functioning and Response	8	7	7	-10%
Coastal, Marine Transport	6	5	6	-3%
Basic Energy Sciences, Geosciences	18	17	19	5%
Fish and Wildlife, Bonneville Power[2]	1	1	1	0%
Subtotal	104	118	129	11%
SOCIAL SCIENCES				
CO_2, Human Interactions	2	0	1	-33%
ENGINEERING AND RELATED R&D				
Atomic Defense Env. Waste and Restor.	187	169	224	9%
Clean Coal Technologies	525	361	390	-14%
Coal Flue Gas Cleanup	17	17	18	2%
Coal Gas Stream Cleanup	17	17	17	1%
Coal Preparation Control Technologies	20	16	12	-22%
Coal Advanced R&D Control Technologies	2	3	2	3%
Coal Waste Management	2	2	2	-4%
Oil & Gas Environmental Research	2	1	2	-3%
Oil Shale Environmental Mitigation	2	2	0	-100%
Subtotal	774	590	667	-7%
INFORMATION AND DATA R&D				
CO_2, Information	2	2	2	0%
Total, DOE Environmental R&D	882	709	799	-5%

Source: Authors' estimates based on agency budget justification.

[1]Computer Hardware, Advanced Math and Model Physics program.

[2]Represents obligational authority financed by Bonneville Power Administration ratepayers.

Table II-3. Estimate of Environmental R&D at the Department of Defense
(budget authority in millions of dollars, by fiscal year)[1]

	FY 1990	FY 1991	FY 1992	Avg. Annual % Change FY 1990-92
ENVIRONMENTAL SCIENCES				
Navy	137	152	149	4%
Air Force	67	79	72	4%
Army	40	37	38	-3%
Services, Global Change	0	0	6	--
Strategic Defense Initiative, Env. Sciences	120	108	112	-3%
Defense Advanced Research Projects Agency	29	28	29	2%
Defense Nuclear Agency	10	9	12	8%
Office of the Secretary of Defense	11	37	13	10%
Subtotal	413	450	432	2%
ENGINEERING AND RELATED R&D				
Strategic Environmental Research Program	0	100	10	--
START Treaty: Rocket Motor Demilitarization	0	3	27	--
Defense Agencies, Special Env. Project	0	0	20	--
Navy, Environmental Quality	0	0	26	--
Installation Restoration	10	18	23	52%
Noise Abatement	4	4	4	-1%
Pollution Prevention	10	14	19	41%
Terrestrial & Aquatic Assessment	4	5	7	39%
Global Marine Compliance	1	1	1	20%
Atmospheric Compliance	1	1	2	18%
Base Support Operations	2	3	6	63%
Subtotal	32	149	146	114%
Total, DOD Environmental R&D	445	599	577	14%

Source: Authors' estimates based on agency cross-cuts for environmental R&D.

[1]Most of DOD's programatic estimates are given in terms of total obligational authority, which does not always correspond precisely with budget authority.

Table II-4. Estimate of Environmental R&D at the National Science Foundation [1]
(budget authority in millions of dollars, by fiscal year)

	FY 1990	FY 1991	FY 1992	Avg. Annual % Change FY 1990-92
ENVIRONMENTAL SCIENCES				
BIO, Ecological Studies	23	25	26	5%
BIO, Systematic & Population Biology	22	24	25	6%
BIO, Long-term Environmental Biology	20	21	24	9%
BIO, Science & Technology Center	2	2	2	6%
BIO, Marine Laboratories	2	2	2	-13%
GEO, Ocean Sciences Research Support	73	82	91	12%
GEO, Oceanographic Centers and Facilities	42	48	52	10%
GEO, Ocean Drilling Program	32	35	36	7%
GEO, Atmospheric Sciences Project Support	53	61	69	14%
GEO, Nat'l Center for Atmospheric Research	45	49	51	6%
GEO, Upper Atmospheric Facilities	6	6	7	9%
GEO, Earth Sciences Project Support	41	46	50	10%
GEO, Instrumentation and Facilities	14	18	20	20%
GEO, Continental Dynamics	6	6	7	10%
GEO, Arctic Research Projects	12	14	20	32%
GEO, Arctic Research Commission	0	1	1	4%
GEO, Science & Technology Centers	1	1	1	-3%
Antarctic Research	17	19	22	14%
Subtotal	411	460	503	11%
SOCIAL SCIENCES				
SBE, Global Change	1	4	7	138%
SBE, Other Environment Related R&D	2	1	*	-48%
GEO, Arctic Research Projects	*	1	1	83%
Education & Human Resources Research	2	3	4	50%
Subtotal	5	8	12	60%
ENGINEERING AND RELATED R&D				
ENG, Natural and Man-made Hazard Mitigation	3	3	3	3%
ENG, Environmental and Ocean Systems	5	6	6	8%
ENG, Earthquake & Ocean Systems	16	15	17	1%
Subtotal	24	24	26	3%
Total, NSF Environmental R&D	440	492	541	11%

Source: Authors' estimates based on agency budget data and cross-cut by field of science.

*Less than $500,000.

[1] NSF Directorates include Biological Sciences (BIO), Geosciences (GEO), Social and Behavioral Sciences (SBE) and Engineering (ENG).

Table II-5. Estimate of Environmental R&D at the Department of Interior, U.S. Geological Survey (budget authority in millions of dollars, by fiscal year)

	FY 1990	FY 1991	FY 1992	Avg. Annual % Change FY 1990-92
ENVIRONMENTAL SCIENCES				
Coastal and Wetlands Processes	7	10	9	12%
Climate Change/Global Change	3	10	11	92%
Landslide Hazards	2	2	2	5%
Volcano Hazards	18	16	15	-7%
Earthquake Hazards Reduction	46	50	50	4%
Nat'l Water Quality Assessment Program	6	16	26	102%
Core Hydrologic Research	8	10	10	10%
Toxic Substances Hydrology	13	13	13	3%
Regional Aquifer System Analysis	10	10	8	-11%
Climate Change Hydrology	2	7	8	99%
Nuclear Waste Hydrology	3	4	3	-2%
Acid Rain	1	1	1	6%
Water Research, Improved Instrumentation	1	1	1	3%
Water Resources Assessment	1	1	1	11%
Tuckee-Carson Water Resource Program	0	0	*	--
Federal-State Water Pgm., Coal Hydrology	2	0	0	-100%
Federal-State Water Program, Water Use	*	*	*	2%
Water Research Grants	4	4	2	-36%
Water Research, State Institutes	6	6	6	0%
Mineral Resource Surveys	47	49	50	4%
Offshore Geologic Framework	26	27	28	3%
Energy Geologic Surveys	27	30	30	5%
Deep Continental Studies	3	3	3	4%
Geomagnetism	2	2	2	3%
Subtotal	239	273	281	8%
INFORMATION AND DATA R&D				
National Geologic Mapping	19	21	22	7%
Cartographic & Geographic Researc	6	6	6	-2%
Advanced Cartographic Systems	10	15	14	17%
Geographic & Spacial Information Analysis	6	6	6	-5%
Global Change Data Systems	1	7	7	174%
Nat'l Cartographic Requirements/Standards	*	*	*	9%
Water Data Collection & Analysis	7	7	7	-2%
Federal-State Water Pgm., Data & Analysis	20	25	24	9%
Subtotal	71	86	86	11%
Total, USGS Environmental R&D	310	360	367	9%

Source: Authors' estimates based on agency R&D cross-cut.

*Less than $500,000.

Table II-6. Estimate of Environmental R&D at the Department of Interior, Fish and Wildlife Service (budget authority in millions of dollars, by fiscal year)

	FY 1990	FY 1991	FY 1992	Avg. Annual % Change FY 1990-92
ENVIRONMENTAL SCIENCES				
Contaminants Research	13	14	14	1%
Wildlife Research	18	19	18	1%
Fishery Research	16	19	19	9%
Endangered Species Research	6	8	8	13%
Technical Development	7	11	14	44%
Cooperative Units	7	8	8	8%
Research Center Maintenance	4	4	5	3%
Total, FWS Environmental R&D	71	82	85	9%

Source: Authors' estimates based on agency budget justification.

Table II-7. Estimate of Environmental R&D at the Department of Interior, Other Agencies (budget authority in millions of dollars, by fiscal year)

	FY 1990	FY 1991	FY 1992	Avg. Annual % Change FY 1990-92
ENVIRONMENTAL SCIENCES				
National Park Service	18	23	27	22%
Bureau of Land Management	3	5	11	89%
Bureau of Reclamation, Global Change	0	2	3	--
Subtotal	21	31	40	39%
SOCIAL SCIENCES				
Office of the Secretary	*	*	*	--
National Park Service[1]	NA	NA	NA	--
Subtotal	*	*	*	--
ENGINEERING AND RELATED R&D				
Bureau of Mines	15	19	18	10%
Minerals Management Service	2	3	6	73%
Bureau of Reclamation	3	4	5	32%
Surface Mining Reclamation & Enforcement	1	1	1	18%
Office of the Secretary	1	1	1	13%
Subtotal	22	28	31	20%
Total, DOI, Other Agencies' Env. R&D	42	59	72	30%

Source: Authors' estimates based on agency budget justifications and R&D cross-cuts.

*Less than $500,000.

[1]The National Park Service conducts social sciences R&D related to environmental management, but data were not available on the amounts involved.

Table II-8. Estimate of Environmental R&D at the Department of Agriculture, Agricultural Research Service (budget authority in millions of dollars, by fiscal year)[1]

	FY 1990	FY 1991	FY 1992	Avg. Annual % Change FY 1990-92
ENVIRONMENTAL SCIENCES				
Soil, Plant, Water, Nutrient Relation.	24	26	30	11%
Improvement of Range Resources	6	6	6	5%
Wildlife & Fish Ecology	4	6	6	20%
Saline and Sodic Soils and Salinity Mgt.	2	2	3	9%
Appraisal of Soil Resources	1	1	1	5%
Alternative Uses of Land	*	*	*	3%
Env. Biology supporting other goals	49	54	57	8%
Physical Env. Sciences for other goals	13	15	17	16%
Subtotal	100	110	121	10%
ENGINEERING AND RELATED R&D				
Alleviation of Pollution	9	12	13	19%
Watershed Protection and Management	8	8	9	9%
Conservation & Efficient Use of Water	7	8	9	16%
Efficient Drainage & Irrigation Systems	3	4	4	21%
Protection from Pollution	2	2	2	14%
Subtotal	29	33	38	15%
INFORMATION AND DATA R&D				
Remote Sensing	1	1	1	10%
CIESIN/Global Change [2]	0	0	1	--
Subtotal	1	1	2	46%
Total, ARS Environmental R&D	129	144	162	12%

Source: Authors' estimates.

*Less than $500,000.

[1]ARS budget authority has been projected based on 1990 data in the Current Research Information System for goals pertaining to the environment and for environmental sciences research supporting other goals.

[2]Consortium for International Earth Science Information Networks.

Table II-9. Estimate of Environmental R&D at the Department of Agriculture, Cooperative State Research Service (budget authority in millions of dollars, by fiscal year)[1]

	FY 1990	FY 1991	FY 1992	Avg. Annual % Change FY 1990-92
ENVIRONMENTAL SCIENCES				
Hatch/Agricultural Experiment Stations	35	37	38	4%
Cooperative Forestry	10	11	11	3%
1890 Colleges & Tuskegee	8	9	9	5%
Special Grants	15	17	20	15%
Competitive Grants	2	17	21	193%
Animal Health	*	*	*	1%
Subtotal	71	90	99	18%
SOCIAL SCIENCES				
Hatch/Agricultural Experiment Stations	1	1	1	4%
Cooperative Forestry	1	1	1	3%
1890 Colleges & Tuskegee	*	*	*	5%
Special Grants	1	1	1	15%
Subtotal	3	3	3	8%
ENGINEERING AND RELATED R&D				
Hatch/Agricultural Experiment Stations	6	7	7	4%
Cooperative Forestry	*	*	*	3%
1890 Colleges & Tuskegee	1	1	1	5%
Special Grants	6	6	8	15%
Competitive Grants	1	1	1	51%
Animal Health	*	*	*	1%
Subtotal	13	15	17	11%
Total, CSRS Environmental R&D	87	107	119	17%

Source: Authors' estimates.

*Less than $500,000.

[1]CSRS budget authority has been projected based on 1990 data in the Current Research Information System for goals pertaining to the environment and for environmental sciences research supporting other goals.

Table II-10. Estimate of Environmental R&D at the Department of Agriculture, Economic Research Service (budget authority in millions of dollars, by fiscal year)[1]

	FY 1990	FY 1991	FY 1992	Avg. Annual % Change FY 1990-92
SOCIAL SCIENCES				
Natural Resource Management	6	6	7	8%
Global Change/Economics	0	0	1	--
Total, ERS Environmental R&D	6	6	7	14%

Source: Authors' estimates.

[1]ERS budget authority has been projected based on 1990 data in the Current Research Information System for goals pertaining to the environment and for environmental sciences research supporting other goals.

Table II-11. Estimate of Environmental R&D at the Department of Agriculture, Forest Service (budget authority in millions of dollars, by fiscal year)[1]

	FY 1990	FY 1991	FY 1992	Avg. Annual % Change FY 1990-92
ENVIRONMENTAL SCIENCES				
Global Change	16	23	21	13%
New Perspectives: Ecological Management	7	13	14	42%
Threatened, Endang. and Sensitive Species	5	7	7	16%
Tropical Forestry	4	4	6	27%
Wetlands	0	*	3	--
Forest Health Monitoring	1	1	2	30%
Resource Management: Other Life Sciences	39	39	40	2%
Resource Management: Physical Env. Sci.	10	10	10	2%
Subtotal	82	97	103	12%
SOCIAL SCIENCES				
Economics of Timber Production	2	2	2	9%
Global Change/Human Interactions	0	0	3	--
Subtotal	2	2	5	62%
ENGINEERING AND RELATED R&D				
Protection from Pollution	3	3	3	2%
Alleviation of Pollution	2	2	2	2%
Watershed Protection and Management	1	1	1	2%
Subtotal	6	6	6	2%
INFORMATION AND DATA R&D				
Geographic Information System	0	1	1	--
Total, FS Environmental R&D	89	106	115	13%

Source: Authors' estimates.

*Less than $500,000.

[1]FS budget authority has been projected based on 1990 data in the Current Research Information System for goals pertaining to the environment and for environmental sciences research supporting other goals.

Table II-12. Estimate of Environmental R&D at the Environmental Protection Agency
(budget authority in millions of dollars, by fiscal year)

	FY 1990	FY 1991	FY 1992	Avg. Annual % Change FY 1990-92
ENVIRONMENTAL SCIENCES				
Environmental Processes & Effects	52	53	65	12%
Env. Sciences w/in Monitoring Systems	25	28	49	40%
Env. Monitoring & Assessment Pgm. (EMAP)	17	16	21	11%
Stratospheric Modification Program	15	22	23	25%
Exploratory Research	13	17	18	17%
Acid Deposition	23	8	8	-40%
Ecological Risk Uncertainty	3	3	3	-1%
Env. Sciences in Technical Liason	2	3	3	4%
Subtotal	150	149	190	12%
SOCIAL SCIENCES				
Socio-Economic Research	0	1	1	--
ENGINEERING AND RELATED R&D				
Environmental Engineering & Technology	97	96	94	-1%
Monitoring Systems	32	31	29	-5%
Exploratory Research	13	16	14	7%
Acid Deposition	7	2	3	-40%
Technical Liason	8	9	9	4%
Subtotal	157	154	149	-2%
INFORMATION AND DATA R&D				
High Performance Computing	0	*	4	--
EMAP/Computer Sciences	0	4	5	--
Subtotal	0	4	8	--
Total, EPA Environmental R&D[1]	307	307	347	6%

Source: Authors' estimates based on agency budget justification and cross-cut by field of science.

*Less than $500,000.

[1]Including administrative overhead and R&D related to environmental health, EPA's R&D funding totaled $424 million in FY 1990, $440 million in FY 1991, and $502 million in FY 1992.

Table II-13. Estimate of Environmental R&D at the National Oceanic and Atmospheric Administration[1] (budget authority in millions of dollars, by fiscal year)

	FY 1990	FY 1991	FY 1992	Avg. Annual % Change FY 1990-92
ENVIRONMENTAL SCIENCES				
OAR, Climate & Global Change	17	44	44	63%
OAR, Weather Research	28	29	28	1%
OAR, Long-term Climate & Air Quality	17	18	18	3%
OAR, Acid Rain/Oxidants/Ozone	4	4	4	0%
OAR, Interannual/Seasonal	7	7	7	0%
OAR, National Climate Program	2	2	0	-100%
OAR, Marine Prediction/Great Lakes	16	18	21	13%
OAR, Sea Grant/Environmental Sciences	12	12	12	1%
OAR, Undersea Research	8	10	11	13%
NWS, National Weather Service	15	12	28	37%
NOS, Estuarine & Coastal Assessment	12	11	14	10%
NOS, Coastal Ocean Science	6	10	12	34%
NOS, Observation & Prediction	2	2	2	-1%
NOS, Estuarine and Marine Sanctuaries	1	1	1	-14%
NOS, Ocean minerals & energy	1	1	1	1%
NMFS, Fisheries Resource Information	75	79	91	10%
NMFS, Fishery Information Analysis	6	6	6	0%
NMFS, Marine Fisheries Grants to States	5	5	5	5%
NMFS, Fishery Management Programs	0	*	2	--
NMFS, Protected Species Management	0	0	*	--
NMFS, Promote & Develop Fisheries R&D	5	7	1	-68%
Subtotal	237	277	305	13%
SOCIAL SCIENCES				
OAR, Sea Grant	1	1	1	34%
NMFS, Fishery Industry Information	*	*	*	0%
Subtotal	1	1	1	26%
ENGINEERING AND RELATED R&D				
OAR, Sea Grant	2	2	2	3%
INFORMATION AND DATA R&D				
NOS, Geodesy	1	2	2	17%
NOS, Mapping and Charting	*	*	*	2%
National Env. Data & Information Service	8	8	9	3%
Subtotal	10	11	11	5%
Total, NOAA Environmental R&D	250	290	319	13%

Source: Authors' estimates based on agency R&D cross-cut and budget justification.

*Less than $500,000.

[1]Agencies within NOAA include Oceanic and Atmospheric Research (OAR), National Weather Service (NWS), National Ocean Service (NOS), National Marine Fisheries Service (NMFS), and the National Environmental Data and Information Service.

Table II-14. Estimate of Environmental R&D at the Agency for International Development (budget authority in millions of dollars, by fiscal year)

	FY 1990	FY 1991	FY 1992	Avg Annual % Change FY1990-92
ENVIRONMENTAL SCIENCES				
Env. Science in Support of Agriculture	15	15	16	3%
Natural Resource Conservation/Management	11	7	7	-18%
Innovative Science	4	3	3	-11%
Marine/Coastal Resource Management	3	2	4	20%
Watershed Management	1	1	1	4%
Subtotal	34	29	31	-4%
SOCIAL SCIENCES				
Economics	3	3	4	29%
Social Sciences	4	4	4	0%
Subtotal	7	7	9	11%
ENGINEERING AND RELATED R&D				
Pollution Prevention	0	0	3	--
Other	2	2	2	5%
Subtotal	2	2	5	56%
Total, AID Environmental R&D	43	38	45	2%

Source: Authors' estimates based on agency budget justification and R&D cross-cut.

Table II-15. Estimate of Environmental R&D at the Smithsonian Institution
(budget authority in millions of dollars, by fiscal year)

	FY 1990	FY 1991	FY 1992	Avg. Annual % Change FY 1990-92
ENVIRONMENTAL SCIENCES				
Natural History Museum	16	17	19	10%
Tropical Research Institute	6	6	7	8%
Environmental Research Center	2	2	2	12%
International Environmental Science	1	1	1	11%
National Zoo Park	2	3	3	8%
Astrophysical Observatory/Global Change	0	*	*	--
Nat'l Air and Space Museum/Global Change	0	*	*	--
Subtotal	27	30	33	11%
SOCIAL SCIENCES				
Human Ecology History/Global Change	0	*	*	--
Total, SI Environmental R&D	27	31	33	11%

Source: Authors' estimates based on agency R&D cross-cut by field of science.

*Less than $500,000.

Table II-16. Estimate of Environmental R&D at the Tennessee Valley Authority
(budget authority in millions of dollars, by fiscal year)

	FY 1990	FY 1991	FY 1992	Avg. Annual % Change FY 1990-92
ENGINEERING AND RELATED R&D				
Nutrients and Water Quality	9	13	13	23%
Agricultural, Municipal & Indust. Wastes	7	9	9	15%
Electric Power Program[1]	3	3	8	57%
Total, TVA Environmental R&D	19	25	31	27%

Source: Authors' estimates based on agency budget materials and communications.

[1]Represents obligational authority financed by TVA ratepayers.

Table II-17. Estimate of Environmental R&D at the Corps of Engineers (budget authority in millions of dollars, by fiscal year)

	FY 1990	FY 1991	FY 1992	Avg. Annual % Change FY 1990-92
ENVIRONMENTAL SCIENCES				
Long-Term Env. Effects of Dredging	1	1	1	0%
Subtotal	1	1	1	0%
SOCIAL SCIENCES				
Economic Impact of Global Warming	0	0	*	--
Water Investment Risk Analysis	0	0	1	--
Subtotal	0	0	1	--
ENGINEERING AND RELATED R&D				
Wetlands Research	1	3	7	182%
Coastal Engineering	5	6	6	4%
Aquatic Plant Control	3	3	4	4%
Flood Control and Related R&D	3	3	3	2%
Environmental Quality	2	2	1	-8%
Water Resource Plan Studies	2	2	1	-25%
Natural Resources	*	1	1	27%
Zebra Mussel Control	0	0	1	--
Subtotal	17	19	23	17%
INFORMATION AND DATA R&D				
Surveying, Mapping, & Remote Sensing	1	2	2	22%
Total, CE Environmental R&D	19	22	27	20%

Source: Authors' estimates based on agency budget justification and related data.

*Less than $500,000.

Table II-18. Estimate of Federal Environmental R&D at the Department of Transportation (budget authority in millions of dollars, by fiscal year)

	FY 1990	FY 1991	FY 1992	Avg. Annual % Change FY 1990-92
ENGINEERING AND RELATED R&D				
Coast Guard	4	4	6	23%
Federal Highway Administration	2	3	6	64%
Federal Aviation Administration	2	2	4	42%
Total, DOT Environmental R&D	9	10	17	40%

Source: Authors' estimates based on agency budget justifications and R&D cross-cuts.

APPENDIX B

From: Intersociety Working Group, *AAAS Report XX: Research and Development, FY 1996* (Washington, D.C.: American Association for the Advancement of Science, 1995), pp. 123-126.

ENVIRONMENTAL PROTECTION AGENCY

Over the last five years, EPA has made a concerted effort to steer its R&D resources toward integrated, multimedia studies. Prior to this shift, the agency's science agenda was largely reactive, focusing on particular contaminants or remedial technologies dictated by statutes. EPA has advocated, with considerable success, that its research program support both the near-term studies needed by regulatory offices and the longer-term research needed to resolve scientific uncertainties about the interrelationships among environmental problems and their effects on ecological and human communities. Striking a balance between the two is expected to strengthen the scientific rationale for assessing and managing environmental risks.

EPA is also taking steps to improve the quality of the research it funds. Acting on the recommendations of a 1994 study of its operations, the agency is consolidating and restructuring its research labs and offices to eliminate overhead and focus on risk-based management. The agency also is in the process of expanding the use of competitive, peer-review grants for its research and is working with NSF to develop award protocols and project reviews. The move toward competitive awards is expected to shift some activity from contract to academic researchers. Following another study recommendation, EPA is initiating a fellowship program for doctoral and masters students, beginning with 100 fellowships in FY 1995. Agency plans also call for a mentorship program, in which top scientists from industry and academia would take temporary assignments to work with EPA researchers.

The net result of these changes is that over half of the agency's $619.2 million budget for R&D in FY 1995 supports multimedia research, more than double its 20 percent share in FY 1990. Much of this growth was achieved by shifting resources from media-specific studies over this period. For example, funding for R&D related exclusively to water quality problems and Superfund sites are each about 25 percent lower than they were five years ago in constant dollar terms. The level of support for hazardous waste and toxic substances have each declined by over 40 percent in the same period.

For the coming year, EPA has requested an additional $62.4 million for R&D, which would bring the total to $681.6 million for FY 1996 (up 10.1 percent over current levels). As shown in Table II-17, funding would continue to flow to the more long-term, multimedia research. The Administration would devote the lion's share of the new money to an expansion of the Environmental Technology Initiative (ETI) under the Multimedia program, proposing $119.8 million for the coming year, $51.8 million (or 76.2 percent) more than FY 1995. Under the ETI program, EPA supports collaborative research to advance the efficiency and cost-effectiveness of technologies related to the climate action plan (e.g., alternatives to greenhouse gases and ozone-depleting substances), advanced manufacturing, and pollution prevention. About two-thirds of the ETI R&D is administered by the Office of Research and Development (ORD).

At $76 million, ecosystems protection research would remain the second largest issue area despite relatively flat funding in FY 1996.[2] These funds are used to develop scientific profiles, indicators, assessments, and strategies for ecosystems at three levels—national (e.g., characterizing esturine, forestry, rangeland, and other ecosystems under the Environmental Monitoring and Assessment Program), regional (e.g., developing a scientific understanding of regional ecological problems and solutions), and watershed (e.g., predicting and understanding processes affecting watershed resources). Apart from watershed research, which accounts for most of the funding for Water Quality R&D, these projects are funded under the Multimedia budget.

The agency wants to boost funding for another priority issue area, criteria air pollutants R&D (up 10.5 percent to $46 million). This program, which is aimed at developing and improving the scientific basis of air quality standards for particular pollutants, has requested an additional $4.4 million for the criteria air pollutants program to address shortcomings identified by the National Academy of Sciences in the current strategies for tropospheric ozone. Following a series of meetings convened by EPA, various public and private organizations have formed a research consortium to fund and conduct research on this issue at a ratio of 2:1 of non-EPA to EPA money. In a related area, EPA is also seeking more money for R&D on air toxics (up $4.4 million to $14.4 million) to carry out congressionally mandated studies on urban toxics and the deposition of toxics in the Great Waters region.

Though much smaller in scale, the Administration has targeted selected health issues for significant increases in FY 1996. The $13 million requested for health effects research in FY 1996 (up 72 percent) assumes an infusion of $5.5 million primarily for studies on the effects of environmental exposure to chemicals that interact with the endocrine system. Funding for health risk assessments would rise to $19 million (up 42 percent) if Congress approves the $6 million increase requested to improve the methods and data upon which risk assessments are made, as recommended by the National Academy of Sciences. Agency-wide support for research on human exposure would fall to $15.6 million in FY 1996 (down $1.9 million or 10.9 percent), because EPA does not envision any follow-on ac-

---

[2] EPA's budget cross-cuts by issue do not allocate all of its intramural and support expenses to particular issue areas. Thus, the figures given here may provide an incomplete tally of the total resources devoted to each issue. However, the data provide reasonable indications of the magnitude and trends in funding.

tivities to the completed national human exposure assessment survey (saving $4.5 million). However, the agency has requested an additional $2 million under the Pesticides program to study special issues associated with the exposure of infants and children.

EPA's effort to enhance its scientific capabilities is apparent in the increases proposed for environmental education and its "Cross Program" (components that cut across issues and programs, such as quality assurance). The environmental education budget would gain $3.2 million (for a total of $11.8 million) to double the number of fellowships from the 100 proposed for this year to 200 in FY 1996. Similarly, about half of the $4 million increase proposed for cross programs would be devoted to expand the number of grants, support the appointment of post-doctoral scientists, and fund the mentorship exchange.

As suggested by the reductions apparent in Table II-17, EPA is proposing to reduce funding for these projects to offset the cost of some of these initiatives. Perhaps the largest single source of savings is the agency's policy of halting funding on most of the "congressionally-directed" projects, which totalled $27.4 million in FY 1995. The effect of this policy is particularly pronounced on funding for exploratory grants and centers, which would drop 21 percent agencywide to $35 million because of earmarks totalling $3.6 million under Multimedia grants (e.g., EPSCoR, oil spill remediation) and $5.8 million under Superfund for the Gulf Coast and Clark Atlanta centers.

Finally, it should be noted that EPA is changing its accounting of certain working capital costs attributable to research beginning in FY 1996. Data processing, mail postage, supercomputers, and other support costs that previously were funded under the Office of Administration and Research Management will now be funded directly by ORD. Although this $35 million intra-agency adjustment does not affect the agency's overall budget or level of support for R&D, it could create the appearance of a surge in R&D funding unless prior-year budgets are presented on a comparable basis. AAAS has therefore revised the estimates of EPA's R&D for FY 1994 and 1995 upward by about $30 million each year so that the year-to-year trends shown in Table II-17 are not distorted by account changes.

APPENDIX C

From: Intersociety Working Group, *AAAS Report XX: Research and Development, FY 1996* (Washington, DC: American Association for the Advencement of Science, 1995), pp. 129.

Table II-1. R&D in the FY 1996 Budget by Agency (budget authority in millions of dollars)

	FY1994 Actual	FY1995 Estimate	FY1996 Budget	% Change FY95-96 Current $	Constant $
Total R&D (Conduct and Facilities)					
Defense (military)	35,509.6	36,272.2	35,161.2	-3.1%	-5.9%
NASA	9,405.5	9,874.2	9,517.1	-3.6%	-6.4%
Energy	6,771.2	6,534.4	7,012.9	7.3%	4.2%
HHS	11,323.5	11,726.9]	12,157.1	3.7%	0.6%
NIH	[10,473.5]	[10,840.2]	[11,293.3]	4.2%	1.1%
NSF	2,242.7	2,543.6	2,540.0	-0.1%	-3.1%
Agriculture	1,528.3	1,539.8	1,483.4	-3.7%	-6.5%
Interior	707.6	686.1	679.3	-1.0%	-3.9%
Transportation	640.8	687.0	619.5	-9.8%	-12.5%
EPA	588.1	619.2	681.6	10.1%	6.9%
Commerce	1,021.9	1,284.1	1,403.7	9.3%	6.1%
Education	175.3	174.8	181.8	4.0%	1.0%
AID	315.6	314.0	255.0	-18.8%	-21.2%
Veterans Affairs	276.5	296.9	272.8	-8.1%	-10.8%
Nuclear Reg. Comm.	90.7	82.0	81.8	-0.3%	-3.2%
Smithsonian	133.9	134.9	138.9	3.0%	-0.0%
Tennessee Valley Auth.	96.0	88.6	98.8	11.5%	8.2%
Corps of Engineers	51.6	54.6	55.4	1.5%	-1.4%
Labor	62.6	61.8	94.0	52.1%	47.7%
HUD	35.6	40.5	40.8	0.7%	-2.2%
Justice	45.8	53.7	55.1	2.7%	-0.3%
Postal Service	51.0	70.0	72.0	2.9%	-0.1%
TOTAL R&D	71,073.9	73,139.6	72,602.0	-0.7%	-3.6%
Defense	38,299.1	38,837.5	37,929.9	-2.3%	-5.2%
Nondefense	32,774.8	34,302.2	34,672.1	1.1%	-1.9%

PART III

KEYNOTE ADDRESSES AND PRESENTATIONS

D. James Baker

Undersecretary for Oceans and Atmosphere,
National Oceanic and Atmospheric Administration

Thank you for inviting me to speak to the forum today. It's a formidable task to try to develop a set of science and technology goals that can help across the board and be relevant to all the different areas of the country. One of the things that impressed me when I took my job in May of 1993 as head of NOAA was the existence of so many reports that laid out goals for the agency. The Academy, the Carnegie Commission, and others had done a very nice job of laying out a series of recommendations about what the new Administration should do, how it should do it, and how we should be organized. We've made some significant progress there, but not as much as we all would like.

The subtitle of this forum, "What Have We Learned in the Last 25 Years, and Where Do We Go from Here?" is particularly apt for NOAA which was formed in 1970. Bob White, who is here and just retired as president of the National Academy of Engineering, created NOAA from a few disparate agencies into a whole new organization. Not many people have had the opportunity to start from scratch and start a whole new agency. It is a tribute to Bob that NOAA has remained intact and become a powerful force for environmental stewardship for the United States.

Today, I'm going to talk about what NOAA is trying to do, how we are trying to follow through with goals and objectives, and how we measure our success. NOAA was formed in 1970, but I think one must look back a bit further to find the foundation for the things that we do. The year 1960 was very important because that was when we launched the first weather satellite. Weather satellites have been a key element for NOAA. Looking back to the 1950s, we have the development of other important elements: the transistor, and then in the 1960s, the development of integrated circuits. Microelectronics has been a key to developing our current capabilities in computing, telecommunications, satellites, and all kinds of automatic and unattended instrumentation. I didn't realize when I started out as a post-doc in oceanography that I would spend considerable time using microelectronics to develop low-powered instruments for unattended long-term observations in the ocean. This new technology has allowed us to learn much more about nature.

As we look to the future, we're looking toward a world where this new information allows us to connect societal decisions and environmental decisions. We would like to see a world where economic growth is coupled with the sustainable use of natural resources. We recognize that these must go hand-in-hand if we want to provide an increasing standard of living for everyone in both the developing and the developed world. We all know that it's very important to have a healthy economy. At the same time, we know that a healthy economy depends on having a healthy environment. Probably the best contrary example of that link now is the situation in Russia, where there was an attempt to have massive development without worrying about the environment. And now they are paying the price.

In NOAA, our mission is to describe and forecast changes in the environment and to manage and conserve natural resources. Our goal is to make those measurements and then provide decision makers with that information. We've had many important technical advances in the last 25 years, which have allowed us to do a much better job of forecasting and understanding the problems. Let me just give you a few examples. First, we are responsible for providing information on the natural changes in the atmosphere on short time scales, otherwise known as weather forecasting. This activity requires about half of NOAA's budget, if we include both weather forecasting and the cost of operating weather satellites. Short-term weather events can have huge impacts. About 85% of the presidentially declared disasters every year are weather related. Hurricanes, tornados, and flash floods are major issues. We see the economic impacts of these events through insurance companies, which are increasingly having to pay more because of the vulnerability of growing populations in regions where natural disasters are likely to occur.

The second major issue for us is to sustain marine fisheries. Marine fisheries are a major source of protein to the world today. And yet in the last couple of years we have seen the total fish catch in the world start to decline. We haven't seen this before. In past years, when particular fish species were depleted, we could always move to another species. But now we've seen the total fish catch start to decline. Even in the United States about half of our fisheries resources are currently overfished. So we have a real challenge if we are to maintain those fishery populations and have both a healthy fishery and a healthy fishery industry.

In terms of longer-term forecasts we have been very successful in the last few years in predicting El Niño, the tropical warming that occurs in the Pacific Ocean and has global impacts. The warming is associated with weakening of the western Pacific trade winds and with global changes in the atmosphere. In the past few years, we have begun to understand the physics of what happens between the ocean and the atmosphere during an El Niño so that we can provide significantly improved forecasts. This is a new and exciting forecasting ability, which has been based on a better understanding of how the Earth system works.

We are also responsible for coastal zone management. In the United States today, we have a relatively slow-growing population, but it is migrating toward the coasts; about half of our population lives near the coast. It is important to ensure that as development occurs, states protect fishery habitats and maintain clean beaches. The Coastal Zone Management Act is a voluntary program between the federal government and the states. We set federal standards that must be incorporated in state plans. It has been a very successful activity.

NOAA is also responsible for all of the mapping and charting of the ocean bottom in U.S. coastal waters and for producing aeronautical charts for aviation. The Coast and Geodetic Survey, the oldest part of NOAA, was established in 1807 by Thomas Jefferson. We continue to maintain our mapping and charting activities, including all those brass plaques that you see around the country. We are now augmenting all those plaques with a satellite-referenced global positioning system. I just visited an office of the National Geodetic Survey in Long Beach last week. They demonstrated how they monitor the ground level as they pump oil from the ground. They try to compensate for the extraction of oil by pumping water into the ground, but there's never an exact balance. They monitor the change in reference to a small brass plaque on the end of a pier near San Pedro. But while the oil pumping changes the ground level locally, on a much larger scale the North Pacific tectonic plate is uplifting and moving north. And all of these changes are about the same order of magnitude. It's wonderful to hear these surveyors discuss the movement and uplift of the North Atlantic plate and the brass plaques and the transition to a satellite-based system. The science and technology come together very nicely.

Let me speak next about goals. There is an old Chinese saying: "Without a long-range plan one is in immediate danger." When I became administrator of NOAA, the first thing I was asked to do was to give a priority ranking to a list of about 150 budget items from the Office of Management and Budget (OMB)—and to do it in two days! We realized that we didn't have a good planning context and we decided to develop a agency-wide strategic plan. We put together a set of strategic planning teams to look across the agency at our different goals. Our strategic plan was developed and is now widely distributed, including on Internet. The NOAA home page shows our seven strategic planning themes: warnings and forecasts for the short-term; seasonal-to-interannual forecasts (El Niño); long-term global change (including ozone depletion and changes in climate, caused by increased radiative gasses); navigation and positioning; rebuilding fisheries; recovering protected species and fulfilling our responsibilities under the Endangered Species Act and the Marine Mammal Protection Act; and coastal ecosystems health. We have a number of national capabilities that undergird these goals: weather satellites, our fleet of ships, environmental data centers, and high-speed computing and communications.

We found that it was a very valuable thing to have this set of seven strategic themes. From these themes we have developed objectives. And then from each of our objectives, we developed measures of performance. Without measures of performance, we really can't say how well we are doing in accomplishing our objectives 10, 15, or 25 years from now. This was a very important and new step. Performance measures allow the government to operate more like a business. We now have the Government Performance and Results Act (GPRA), which provides the context for developing these measures of performance.

I'd like to discuss just three areas of our strategic plan in which we are trying to measure our success. We started with our seven themes and developed our budget based on these themes. As an aside, I might note that there was an initial reaction against this by those who were used to working with the budget based on line and program offices rather than on strategic planning themes. Others thought that this budget format made a lot of sense and that we should reorganize the whole agency around our strategic planning themes. Well, we haven't gone that far. But at least the strategic plan provides a good framework for focusing our activities.

Let me use short-term warnings and forecasts as the first of the three points that I wanted to make. On Palm Sunday in 1994, a tornado struck a church in Alabama and killed a number of people. This event reminded us of the difficulty of disseminating warnings. Tornados, like other explosive storms, are very hard to forecast. We have some new observing systems—Doppler radar systems—which are very good at giving initial indications, and we have already seen improvements in forecasting. Floods are another weather-related problem. The Mississippi River flood of 1993 caused extensive damage and dislocation. There is a major flood in some part of the country every week. We're doing a much better job of forecasting floods these days, but how do we quantify the improvements?

We've spent a lot of time developing specific measures of performance. We asked each of our strategic planning teams questions like, Given that we have a lead time of 15 minutes for a flash flood today, what do you think you can do in 10 years? What do you think you can do in two years, three years, or four years? Given our scientific knowledge and expected improvements from our observational and computer modeling systems, we think that we could improve the lead time for flash floods to 40 minutes in 10 years. For tornados, the average lead time today is 13 minutes. We think we can eventually provide 25-minute lead times. That would be good, since tornados can grow from nothing to full magnitude in about 30 minutes. We're also improving lead times for predicting severe storms, temperatures, and snowfalls, and for issuing aviation forecasts. These numbers give an indication of the incremental improvements that we think we can deliver for the public and for policy makers as they make decisions. Insurance companies are very interested in this as they develop their actuarial tables. They want to know how much better our forecasts can be.

Forecasting lead times are one type of measure. Another type of measure is climate extremes. A major question today is what will happen with the increase of radiatively important gases in the atmosphere. What changes in climate are likely? We have models that give indications, but the models are not yet fully accurate in representing the physical interactions. We do know that the trace gases are continuing to increase. We're studying the data to see what they tell us. NOAA has developed a climate extremes index, showing extremes in the United States of temperature, droughts, and rainfall. Have climate extremes changed? Since about 1976, there is no doubt that the atmospheric circulation strength over the Pacific and North America has changed. But the increase that we've seen over the last 15 to 20 years is really not sufficient for us to say that there has been a change in climate. The greenhouse climate response index is another measure of the state of the environment. It involves looking at temperature, rainfall in cold regions, drought in warm regions, and extreme rain. These are the parameters that would be expected to change with increasing greenhouse gases. We would expect this value to be up about 10%. And since 1976, it's been up about 12%. This points toward global warming, but it is not yet statistically large enough to reject the idea that we're simply seeing a stable climate.

My third and final example of measures relates to fish. Salmon are particularly interesting as an indicator for a total ecosystem. If the salmon are healthy, the whole ecosystem is healthy, because the salmon live in the ocean and in the fresh water. Dams are a particular problem, since they make it difficult for salmon to reach their spawning grounds. The United States has decided that it is beneficial to build dams on rivers to produce clean hydropower, but newly spawned salmon find it difficult to travel past the dams. They are either crushed by the turbines or eaten by other fish. In addition, their habitat is degraded because of agriculture and logging. So we have a real problem in trying to maintain a healthy salmon population. We also have problems in maintaining fish populations in general because of the conflict between the fishermen who would like to catch as much fish as they can, and the need to maintain healthy levels of stocks.

In measuring the health of fisheries we can try to assess fish stocks. Based on these assessments, we then develop models of management. How many fish can be taken from a particular stock in a sustainable way? Can we take 1% or 5% of the fish and still maintain healthy fish stocks? Knowing how much fish we can harvest is a key to building sustainable fisheries.

Fishery management plans are a key to sustaining healthy fisheries. Today we have a successful process for involving all fisheries stakeholders through the Regional Fishery Management Councils. The councils were established in 1972 under the Magnuson Fishery Conservation and Management Act. They bring together commercial fishermen, scientists, and representatives of regional, state, and federal governments to develop fishery management plans. NOAA then implements these fishery management plans.

We need not only scientific data, but also economic and social data. We need to know about fishermen in the fishing communities, and how they interact. We are trying to reduce the amount of bycatch—nontarget species that are caught and discarded. Unfortunately, this bycatch is not a small fraction of the total amount of fish caught. In the Gulf of Mexico, about seven pounds of fish are caught for every pound of shrimp, but only the shrimp are harvested; the other fish are discarded. Bycatch is a real problem, and we are trying to reduce it.

Returning to the broader theme of strategic planning, in the process of developing an agency plan, we are working not only within NOAA but also with a wide range of constituents. We also work very closely with Congress. NOAA has always received strong bipartisan support and continues to receive good input from Congress and the public on its products and services. Almost everything we do touches most Americans—from forecasting the weather to managing fisheries to promoting safe navigation. The strategic plan has allowed us to show the Administration, Congress, and the public that we have examined our mission and established goals and priorities. In 1996, we were pleased to be one of the few agencies for which the President proposed a budget increase. Even with all of the financial constraints, I think there is an understanding that the things we do are important. We're faced with some real challenges, as are all federal agencies. We're faced with the challenges of downsizing and of living within the budget limits. Both the Administration and the Congress have agreed to balance the federal budget, so we must learn to operate in this mode. The strategic planning process with goals, objectives, and measures of performance is the best way that we can meet the challenges. One of the things that you can do during this conference is to look even more broadly at the science and technology activities that are important to this Nation. In the end this will help us all achieve our goals.

Thomas Grumbly*

Assistant Secretary for Environmental Management,
U.S. Department of Energy

Good morning. I've been asked to give my thoughts on what the nation's environmental goals should be for the next 25 years, now that we're 25 years down the road from Earth Day. To suggest the country's environmental goals for the next quarter century, we first need to review past practices and policies, see what lessons we've learned, and move forward to fix what's broken. Although progress in improving environmental quality has been uneven and unquestionably expensive, I think most people would agree that the state of our environment in this country is, overall, in healthier condition than in 1970: Most, though not all, of the pollutants in our air have been reduced; many species formerly on the endangered species list in this country are coming back; improvements in water quality are sketchy but still moving forward.

BACKGROUND

The nation, through its elected representatives, has laid a groundwork to protect the environment in numerous environmental statutes since 1970: the various clean air and water acts; our hazardous waste and toxic substances laws; the variety of all the other statutes that are on the books. In spite of all of this legislation, most public policy observers agree that improvements in environmental quality and protection could have been achieved at much lower cost, with less cumbersome and more intelligent regulations and, consequently, with much less industry hostility. We seem to have been particularly good in changing environmental regulations from the kind of consensus-based policy making in the beginning of

*Via video link from Washington, D.C.

the environmental movement into what currently has become an adversarial process that pits industry against government and parts of government against itself.

However, one major tenet remains that nearly everyone can agree on: While the general ends of these policies are not in dispute, the means to achieve them are. Many public opinion analysts would agree that environmentalism has become an idea fixed in the composite of values that define America's basic political beliefs and, thus, now competes for its "market share" of government resources at all levels. In short, our environmental goals should aim for what most Americans want: the protection, preservation, and restoration of our environment. If we focus on protection and preservation in the right way, then restoration needs will diminish over time.

In the few minutes I have on the satellite, I would like to briefly discuss three principles that must guide and inform our goals in environmental policy making: (1) applying risk management to our environmental programs to ensure we're spending our money in the most optimal way; (2) using market forces to the maximum extent feasible to alter the incentives of environmental protection; and (3) recognizing and acting upon our moral obligations to others and to future generations in exercising environmental stewardship. I believe that these three values need to be utilized within the legal, political, and institutional foundations and processes that determine our environmental goals.

ENVIRONMENTAL GOALS

Let me begin with a basic environmental premise: If you pick something up in the universe, you find it, connected to everything else. Coherently integrating the complex web within ecological systems to the social, economic, and political sectors of our society bespeaks of the difficulty in formulating environmental goals. William Pederson has put forth a relevant, but critical, perspective on this point in a recent *Loyola Law Review* article[1] : We have been making environmental laws but not environmental goals for the past quarter century. He states, for example, that in lieu of a clean water policy, we have a Clean Water Act. Although the "goal" in the Clean Water Act of 1972 was fishable and swimmable waters in the United States by 1980, the government pursued many activities that adversely affect water quality: agricultural and transportation subsidies, flood insurance, timber leasing, and water pricing. Fixing this compartmentalization of governmental activity should be one step in establishing environmental goals.

Another example of the duplicity and compartmentalization of environmental policy lies in the patchwork regulation of my own program. The Department of Energy's Environmental Management program is charged with the mission of cleaning up the legacy of 50 years of nuclear weapons production at our complex

[1]William F. Pederson, *Loyola Law Review* "Protecting the Environment—What Does That Mean?" (April 1994).

of sites across the country. We must deal with millions of cubic meters of radioactive, hazardous, and mixed wastes in the soil, groundwater, in the ducts and pipes in our buildings, and in tens of thousands of tanks and canisters. No fewer than 10 major federal laws govern our program activities, with an additional host of state laws, and scores of consent agreements. How can we sensibly formulate a specific set of strategies and goals with over 133 sites in 33 states and Puerto Rico, with over 7,000 buildings, all in an area larger than Rhode Island and Delaware combined?

RISK MANAGEMENT

One way is to prioritize the risks present in the system, and here I may be preaching to the choir. For those who don't know, I requested in July 1993, two months after assuming my current responsibilities, that the National Academy of Sciences-National Research Council advise the Department on whether and how risk and risk-based decisions could be incorporated into the Environmental Management program. My request to the National Research Council resulted in the January 1994 report, *Building Consensus Through Risk Assessment and Management of the Department of Energy's Environmental Remediation Program*. In the report, the Council identified the major obstacles, issues, and barriers to implementing a risk-based management approach. Nevertheless, the report concluded that the use of risk assessments and a risk-based approach would be feasible and successful, provided its purposes and limitations are clearly defined; early and full public involvement is gained; and consideration is given to cultural, socioeconomic, historic, and religious values.

In turn, we used these principles in studying our sites and coming up with our own Risk Report. Our report *Risk and the Risk Debate: Searching for Common Ground* represents the first step toward a consistent approach to evaluating the risks to human health, worker safety, and the environment posed by conditions at Department of Energy sites and facilities. While this first draft is not perfect, it provides three outcomes in my program that should be applied in other environmental regulatory policy and goal-setting areas. First and foremost, it articulates one, consistent analytical approach that captures the spectrum of risks associated with our program activities across the nuclear weapons complex; secondly, it assesses the degree to which risks are addressed in the patchwork of regulations and compliance agreements that govern our activities; finally, it links the risks in a qualitative fashion to regulatory program performance and the budget.

A properly structured risk assessment program can have significant benefits, for the Environmental Management program and for other environment protection efforts. Incorporating risk management into the policy-making and goal-setting processes forces institutions to pose the question: How much risk reduction at what cost? Most important in this process is allowing society to become involved in the debate as to how much should be spent to address specific risks.

Allowing this debate to occur in the marketplace can provide powerful incentives for both the producer and the consumer to shape our environmental priorities. I'll expand on that thought in just a moment, but to sum up my thoughts on risk management, I believe an important goal for the next quarter century should be incorporating risk assessment and risk management into the political, legal, and institutional processes that themselves lead to the establishment of environmental priorities and goals.

ENVIRONMENTALISM IN THE MARKET PLACE

In his book *Earth in the Balance*, Vice President Gore writes, "Free market capitalist economics is arguably the most Powerful tool ever used by civilization." He further described how to conflate classical economic and environmental principles; one way is through market-based environmental policies. Harnessing the power of market forces can provide more cost-effective and far-reaching solutions for many environmental problems than can regulation. The command-and-control regulations that have dominated environmental protection in this country suffer from the same limitations that have led to the collapse of command-and-control economics around the world, They are inefficient, they stifle innovators, and they lack the flexibility to account for important differences among individuals, firms, and regions in the market. In contrast, market-based policies motivate both producers and consumers to seek out the best methods of environmental protection by changing the financial incentives that drive what to consume, how to produce, and where to dispose. The results are a lower cost of compliance for industry and, by extension, for the consumer, Clearly, a long-range national environmental goal must be a greater reliance on market-based strategies—we've simply got to get away from command-and-control regulations,

By the year 2020, several important environmental changes within the private sector should be firmly established as well. An ethic of environmental responsibility must be ingrained in the management culture of every private company, large and small. We need to have the environmental equivalent of the Malcolm Baldrige Award in the major corporations around the United States. When I spent the last seven years in the private sector before coming to this job, it was amazing to me to see the competition for the Baldrige Award really changed the way CEOs thought about their company. You can see CEOs competing with each other about who could be the best company in the marketplace. And we must do the same thing in the environmental arena. Corporate America must proactively protect the environment, rather than react to regulatory or public pressure. The new generation of middle and upper management must recognize the marketplace value of environmental stewardship. That kind of heightened sensitivity needs to extend to the citizenry and to government managers, as well. These "sea changes" in the marketplace, along with risk management, are an important part, in my view, of where the nation should be 25 years from now. But neither of these elements can work without a healthy dose of morality. Let me take a few minutes now to elaborate.

OUR MORAL OBLIGATIONS

The right kind of environmental stewardship for our nation, and indeed for the world, has to include an acknowledgment of our moral obligations. How, for example, do we treat Native Americans who have been exposed to low risks from our nuclear weapons production plants? The answer to questions like this will not come from the marketplace. People have already been compromised, and the federal government, acting for all of its citizens, cannot ignore those citizens who have been inequitably exposed to risk. Thus, our goals and our environmental policy must be guided by our collective moral compass. In our age, human activities are transforming the planet in profound ways. Our ability to change the environment is magnified along with our growing numbers, our continual quest for a higher quality of life, and our technological and institutional capabilities. Our moral obligation to future generations compels us to ask how these factors can provide means not just to alleviate environmental damage but to advance the state of the environment.

The question that we should build into our daily routines is: What are we doing today that will prompt another generation to say, "How could these people not have seen the consequences of their actions?" This question could be directed at our waste disposal practices, our transportation methods, even our eating habits. No one, of course, knows what these future questions will be, much less the correct answers. Nonetheless, part of the inheritance of my kids and yours in the year 2020, should be a desire to look to the future and anticipate these questions now.

CONCLUSION

This morning I've given you a very brief overview of what I see as the direction we need to take in establishing our environmental goals for the next quarter century. Ideally, all major environmental risks will be eliminated by the year 2020, in keeping with the American public's desire to continue to protect, preserve, and restore our environment and our natural resources. Over the past quarter century, the means to achieve environmental protection has been the "stick" rather than the "carrot" and has, as a result, not been either cost-effective or conducive to the kinds of technological innovation we need. Aggressive risk management and marketplace incentives, informed by our moral obligations, could yield more sensible, powerful, and effective results.

I'd like to conclude by noting that the ultimate success of our nation's environmental policies and goals—in fact, all of our public policymaking—rests on the two pillars of representative democracy: the participation and education of the citizenry. It seems to me that we are in the middle of a revolution about how people want to conduct policy making in the United States. The end of the Cold War has essentially led to the end of the supposition that the federal government is able to solve many or most of the problems we have. States, local governments, individual citizen groups want to take back the power to themselves. One of the major challenges for environmental policy making in the next 25 years, particularly policy making that requires national action, will be how to formulate some kind of new national policy that includes the many factions of our national politics. I personally happen to be a Hamiltonian, and by that I mean that if the United States is to maintain its position in the twenty-first century, we must have a reasonably strong central government. So ladies and gentlemen, the huge problem of the moment is that the average citizen and public policy maker at the state and local level simply do not want the federal government to deal with the problem anymore. Putting all of these interests at the local level together so that we can develop a coherent national strategy against both our domestic environmental and our international environmental problems will be a major issue in the years ahead.

With respect to education, all of us understand that the level of literacy in society is not at the level that it should be if we are to have the necessary informed dialogue, but it seems to me that we cannot engage in talking down to people or being paternalistic. We must work from the ground up again to get people to participate in the development of policy; we are going to have to work to educate folks to the level that they want to be educated to. By way of combining participation and education, I think EM's site-specific advisory boards have done a very good job in grasping the essentials of economics and science, and then making decisions and recommendations based on that.

Nurturing these two democratic fundamentals is as important as the policy-making itself—the environmental progress that has been achieved in the past 25 years underscores their importance. So, we must continue the hard work and challenge of maintaining the basics of our political system while continuing to engender environmental protection as a social value. In the end, we need to re-create a country in which every private citizen has a public responsibility and feels that responsibility in a direct way.

Barry Gold

Chief, Scientific Planning and Coordination,
National Biological Service, U.S. Department of the Interior

First let me commend the NRC for convening this forum on linkages between science and technology and societal goals for the environment. It's important because it acknowledges the legitimate role that societal goals must play in shaping environmental research priorities, as well as the role that research results play in shaping society's goals for the environment. Dialogue such as this one can help ensure that environmental science and technology activities are relevant to society's needs and aspirations. In fact, at the National Biological Service, we've initiated a similar series of dialogues with our sister bureaus at the Department of the Interior, with representatives of state agencies, private corporations, nongovernmental organizations, and the academic community.

The emphasis of my talk will be on the second of the two questions we were asked to address. How can science and technology contribute to meeting future societal goals for the environment? I want to begin by stating a few personal beliefs. First is my belief that environmental challenges confronting the United States are some of the most critical issues we face today. Others have argued that ecological issues are one of the key defining forces in the still emerging New World Order. I think current and emerging environmental problems are different in scale and kind from those that led to the passage of the National Environmental Protection Act and the creation of the U.S. Environmental Protection Agency in 1970. In the 1970s, our focus was primarily on human health issues. Today, environmental concerns result from less tangible causes, may be less amenable to treatment, and may have potentially more cumulative irreversible effects than previously thought. In addition, many of the new and emerging environmental problems are fundamentally ecological in nature. These include the disposal of solid and toxic wastes, deforestation, watershed destruction, loss of biodiversity, and climate change. As a result, environmental research has increasingly focused on the health of the environment, the value of ecosystems services, and the ability of the biosphere to provide—on a sustainable basis—the goods and services demanded by the needs and aspirations of an increasing human population. In many ways, the situation we face at the National Biological Service with respect to natural resource issues is analogous to the one faced in the pollution and human health arena. Initially, we rallied public support in support of saving bald eagles, gray whales, and the like. We focused on individual species and on case-by-case responses. Now, we recognize that fungi and mussels and issues such as habitat fragmentation are at least as important, if not more critical.

The nation's biological resources are the basis for much of our current prosperity and social well-being, essential parts of the wealth that we will pass on to future generations. Our very existence is dependent on the plant and animal products that provide us food, fuel, fiber, shelter, and pharmaceuticals. In addition to these biological products that enter our market economy, we depend on healthy ecological systems for critical services such as clean air, clean water, and fertile soil. Recently, Tim Worth, Undersecretary of State for the Environment, has asserted "that the economy is a wholly owned subsidiary of the environment." Since biological resources and ecological systems are an essential part of our nation's wealth, like other forms of wealth, they should be managed wisely. Human population growth and the pursuit of an improved quality of life have produced unintended threats to the health of biological resources and the integrity of the ecological systems.

Increasingly, the private sector will determine the boundaries in which the individuals and households make environmental decisions. Ecosystem health and integrity are concepts that are widely used when talking about biological resources and environmental quality. Yet they are still without precise scientific definition. Healthy ecological systems provide goods in the form of food, raw materials, sources of energy, medicinal plants, and genetic resources among others. They also provide services, such as maintaining hydrological cycles, cleansing water and air, pollinating crops and other important plants, storing and cycling essential nutrients, providing sites for tourism and recreation, regulating climate, maintaining the composition of the atmosphere, generating and maintaining soils and reefs, naturally controlling disease vectors, and absorbing detoxifying pollutants. And I could go on. This growing awareness of the inescapable interdependence of human health and welfare and the integrity of ecological systems has led to a recognition of the need for strengthening the research on and monitoring of the nation's biological resources and the ecological systems for which they are imbedded. The extent to which biological resources and ecological systems are sustained will, in fact, determine the variety of socially and economically viable management options that can be retained for the future.

Let me turn to the National Biological Service. Why was NBS created? The mission of the National Biological Service is to work with others, to provide the information and technologies needed to manage and conserve the nation's biological resources. To achieve a goal this large, NBS must first work in partnership with others and second take a more integrated ecological systems approach to problem solving. NBS was created out of the recognition by Secretary Babbit, that ecological train wrecks, as he called them, come from taking a single resource or a single species approach when there are many competing interests, and from focusing on current needs and not looking ahead. Based on this analysis, he concluded that the most important step to take in avoiding these crises involves integrating research efforts to create a unified attempt to understand whole systems. It is his view that such an enhanced scientific understanding will help us anticipate and plan for the future, rather that simply react to it. If NBS is to help prevent train wrecks, it must address multiple resource and multiple species issues at ecosystem to regional scales. To do this, NBS must be more strategic and more proactive in what research it undertakes. NBS research must expand our understanding of how ecological systems work and use that understanding to both project trends into the future and address real problems facing resource managers. This approach has required NBS to rethink ongoing projects and to ask how funds can be redirected to link together projects that once were viewed as unrelated.

Where did NBS come from? NBS is the newest bureau of the Department of the Interior. It was created in 1993 by merging the biological research, inventory, monitoring, and information technology related capabilities of seven Department of the Interior bureaus. NBS has no regulatory or resource management responsibility—thus lessening the chance that our findings would be unduly influenced by policy or enforcement issues or that they would be tainted by the suspicion that science was made subservient to the needs of policy makers. NBS's job is simply to provide better science relevant to the needs of resource managers. Let me say a bit about what NBS is not—both because it's worth examining the kind of misinformation that has been spread about NBS and because it raises a key aspect of how NBS will do its business. NBS has been accused of being a single giant survey that will sweep the nation, "An army of intrusive agenda-toting environmentalists in science disguise." This is not what NBS is. Most of NBS's time is devoted to research on the causes of environmental change and how management decisions affect biological resources.

It is certainly true that we need to know more, but a critical element of that is making sense of the myriad efforts that are already going on. Federal agencies, state agencies, museums and universities, private corporations, and nongovernmental organizations all spend money to collect data and conduct research on biological resources and ecological systems. These efforts do not add up to coherent views of the biota of even places we've studied intensively, like South Florida, yet alone the entire United States. Thus, Secretary Babbit requested advice from the National Research Council in the establishment of NBS. In its 1993 report, the NRC encouraged the development of a new national multisector cooperative program of federal, state, and local agencies; museums; academic institutions; and private organizations to collect, house, access, and provide access to the scientific information needed to understand the current state of the nation's biological resources. NBS is embarking on creating such a national partnership.

Now let me turn to my first recommendation for the forum, and that is: Don't ignore how we do environmental science and technology. We need to focus not only on what we study, but how we study it. In trying to bring about this national partnership, NBS can play an important role in ensuring that there is more and better information on the status and trends of the nation's biological resources. First, NBS can work with DOI land management agencies to ensure that there is adequate inventory and monitoring. Secondly, NBS can work with federal agencies, states, and the private sector to ensure that more rigorous standards and protocols are developed. Here the major NBS role is not to conduct new surveys, but rather to help provide a common architecture that ensures the reliability of the information collected by others. Third, NBS can play a critical role in making certain that both existing and new information are fully available to all decision makers. Providing the science to guide the resource management requires that we understand and are able to predict the major causes of biological resource degradation and the effects of various drivers of change in ecological systems. Key driving forces generating change in ecosystem structure and function include natural and environmental perturbations and anthropogenic stresses, such as extractive and nonextractive uses, habitat destruction, and impacts from contaminants and pollutants.

The NBS both conducts tactical research on issues already identified as serious problems and undertakes strategic research aimed at addressing emerging and future problems. In moving toward a more integrated research approach, NBS has established the following general principles to guide the development of more integrated research. First, to co-locate seemingly unrelated projects on upland, bird, stream, fishes, and wetland plants on the same site and, then, share information on land use, change, weather patterns, contaminant exposure, etc. We believe this can avoid duplicated efforts and eventually lead to an integrated understanding of multispecies responses to management practices. To coordinate research protocols and methods so that studies in different area or by different agencies, but on the same species or on the same contaminants, when possible use the same methods. This would prevent us from duplicating efforts and will allow more sharing of data and more across-sight comparisons. To emphasize studies that link biological trends with physical forces that drive those trends. For example, studies of habitat suitability must be linked to an understanding of the physical and anthropogenic factors that create and destroy habitat or change habitat suitability. To increase our effort on establishing cause-and-effect relationships between environmental factors and trends in biota. For example, we need to establish cause and links between contaminant exposure and population declines.

To begin to link our biological and ecological studies to the social and economic driving forces of environmental change and to project simultaneous ecological and economic trends. To place greater emphases on predictive studies that will allow us to forecast the impact of alternative management and policy decisions on populations and ecosystems. And to develop more taxonomic breadth and devote more funds to studies of species other than birds, mammals, and fish. It is our belief that these steps can lead to a more proactive ecosystem approach that places individual studies in a broader context and enhances our ability to predict the consequences of management decisions for a wide range of biological resources and ecological systems. So my final recommendations to the forum are: Recognize the increasingly ecological nature of environmental problems, foster integrated ecological research where appropriate, and don't ignore the role of the private sector in establishing the boundaries within which individuals and households will make environmental decisions.

Keynote Address

Harlan Watson

Staff Director,
House Committee on Science

I want to thank the sponsors of this forum for the opportunity to be here this morning. First of all, I want to make clear that any comments I make represent my own opinions and views, and do not necessarily represent those of either the Committee on Science or any of its individual Members.

Looking ahead 25 years is a strange topic for someone from the U.S. House of Representatives to address, since we're used to thinking about one two-year term between elections. Presumably my colleague from the Senate, Dave Garman, who is accustomed to six-year terms will provide a more reliable long-term view. However, there is still the issue every two years of which political party is going to be in power.

I did have some learned remarks prepared and some great recommendations, but unfortunately Tom Grumbly stole them all. I do want to note, however, that I would agree and I believe most Members of the new House of Representatives would agree with what Tom said about risk assessment and analysis, and the need for sound science and movement to free-market economics. I won't dwell on those areas since he went into them in some detail, but what he said is certainly in broad agreement with the current majority of the House.

There has been, as you know, a sea change in the makeup of this Congress as a result of the 1994 election, particularly in the U.S. House of Representatives, which attained a Republican majority after 40 years of one-party rule. The new Members of this Congress—who total over 70—are for most part very conservative. In fact, as a class they are probably more conservative than the leadership of the House, which includes House Speaker Gingrich and Majority Leader Armey. And they can be, by and large, characterized by a belief in "minimalism"—that is, a belief in minimal governmental regulation, in minimal government intrusion into the lives of individuals, and in minimal government taxing and spending.

The focus of Congress for the next seven to ten years will be, I believe, on getting control of the deficit—and it will remain a focus whichever political party, Republican or Democrat, has the majority in Congress or has the presidency. What we have right now is agreement between Congress and the Administration on balancing the budget; the only disagreements are whether it should be balanced in seven years or ten years, and which should be the spending priorities. However, the important thing will be the "bottom line."

This circumstance pits spending on the environment and environmental issues in general—which are not registering very high with the U.S. populace in the current polls—against spending on many other important areas such as Social Security, Medicare, crime, and so on. In fact, the Ladd-Bowman paper included in this forum's agenda book, noted that a January 1995 Gallup poll found that the environment was mentioned as the "most important problem" by only 1 percent of those surveyed. And Congress will tend first to respond to and address those issues in which there is strong public interest.

The near-term focus on balancing the budget—that is, the near-term focus on the bottom line—will make it difficult to concentrate on the long term. And it will also make all environmental agencies budget targets. The Environmental Protection Agency budget has been cut at this point, and the National Biological Survey's existence as a stand-alone agency is uncertain—it may become part of the U.S. Geological Survey or downsized considerably if it survives. And, although several forces are at work here, financial considerations are the overwhelming driver.

Tom Grumbly also brought up another good issue—namely, that the old environmental consensus created in the late 1960s and early 1970s has broken down—and I would like to take a few minutes to address this topic.

That old environmental consensus remained largely intact throughout the 1980s, although there were controversies during the Reagan years. On the whole, however, environmental legislation was reauthorized, environmental funding grew, and things did move along. One could argue about the pace, but whichever side of a given environmental issue you were on, there was a broad consensus for a "command-and-control" approach to the environment—with the government either setting standards to be followed by the private sector or prescribing in detail the technology that the private sector must use to combat a particular environmental problem.

This is not to say that there were not complaints about the costs of environmental compliance, particularly from individual industries or companies that were impacted by a given environmental regulation. Those early costs, however, tended to be localized and tended to reap relatively large environmental benefits.

As we advanced into the late 1980s, however, further improvements to many areas of the environment became much more expensive, with fewer and fewer apparent benefits. In addition, many corporations started "right-sizing" with resultant job losses, and there arose widespread concern about the phenomenon of wage stagnation, or "stagflation," with the slump in the economy, and public attention and concern increased about the costs of environmental regulation and their contribution to these problems.

In addition, many of the highly localized "end-of-the-pipe" problems had been largely dealt with by early regulations, and regulators began to extend their reach to address nonpoint sources, with far-reaching impacts over broad areas. Furthermore, the implementation of wetlands regulations authorized by the early 1980s extension of the Clean Water Act and Endangered Species Act regulations have had widespread impacts, particularly in the West. And with these increased impacts have come increased controversy, political polarization and adversity, which have strained and—in many cases—broken the old environmental consensus.

As an example of the current tenor of the debate, let me quote Carl Pope, executive director of the Sierra Club, from an article he authored in *The Washington Times* on July 27, 1995, as part of a special section entitled "Debate on the Environment: Which Way for Hill Reforms?" "The 104th Congress," said Carl, "has launched a fundamental assault on every American's right to a safe and healthy environment. This assault is cloaked in populist rhetoric, yet is funded by billion-dollar business interests that want to increase pollution of our air and water back to the levels of the 1960s, to cut down our national forests and obtain the right to wipe out wildlife on our public lands."

Now I could have quoted from other articles in this same special section from the opposing point of view that would have been just as vociferous against the environmental movement. But the point is that this sort of dialogue illustrates that indeed the old environmental consensus has broken down.

And so my principal recommendation—or perhaps plea—to the Academy, in addition to addressing those important technical issues raised by Tom Grumbly about risk assessment and the need to establish risk-based analysis, is to help devise ways and means to forge a new environmental consensus, which is beyond the ability of science and technology alone to answer but more appropriately within the realm of the social sciences. Whatever the final form of such a new environmental consensus, it will have to give results that are "cheaper, faster, better," that rely on much less government intrusion than in the past, and that rely largely on market-based solutions.

One of my favorite American philosophers that I like to quote is Groucho Marx, who once said that "[p]olitics is the art of looking for trouble, finding it everywhere, diagnosing it incorrectly, and applying the wrong remedies." Unfortunately, in the past that's often been the case. However, the new budget climate, with the increased requirements for accountability and with fewer resources, means that we have to do it better. We can no longer look for trouble and find it everywhere—we can't afford it because we're going to find trouble practically everywhere we look. And we can no longer afford incorrect diagnoses, nor can we afford wrong remedies. So, as I said earlier, if there is any way the Academy—realizing it is beyond the realm of science and technology alone—can address ways and means to forge a new environmental consensus, it would be extraordinarily helpful.

Congress also has to get its act together to properly address long-term problems, such as the environment. In spite of the downsizing of the House, which cut several committees and staff by one-third, there is still an enormously complex and laborious system in place, with something like 21 committees in both the House and the Senate, four joint committees, and some 187 subcommittees. It's very easy to get bogged down, particularly in environmental issues where jurisdictions are spread across a number of committees and subcommittees in the House and Senate.

At this point, I think I'll bring my remarks to a close and turn the floor over to my colleague from the Senate, Dave Garman. I look forward to participating in the remainder of this forum and continuing dialogue on this important topic.

David Garman

Professional Staff Member,
Senate Committee on Energy and Natural Resources

In thinking about how to establish our environmental goals for the next 25 years, I settled on **some core concepts** which I believe are fundamental to the exercise. I'll briefly outline these core concepts, lay the groundwork for their defense, then get on with an outline of some goals I believe we ought to pursue in the next 25 years.

The first core concept is one that seems obvious to me, but is often lost to the activists of orthodox environmentalism:

- Environmental values are important, but they are neither absolute nor intrinsically superior to all other values, and they cannot and should not be pursued at the expense of all others. As Lynn Scarlett writes in a paper that will soon be published by the Reason Foundation: "The challenge we face is how to give robust expression to environmental values among the many values individuals hold and pursue in their daily choices about how to allocate their (and society's) scarce resources . . . and how to allocate rights and responsibilities to enhance overall quality of life, including environmental quality."

My second core concept:

- The "spaceship earth" paradigm that holds that we are rapidly running out of key natural resources—conventional wisdom at the very foundation of the original Earth Day in 1970—is wrongheaded and detrimental to our efforts to achieve a cleaner, healthier environment for the majority of the planet's inhabitants.
- Let me say it again and say it bluntly: The Club of Rome-Limits to

Growth-End of Affluence-Small Is Beautiful world view that we are running out of resources is not only wrong, it stands in the way of environmental improvement in most areas of the world.

The third core concept:

- Technology, long held by orthodox environmentalism to be the **curse** through which man has despoiled the "natural" environment (whatever that is!), has instead turned out to be the **blessing** through which environmental gains have been and will be achieved.
- Moreover, the wellspring of environment-friendly clean technology will be the old "enemies" of orthodox environmentalism—industry and market mechanisms.

The fourth core concept:

- Market mechanisms can't do it all, and environmental regulation will necessarily continue to play a key role. Many environmental regulations are reasonable allocations of responsibilities, but we have to use them cautiously. They are like "headless nails," easy to put into place but nearly impossible to remove.
- Regulation can also result in gross misallocations of society's resources. (The $8 billion "cleanup" at Hanford that has yet to really clean anything up and hazardous waste cleanups under the Superfund program come to mind.)

The fifth core concept:

- We must recognize—and celebrate—the environmental successes of the first 25 years since the first Earth Day.
- We have come a long way. The air we breathe and our lakes and rivers are far cleaner than they were at the time of the first Earth Day in 1970.
- This success has left us with a new and exciting fundamental choice: We can continue to allocate political and financial resources in the pursuit of marginally diminishing levels of "environmental success" here in the United States, or we can adopt a broader, more global view that achieves a greater good for the health and safety of a majority of the Earth's inhabitants.

And finally, my sixth core concept:

- Perhaps the greatest tools that we can employ for the sake of the global environment are democracy and the promotion of economic prosperity, even if new economic prosperity in the Third World is accompanied by increases in resource consumption.

Now let me highlight and expand on just a few of these key concepts. Because time is short I'll skip directly to key concept number two:

- Earth Day 1970 was greatly influenced, perhaps even inspired by, notions that we were running out of key resources.
- This "declinist" point of view was predominant in the 1960s and 1970s, and it continues to have a following among orthodox environmentalists.
- This view holds that natural resource scarcity and rising commodity prices require that we undertake a crash program of government intervention to reallocate and preserve resources for the sake of the environment.
- This view is wrong.
- In their 1976 book *The End of Affluence*, Paul and Susan Ehrlich predicted that mineral supplies would be largely depleted by 1985, and referred to the 1980s as "the catastrophic decade" in terms of resource consumption. In fact, as Harvard's Roger Stavins points out, for nearly every mineral resource predicted by Ehrlich to be gone by the year 2000, "reserves have increased, demand has changed, substitution has occurred, and recycling has been stimulated."

Another example:

- A popular 1972 book entitled *The Energy Crisis* projected 30 years of gas reserves and 20 years of oil reserves. It was not out of line with prior "official" estimates:
  - —In 1939 the Department of the Interior projected that the United States would run out of oil in 1952.
  - —In 1947 the State Department pronounced that no more oil would be discovered on U.S. territory.
  - —In 1951 the State Department declared that global petroleum supplies would be exhausted by 1964.
  - —In 1979 the International Energy Agency predicted that global petroleum reserves of 645 billion barrels would be exhausted by 1985.
- The fact is, proven oil and gas reserves have risen over 700% since 1950. In 1990, we had 1 trillion barrels of global reserve.
- Natural resources are today half as expensive relative to wages as they were in 1980; three times less expensive today than they were 50 years ago; eight times less costly than they were in 1900.
- Jane Shaw of the Political Economy Research Center claims that there "has never been a nonrenewable resource that has actually disappeared, because in a market system people start looking for substitutes when prices rise."
- Gregg Easterbrook, in his recent book *A Moment on the Earth*, generally agrees. He writes that "wood, coal, rubber, oil, copper, tungsten, chromium and platinum have all been subject to pronouncements of eminent exhaustion during the industrial era . . . . All now exist in greater supplies, selling at lower real-dollar prices, than they were when they were supposedly about to exhaust."

We will get to why this is so important a little later on, particularly as we take a more global view.

Expanding a bit on core concept number three:

- Technology, long held by orthodox environmentalism to be the **curse** through which man has despoiled the "natural" environment, has instead turned out to be the **blessing** through which environmental gains have been and will be achieved.
- The fact is, the **real heroes** behind our environmental successes are not the environmental activists in Washington, New York, and San Francisco.
- Instead, they are the engineers in hundreds of businesses and R&D shops in places such as Detroit, Seattle, and Albuquerque.

Why do I say that?

- In 1976, the best selling sedan was a high-emission Chevy Caprice. At 4,424 pounds it got 16 miles to the gallon.
- In 1993, the best selling sedan was a low-emission Ford Taurus. At 3,420 pounds it got 29 miles to the gallon.
- The Boeing 757 I rode out here on burned 30% less fuel per passenger-mile than jetliners of just a decade ago.
- When I finally get to ride on a 777, I will take comfort in the fact that it burns 50% less fuel per passenger-mile.
- The old dilemma of "paper or plastic" we all confront in the grocery line is no dilemma to me: 1,000 paper sacks require 140 pounds of material; 1,000 plastic sacks require only 40.
- New technology and techniques, coupled with market demands and, yes, environmental regulations, are driving a materials revolution where knowledge, rather than brute force, is enabling the manufacture of new materials and products with fewer resource inputs.
- After this conference, I'm off to Sandia National Laboratory to see what I'm told are some new technological breakthroughs in materials development that were achieved through private sector partnerships.
- These materials technologies represent the frontier of our future environmental success, enabling us to enjoy economic growth and the creation of wealth without ever-increasing resource inputs.
- The orthodox environmentalists may be uncomfortable with the fact that some of their old enemies—industry and market mechanisms—have been and will be the wellspring of environment-friendly clean technology, but that's the way it is, and the way it will be.
- The deregulation of electrical generation, for instance, will result in market imperatives to seek least-cost power, achievable in part by minimizing fuel input.
  — A few weeks ago I met with inventors of a patented technology that promises to achieve steam cycle efficiency gains, as well as the attendant lower emissions and fuel savings those gains will bring.

— Unfortunately, they were having problems signing up a utility to demonstrate the technology. Why? Because the utilities had little incentive to take the risk. PUCs would pass any savings back to ratepayers in a regulated environment, so why bother?
— Market forces at work in a deregulated environment will be good for the planet.

Skipping ahead to the fifth core concept:

- We must recognize and celebrate the environmental successes of the first 25 years since the first Earth Day. Here are just a few examples:
  — In 1970, only 25% of U.S. rivers met current Clean Water Act standards for fishing and swimming. That number is up to 56% today.
  — Ozone concentrations in Los Angeles have declined 40% overall since 1970, notwithstanding a tripling in the automobile population in the basin.
  — Los Angeles hasn't violated federal standards for sulfur dioxide emissions since the 1970s. Nationally, sulfur dioxide emissions have declined 20% during the 1980s, even as electrical generation from coal has increased.

These are stunning successes. However, many orthodox environmentalists seem uncomfortable with that success. In fact, many endeavor to deny it:

- Our Vice President says the U.S. environmental situation is "extremely grave . . . the worst crisis our country has ever faced."
- The former Senate Majority Leader, George Mitchell, declared that "we risk turning our world into a lifeless desert."
- The "father" of Earth Day, Gaylord Nelson, said in 1990 that our current environmental problems "are a greater threat to Earth's life-sustaining systems than nuclear war."
- According to Gregg Easterbrook, a self-described "liberal environmentalist," these statements are not only demonstrably wrong, they are fundamentally counterproductive to the environmental movement in the long term. Orthodox environmentalists, according to Easterbrook, are on the "right side of history" but on the "wrong side of the present, risking their credibility by proclaiming emergencies that do not exist."
- The fact is, we can't afford to allocate greater and greater political and financial resources toward a marginally diminishing level of environmental success here in the United States.
- We can't have more dollars chasing fewer results here at home. Instead, we should adopt a more global view that achieves a greater good for the health and safety of a majority of the Earth's inhabitants.

That brings me to the last of my guiding principles:

- Perhaps the greatest tools that we can employ for the sake of the global environment are the promotion of democracy and economic prosperity.

Let's take a look at the pressing environmental problems that affect human health globally:

- 1.3 billion people in the developing world live in areas with "dangerously unsafe" air;
- according to the World Health Organization, in 1993, four million Third World children under age five died from preventable respiratory diseases brought on by air pollution, mainly from the dung and wood smoke of cooking fires—that's roughly the number of people *of all ages* who died *of all causes* in the United States and the European Union in that same year;
- in 1993, another 3.8 million developing world children under the age of five died from diarrheal diseases caused by impure drinking water.

Again, to quote Easterbrook:

> Institutional environmentalism focuses on the real but comparatively minor problems of developed nations in part to support a world view that Western material production is the root of ecological malevolence. The trough of such thinking was reached at the Earth Summit in Rio in 1992. There, having gotten the attention of the world and of its heads of state, what message did institutional environmentalism choose to proclaim? That global warming is a horror. To make Rio a fashionably correct event about Western guilt tripping, the hypothetical prospect of global warming—a troubling but speculative concern that so far has harmed no one and may never harm anyone—was put above palpable, urgent loss of lives from Third World water and smoke pollution.

- Western, orthodox environmentalists oppose zero-emission hydropower in China and India, seemingly indifferent to the benefits to human health that would result if electricity replaced wood and dung fires for cooking.
- They oppose expansion of propane and kerosine, fearful that we will be setting the Third World on the "consumptive pathway."
- In truth, there is an ecological need for *more* resource consumption in the Third World. The "consumptive pathway" will be good for the environment—and good for the humans, particularly if it is done with technology transfers that make industrialization smarter and cleaner than what was experienced in the West.
- Coupled with democracy, industrialization results in citizens who are freer, better educated, less fertile, less sexist, more environmentally aware, and yes, richer. And there is nothing wrong with that.
- And as we've seen, there is no resource crisis that requires that the developing world be denied these advantages, even if we were inclined or empowered to do so.

— China is going to continue down the pathway toward industrialization, and there's nothing we can do about that. But we can assist them in choosing hydro over lignite.

So it's easy to see where I'm headed:

- The activists among us must endeavor to raise the public consciousness about the environmental problems that *really matter* to human health. And they ought to shed the doomsday pronouncements aimed at the proposition that consumption is evil.
- The politicians among us must be more courageous and wise in allocating the dollars where they can best make a difference.
- The scientists among us must endeavor to refute the doomsday exaggerations of the orthodox environmentalists if the science and data don't support it.
- Finally, all citizens should promote the spread of democracy, democratic institutions, and economic prosperity around the world. The absence of democracy and economic prosperity breed environmental indifference.

Thank you for allowing me to share these thoughts with you, and I'm happy to take any questions in the time we have left.

Presentation

John Wise* and Peter Truitt†

Deputy Regional Administrator, Region 9, and*
Manager, National Environmental Goals Project,†
U.S. Environmental Protection Agency

REVIEW OF PROPOSED NATIONAL ENVIRONMENTAL GOALS

John Wise:

While those papers are going around the room [handouts], I will proceed with my remarks. Thank you, Guy [Stever], for the nice introduction. You can in fact read my bio in your folders.

In December of this year, EPA will celebrate our 25th anniversary. This is an opportune time to assess where we have been for the first 25 years and position ourselves for the next 25 years. As a charter member of the EPA, having started in 1971, and a career executive in Region 9, in San Francisco, I want to offer you a perspective on this span of time and specifically on the evolution on environmental policy, which many of our commentaries this morning spoke about. I will be joined in my presentation by Peter Truitt, who is the manager of EPA's National Goals Project, who will outline the agency's work in progress on the national environmental goals.

As you know EPA was created in 1970 by a presidential executive order. The agency was not provided with an organic statute defining our mission and national environmental goals. EPA has fashioned a national environmental agenda out of the sum of many legislative parts. Today EPA administers a portfolio of some 14 major statutes, each one mandating a regulatory and enforcement structure to clean up, abate, control, and remediate pollution releases. Most of these statutes are single purpose and mediaspecific. And interestingly, each statute contains some form of a goal statement. For example, in the Clean Water Act there's a national goal for fishable and swimmable waters.

One can distill EPA's overall mission by looking inside each statute. You will find that we set national standards, we promulgate federal regulations, we issue permits to conduct certain activities, we license and register products, we inspect for compliance, and we enforce where it is necessary. We monitor for results. This is EPA's core regulatory agenda. And of course all of this cascades down through our system of government by delegation of regulatory authorities to state and local entities. Indeed, most environmental regulation in America is now performed by states and localities, not by the federal EPA.

All of this has produced some remarkable accomplishments in the last 25 years. We have substantially reduced mass-loadings of pollutants to the air, the water, and the land. We have installed pollution-control technologies at the end of the pipe or top of the stack. We have provided essential public health protections in the air we breathe, the water we drink, and the food we eat. We have arrested some ecosystem losses, and now we are starting the long journey of ecological restoration. Americans are adopting the ethic of pollution prevention. Prevention is now the environmental strategy of first choice for EPA and American enterprises. We have opened a window for public scrutiny under the public disclosure provisions of the Community Right to Know law. And we are starting the transformation to long-term sustainability and eco-efficiency in our uses of energy, water, and materials.

America has much to be proud of. I want to be so bold as to suggest that environmental protection is one of the most successful governmental interventions in the modern era. When we look at our accomplishments in the first 25 years, I think we should acknowledge and celebrate those successes, which incidentally happens to be Mr. Garman's fifth core concept, to which I heartily subscribe. And yet, a troubling mood of denial and despair seems to have settled over America. The anti-regulation forces seek to constrain or even roll back some of our environmental management system. On the other side, the environmental activists continue to proclaim doomsday. Both sides I suggest are preparing for the wrong battle, for the wrong reasons.

The American public expects continuing environmental quality. They demand equitable enforcement of environmental laws. Every poll I've seen seems to validate this; every public discussion I have engaged in Region 9 communities seems to tell the same story—a continuing expectation for environmental quality. So I suggest it's no longer a question of *whether* we shall have a quality environment, but rather *how* should we proceed?

For the next 25 years, America must fundamentally reorient our environmental agenda by building upon our successes. We need to update our statutory portfolio. Some of our laws are 25 years old, and they need to be updated to address a whole new generation of environmental challenges. The environmental landscape has changed. Our statutes need to be upgraded to vest EPA with a new set of tools to do the job. We need authorization for performance-based and prevention-based approaches, financial incentives and rewards, environmental information-driven decisionmaking, and voluntary and cooperative actions—all of these as opposed to reliance on the prescriptive technology-based controls that currently prevail under existing law. We need to shift our regulatory and enforcement authorities from compliance assurance to compliance assistance. We need to facilitate investment in pollution prevention by businesses, and we need to shift our priorities from strictly human health protection to the long-term sustainability of ecosystems, looking specifically at restoration of the integrity of ecological processes. We need to merge environmental and economic policies for a long run sustainable future. And we need to set ambitious but realistic national targets for environmental improvement and measure our progress towards those goals.

Now I will outline the purpose and history of our National Environmental Goals Project, then ask my colleague Peter Truitt to provide a more detailed overview of the proposed goals. We've handed out a piece that will guide you through that dialogue.

First of all, why are we doing this? The purposes of the National Goals Project are twofold. Number one, to strengthen understanding and support for the national environmental agenda by describing the expected real-world improvements that will result if we do our jobs well. Number two, to manage better—to tie our plans, budgets, and program evaluations to environmental outcomes so that the investments we make with taxpayer money pay back in terms of measurable and recognizable environmental improvements.

This project didn't start yesterday. Indeed, back in 1992, former Administrator Bill Reilly launched it. The project languished for a while until current Administrator Carol Browner directed her staff to develop the goals with full public participation. That process has ensued over the last couple of years. In 1994, EPA held a series of nine public roundtables around the country, which discussed what the national goals should cover and how they might be expressed. That process was extraordinarily rewarding. When you gather a group of a hundred people into a room and spend a day talking about the future, you begin to sense the different values of the participants, the stakeholders. You see how the dialogue starts to shape a collective future. That kind of public engagement informs the process for articulating national goals. And that public process will, of course, continue through the formulation of the final goals. Because without public involvement these goals are essentially meaningless.

In early 1995, EPA prepared a summary goals report that was reviewed by many government agencies and participants of the roundtables. Some of the guidance that reviewers provided to us was that the goals should be more visionary, which I think means that they should inspire people towards a future and not just be numerical measures of this or that. Reviewers said that we should describe how we will attain these goals, including the costs and who will pay. We also engaged Congress in the process, who came back to us saying, "Which goals and which milestones ought to be priorities? Just because we can attain these targets doesn't mean we should." I think that joins the debate we heard this morning about congressional intentions. But I believe that engaging the public who want a visionary statement, looking at the costs and the benefits, considering who's going to pay, and considering the question of priorities are all substantive issues that the agency is wrestling with.

The revised draft is now nearing completion. We will circulate it for review during October-November. I invite the National Academy to look at our goals and consider them in your own deliberations. Ultimately, if the government and public reviews go well, and we gain some measure of consensus, we plan on releasing the goals report by Earth Day 1996.

With that overview, let me ask Peter Truitt to walk us through the substance of the goals report. And then we'll wrap up and have a brief discussion.

Peter Truitt:

Thanks John. As John mentioned, last February we circulated a goals summary report to government agencies and the 1994 roundtable participants for comment. Reviewers wanted more detailed information on how we propose to attain the 10-year targets and on the economics, and they wanted a more visionary quality to the document. We are doing our best to respond to these concerns.

The current draft begins with an explanation of why EPA is developing the goals, how it is doing so, how the goals and milestones will be used in managing the agency, and how we intend to address the issue of priorities. We then propose a vision statement, which I will go into in a minute. Next, we describe the seven guiding principles that are shaping EPA's operations under the Browner administration: (1) Prevent pollution. (2) Be a partner. (3) Strengthen science and information. (4) Protect all communities equally. (5) Protect ecosystems. (6) Promote environmental accountability. (7) Reduce costs and red tape.

The report then examines the goals in the major laws administered by EPA. The message here is that few of our laws have measurable environmental goals. The 1990 Clean Air Act is an exception, with its targets for cities to meet air quality standards by specific years.

The meat of the report is the presentation of the long-range goals, the milestones for 2005, and the strategies to attain them. There are now 15 goals. They cover clean air, climate change risk reduction, stratospheric ozone layer restoration, clean waters, healthy terrestrial ecosystems, healthy indoor environments, safe drinking water, safe food, safe workplaces, preventing accidental releases (like Chernobyl and Exxon *Valdez*), toxic-free communities through preventing wastes, safe waste management, restoration of contaminated sites, reducing global environmental risks, and better information and education to improve environmental understanding.

For each of these, there is a long-range goal statement and a set of milestones. The milestones include a discussion of what we know about past trends, how the 2005 target level was set, and the data we will use to track progress.

Also, for each goal there is a strategy section providing an overview of the current approaches EPA is taking to achieve the set of milestones.

The report winds up with a section on overall costs and benefits. We had hoped to discuss the economics relating to each goal and its 10-year targets, but we simply don't have the information to do that now.

Now, it's important to understand that we are not announcing these goals. We're proposing them. We're proposing them to federal and state government agencies, Congress, and the American people. After our government review of the proposals, we'll be having discussions with the public. We'll probably have a second set of roundtables. Given the kind of specifics in this report, it will make for a very educated, enlivening discussion about what Americans are trying to accomplish in environmental protection.

John mentioned that earlier public reviewers thought the report should be "visionary." We are proposing the following vision statement for the U.S. environment:

> We envision a 21st century America where healthy and economically secure people sustain and are sustained by a healthy environment. Everyone breathes clean air, drinks clean water, and eats safe food. Homes and workplaces are free from toxic pollutants. We swim and fish in clean waters in the cities and the countryside and enjoy the gifts of nature in our neighborhoods. All Americans have a respect for nature. We use and recycle natural resources efficiently. Clean and plentiful energy fuels a growing economy. The renewal of diverse, thriving communities of plants and animals offers bright prospects for prosperity and fulfillment for generations to come.

The 15 goals and 76 milestones are proposals for policy statements and measures of progress, with targets, for moving toward this vision. The milestones are being developed in a number of ways. Some of them (a very few of them) are actual statutory, regulatory, or treaty requirements, such as the clean air, climate change, and stratospheric ozone targets. But most of the rest are discretionary targets, established by EPA staff using their professional judgment about what we realistically can achieve if we as a nation do a good job protecting the environment. We assumed continuation of existing and planned EPA programs, funded at the 1995 level. Of course the government share of the environmental protection costs is pretty small. Achieving some of the milestones may require substantially more societal expenditures.

John mentioned the statement by a Senate Appropriations Committee staff person that just because we can attain these targets doesn't mean we should. It's a very revealing statement, given the budget situation. I should note that the goals and milestones were not selected mindless of risk. We cover every one of the high risk problems identified by EPA's Science Advisory Board in the Reducing Risk report, which many of you may be familiar with. We also have goals for six subjects that the Science Advisory Board did not consider to be high risk. But many of these, such as waste cleanup, are of high interest to Congress and the

American people. Also, achieving the waste goals will, of course, contribute to achieving many of the other goals such as clean water. They're all linked together, so it is hard to evaluate the risk reduction associated with one milestone.

Aside from the criticisms I mentioned earlier, we have heard very positive comments about the goals project. People are saying, congratulations EPA, at last you are delivering an environmental results management tool that's going to be very useful. It is a strong direction-setting document. It's useful not only for planning and budgeting, but for program evaluation. As time goes by, we will be reporting whether we are above our targets, or below them, and why. So it's an evaluation tool. And it will be useful for communicating with people in terms of results.

There are some issues that concern us. Will our report stand up to scrutiny by scientists and economists? We don't know yet. Is it unrealistic to assume continued societal investments and government programs? We believe the American people are willing to invest in environmental protection that promises results. Are all the goals equally important? Probably not. Perhaps we should be setting higher targets for higher-risk problems and paying for them by easing the targets for lower-risk problems. It makes conceptual sense, but drawing conclusions from the information we currently have is difficult.

With that I'm going to turn it back over to John Wise who will wrap up.

John Wise:

I'll wrap up briefly. As you can see, the scope of this endeavor is truly heroic, and to the extent that we are successful in engaging a public process, that we can withstand the scrutiny of the science community and the economic community, we will actually propose a set of goals for America next year.

How is all of this going to be used? That's an important question. When these goals become generally accepted as part of the country's environmental agenda, we will then craft EPA's strategic plan to chart a course to the milestones. We'll use that strategy with the milestone targets as a base for our annual planning and budgeting, and to develop our performance agreements with state and local agencies—which as I mentioned earlier, carry the majority of the load in terms of environmental protection at a state and local level. We'll also use the goals-based strategic plan to fulfill some of our obligations under the Government Performance and Results Act.

Lastly and most importantly, we'll prepare annual reports for the public that explain the progress that we're making—or not making—and reaffirm our commitment to environmental quality. So with that, let us stop and listen to you.

Judith Espinosa* and Peggy Duxbury†

*Former Secretary of Environment, State of New Mexico, and Member;**
Coordinator, Principles, Goals, and Definitions Task Force;†
President's Council on Sustainable Development

REVIEW OF PROPOSED NATIONAL ENVIRONMENTAL GOALS

Judith Espinosa:

Some of the comments that you all have made just now, plus the questions that you had of EPA, are probably some of the comments that you will want to have of the President's Council on Sustainable Development and our goals. Some of them track. Ours, however, are probably even broader than many of the EPA's goals.

For those of you who have not seen this lovely green and kind of yellowish colored piece of document out on the table, please pick one up. This is the public comments survey for our goals. There are goals, and there are also the five themes that the President's Council developed along with some policy issues and considerations. The goals Peggy is going to talk about, along with the themes and some of the indicators. I am going to talk about some of the background to the Council's work and some of the more controversial points that were brought up during the course of the Council's discussions over the last two years.

It was in June of 1993 that the Council was formed by President Clinton. It is composed of 25 persons, about one-third from industry and business. Large corporations, no small businesses, are represented. About one-third are cabinet members and other governmental agencies, such as Dr. Baker this morning, and one-third are other "nongovernmental organizations." I wonder sometimes how I was selected to be on the Council, but I was to represent state government. At the time, I was Secretary of Environment for the State of New Mexico. Since the elections of 1994, I am no longer that and am back in private business. And so we have come a long way in a couple of years. The Council did have some folks who were critical of the way it got started and of the slowness of the starting, but those of you who have worked with a diverse group composed of 25 people with adversarial viewpoints, in many cases, hopefully can well understand that much of our first six months was taken up with forming relationships and looking at developing a vision statement and actually a definition of sustainability.

We wound up starting with the definition of the Bruntland commission, which was, "To meet the needs of the present without compromising the ability of future generations to meet their needs, as well." Our vision was, "To look for a life-sustaining Earth with peace and dignity and equity for all persons, and the opportunity to have a healthy and safe environment. To live the quality of life that we ourselves would like to live, and to maintain the viability of our natural resource base." Simple visions. Simple statements that you heard from many of the speakers here this morning and that we heard from people all over the country.

We attempted to go outside the 25-member Council by engaging persons outside the Beltway, outside government—and real people who live in communities so that we could get a diversity of information and a diversity of viewpoints for the Council—so that we could provide some outreach and so that we could expand the view of the Council, as well as provide for that kind of, if you will, nebulous thing called "public input." We did this through putting together about eight different task force groups, some of which involved Energy and Transportation. I was on the Energy and Transportation task force. I was the co-lead on the Public Linkage and Dialogue task force, which was to engage grassroots groups and other persons around the country. We also had a Natural Resources task force looking at various issues. Ecosystem task forces looked at industrial processes and product stewardship and the like.

The task force was reaching into the general populace of the American public to attempt to get a broader viewpoint. We had put out our principles and vision statements in the summer of last year, and right now, we are putting out our goals. So if you would please pick them up and, if you cannot comment today, please send comments to us later. We certainly need to hear from as many persons as possible—and certainly those of you who are members of the National Academy of Sciences, those of you who work on the committees, it is important for us to hear from you. I don't need to go into a lot of discussion because much of what has already been said by the speakers are points of contention during the Council's discussions during the Council's debates.

Our meetings were in public forums. We also had separate committee meetings, retreats, to talk about principles and goals and where we wanted to go for the future. Some of these are controversial issues, and I think it's important for to you know this because these issues are now in the public debate. And there are some really, I believe, hostile discussions going on between both extremes, sometimes not too much in the middle. And I think it behooves all of us to understand what those issues are and to see if we cannot come to some middle ground.

The Council is a consensus-building process. Therefore, a lot of what you see in this document was built out of the controversial issues and was brought into the process of consensus. One of the first concerns that we heard this morning, and one of the issues that was brought up time and time again in Council discussions, was risk assessment. Where are we going with risk-based regulations? Where are we going with risk assessment? What is risk assessment? What are the concerns with risk assessment? The debate has been that if you look purely at risk assessment, you are not going to be able to take into consideration some of the social issues and as well as some of the non-economic issues (or the economic issues, as well) that befall a community when it's looking at risk—whether it be risk in toxics or whether it be risk in hazards or clean water issues, as well. You can see, I think, in the goals, that risk assessment is certainly one of those indicators and one of those policy issues that has been put out in the goals. I must say that we came to no consensus in the Council regarding risk assessment. We didn't feel it was our job, and we did not get into the specifics of what would be risk—what is a good risk-based process in our environmental regulations. I think there was some consensus that we need to look at risk factors and that we need to look more in that direction.

The concern, of course, is what I think I heard this morning from one of the panelists from the congressional side—that there will be some risk-based types of regulatory and statutory mandates issued by this Congress. I also heard that Congress doesn't know right now what the exact procedures will be. That kind of concerns me, because I don't think Congress should be dealing with what the exact procedures for risk assessment should be—I believe that it should be groups of scientists and groups of individuals who work in the environment as well as social scientists looking at risk assessment issues.

One of the other concerns, or one of the other debates that went on, was the issue of equity. What is equity? You will also see a goal on equity. Equity sounds good. Environmental justice is wonderful—societal equity, the opportunity for all people. But there was a big debate on that because one of the issues that kept coming up on equity is whether we were talking about distribution of wealth? Are we talking about types of implementation of policies and procedures that are going to be costly? Are these going to be based on non-economic or non-capital types of provisions? Are we talking about socialism? We got very basic about this. So the issue of equity was one that was debated for a year among Council members, and we think that what we drew up relating to equity is the consensus of the Council.

Another issue was better science. I'm not quite sure what that means. Perhaps the National Academy of Sciences can help me out and help all of us out on what better science is. There was a lot of discussion by business people that we need better science in the regulatory process. I don't think anybody disagrees with better science, but I'm not sure that anybody knew what it was. The other concern was that we already have standards. We already have environmental standards that are set by scientific methodologies by modeling and by scientific endeavors. So what do we mean exactly by better science? I'm sure that John and Peter can tell you what type of scientific proof they use before they go out to set their regulatory standards. I know in state government we have to do the same thing. We cannot go before our boards and commissions and ask them to set a regulation without our having some scientific background regarding that regulatory process. So we need to decide what is better science. That is not reflected in the goals. That is one of the challenges that I would pose to the National Academy of Sciences as we look at science and technology and how it relates to not only the environmental goals of the PCSD, but also the EPA, and also to what Bob Watson is going to be talking about here in a bit.

Climate change was also a big issue. I think we decided that we would agree to disagree. So you don't see anything specific in the goals and policies on climate change. Most of the business people really were very strong in their assessment. They did not believe that global climate change presented itself as an environmental factor. Not being able to get over that, we proceeded with other policies and goals. So those of you that have some strong opinions of climate change may want to make some comments on that.

Sustainable agriculture—again, as I mentioned, we talked a lot about safe food sources. We talked a lot about subsidies in the agricultural field, which basically did not fall within the goals of sustainable development and we need to rethink. We need to rethink these government subsidies that do not assist us in reaching our goals of sustainability. We also talked a lot about the export of non-safe chemicals to other countries, thereby contributing to global food safety issues as well as the use of pesticides. We also talked about the use of pesticides in this country. It's a difficult issue, very contentious. There are some policies and indicators on sustainable agriculture, but probably not as strong as, and I will say this personally, as I would like to have seen it. Perhaps you can give some direction to all of the Council on what we ought to look at in sustainable agricultural issues.

One very large issue was economic growth versus economic prosperity. As you can see by the goals, the economic prosperity folks won out. Many people, (and we heard this in our San Francisco meeting from a lot of people) community-based people, local government people, local groups of real people, came to us and said that we no longer need to talk about economic growth in this country, because that is not sustainable. It is not a definition within the term of sustainable development. What we need to talk about is economic prosperity. We cannot continue to grow and grow, and consume and consume at the rates at which this country is doing, and overpopulate ourselves—not only in this country, but around the world—and be able to contribute to sustainable development. So prosperity is what we need to look at. And we need to make sure that people have the quality of life that they're looking for. I don't know that everybody that went away from the Council, or has gone away from the Council, is satisfied with the issues of growth versus prosperity. So that debate continues, but it was a healthy debate and I think it should continue in the mainstream of society.

Energy conservation, renewables, mobile sources versus stationary sources of energy, and energy issues generally were, as usual, a topic of discussion. We had representatives from General Motors, and we had representatives from the electrical utilities, and then we had, of course, all the representatives from the environmental community. So there was a great topic of debate regarding the issues of transportation, mobile sources, how do we look at the automobile, how do we look at emission standards, how do we look at energy conservation in relation to utilities and mobile sources. Those, again, are not reflected in this document. They were discussed, and they may be at the next round of goals and policies.

Population—the issue of population stabilization, as you well know, in general is certainly a big and controversial issue. We had a religious roundtable where we brought in religious leaders from all over the country, who actually have been engaged in sustainable development concerns for far longer than the Council has existed. They gave us some very good points regarding the process of population stabilization—the religious issues as related to environmental stability and environmental protection. It was a very interesting discussion to hear from a group of people who were involved in many issues of quality of life and population growth and environment—and how they view, as religious leaders from all sectors, both Christian and non-Christian, how they view sustainable development in the context of our country. So we did talk about population stabilization. Within the context of the goals, you see those.

Free market mechanisms certainly were debated back and forth. I think our consensus was that we have market mechanisms that we need as tools in the marketplace in order to look at regulatory and environmental goals. We no longer need to look at the typical command-and-control type of system that we have had in place for 25 years. Of course, that isn't all agreed to by everyone. There is a certain lack of trust that corporations, as Tom Grumbly would put it, are going to have the moral ethic and the morality to go forward with sustainable development and good environmental protection goals without the force of law, without the force of regulations, without the command and control that we've known. And it's going to be very difficult, I think, for that trust to be there, in the long term for people. I'm not just talking about "environmentalists"; I'm talking about people in general (because I've worked with grassroots people around the country). People must trust their business and corporate entities to do the right thing environmentally. If that is the case, and if we can come to that corporate ethic in this country, then I believe that we can certainly get to the point where we do not need command and control. We can have the goals, and we can have the programs that are going to reach the goals, and we can have the flexibility.

Certainly as a state regulator I'm kind of tired of having to regulate all the time. When the EPA tells you that it's the states who are doing the environmental enforcement, they're right. And it's nice for them now, because they don't have to worry about it, for the most part. We're the ones that take the flack. We're the ones who go to the legislatures, and we're the ones that have to ask for the budget moneys to be able to do the environmental enforcement. Some of it no longer works the way that it did 25 years ago, and some of it is no longer necessary, and some of it has to be more flexible. And we certainly, as many of the other speakers this morning said, have to get away from the way we've been doing business. But we certainly have to take care of our business. And when the command-and-control issues go, what will be the standards? Are they going to be scientifically based? And do we dismantle the environmental regulations that we have, or do we make sure that we have the good environmental protection we need and go forth with different types of ways of achieving that.

So just to conclude, I would like to leave the National Academy with some challenges and maybe some questions. I might say that as we moved around the country and talked to various people, we found that many communities and many local governments and many grassroots organizations in neighborhoods and in their circle of friends have been doing sustainable development activities now for three or four or five years. They have looked at the quality of life in their communities. They have looked at the necessity for environmental protection within their communities. And they have looked at how to link up those communities, those issues of environment, with issues of how they live day to day in their communities. It was very impressive. I think the Council members, at least for myself and most of the Council members, learned just about as much on sustainable development by talking to real people and getting out and seeing what they've already done—far ahead of all of us, EPA and PCSD—on setting goals for their communities. They have indicators. Seattle has indicators. Du Page County in Illinois has indicators for successful, environmental, achievable, sustainable development goals. They're out there. We don't need to reinvent a lot of this. What we need to do is scientifically work on them to make the science good, if you will, and to make sure that we have the backup for those indicators that we believe truly work in our communities.

So I would ask the National Academy of Sciences how scientific methods and your informed science research and procedures can help the PCSD's goals and policies. How can we measure quantitatively and scientifically some of the indicators that we've set out and some of the policies that we wish to achieve so that we test them—so that we test them in the next five to ten years and know whether they're right or wrong—know whether this country is proceeding in the right fashion. How can the Academy help us with risk assessments? What are the variables that need to go into looking at risk assessments—cultural, religious, social, biodiversity values that are not found in a hard science context? But how can we make a risk assessment work including those variables. I think you heard Tom Grumbly talk about that this morning, as well as others. It is a challenge. How can the scientific community bring trust between the large corporations and the businesses in this country, and the people? How can you make that link, so that when we talk about the science of environment and the science of biological diversity and the importance of that to all of us, we ensure that there's going to be that link between the community and between people and between its institutions? And I do believe that science can help us with that. How can we look at scientific methods to influence and professionalize some of the debate that is going on today among all sectors of our society? How do we make sure that it's professionalized and that we look at it in a way that is not one extreme or another, so that we truly have some real backup to do that. I'm not talking about playing games with numbers, which you all know can be done. We can get one expert or another to come in and say what we want to say, but how can the Academy truly look at the nation as a whole and make sure that we are going in the right direction and that we inform our public institutions in the correct way?

And, lastly, how can science and how can you help us to get an informed citizenry so that we make good, informed decisions and the public can inform their politicians or their government institutions on where they want to go? Because, frankly, as far as I'm concerned, sustainable development is not going to work unless the public buys into it. And it's not going to work unless we have an informed public that really cares about what it is we're talking about. That informed public is going to back up the environmental goals and the economic and equity goals that we seek to set out for this country. Thank you.

Peggy Duxbury:

What I'm hoping to do is to walk through the ten goals and the three sectoral goals that the Council did develop after having had this long process that Judith has described. And, I think, keeping in mind before I do walk through them, that the Council basically did the same thing that we all did this morning or that the speakers this morning did—which was, they attempted to think about what should be, and not just the environmental goals for the next 25 years but what should be the goals of this country for the next 25 years. And I think that there they would answer the questions of what should be the environmental goals by saying they must be developed looking through the lens of sustainable developments.

And as David Garman discussed this morning when he talked about his core concepts, environmental goals, environmental policies, must be looked at with an eye more carefully to the economic and social costs and economic and social values that are involved. Environmental decisions will be better if some of the economic costs are taken into account. But I think what he didn't get into is that this is not necessarily a competition or a zero-sum game, when you're trying to balance the environmental, social, and economic consequences of public policy decision-making. This isn't in competition or this isn't always a conflict. Sometimes it will be and sometimes there will still be trade-offs, but there are also times when you come up with better decisions.

I think one of the other dimensions of sustainable development that is crucial in our developing goals, is one that Bob Watson talked about this morning as well, and that is the issue of generational equity. The whole issue of equity is something that is woven through most of what the Council does—both generational equity and distributional equity. And that's a very hard thing to do, because equity is a broad term, it's a vague term, and it means something different to almost everybody that you ask, "How would you define equity?" But it's clearly one that the Council has spent a long number of hours trying to define, at least for themselves as a council. So with that in mind, one of the things that I was going to go into was the process of how did we get to these ten goals? But I promised Deborah Stine that I wouldn't go through a blow-by-blow process discussion of how we went from basically 300 goals, which is where we were in February, to the 10 goals that you're looking right now, plus the three sectoral goals. So I'm saving you the war stories.

But I think by saving you that, I don't want to dismiss that process as not important. Especially, in looking at this new approach of decision-making—where you're trying to do collaboration and you're trying to bring in more stakeholders that perhaps didn't have a voice in the past—you really do have to have a process that has some integrity, a process that people believe in, and a process that involves trust on all sides. Because if you don't have those things, if you don't have that kind of process, you can't get to the next step, which is actually trying to come up with goals and trying to set goals. And then when you try to implement them later on, if you haven't had that open process, you then run into troubles when you do your implementation.

So with that said, I think you have in your briefing books under, I believe it was Tab 13, the goals that were discussed by the Council; they are still draft goals, but they are closer to final than they are closer to draft, I believe, at this point. But we do have a public comment period in the next three weeks, and certainly this group, any comments that you can make, either right now in the next few minutes or in the written document that we've handed out, would be extremely valuable to us. Obviously, when you go from 200 or 250 goals to 10, you do some prioritizing, and these goals, as Dick Morgenstern pointed out in his paper, are really very broad and very general in their scope.

The Council, like EPA, faces the same tension between very visionary, broad goals that were more inspirational in nature and very specific goals that would tend to be much more quantifiable and probably much more specific, but also could be seen as much more prescriptive in how they were set forth. I think it's obvious where we fell on that tension. We fell toward the broader side. And part of that was the fact that we were 25 people who were trying to come up with a collaboration. And part was that we felt it focused us a little bit, if we just did 10, rather than 200. The first goal is a healthy environment goal. It was from about a hundred different ones. Like all of our indicators, we spent quite a bit of time on indicators of progress and learned a lot about how absolutely important indicators are to a goal-setting process. If you can't measure it, is it really a goal? And in some cases these can't be measured yet. Either our goals can't be or there are not ways yet to measure what it is we're trying measure. Because a lot of what the PCSD is looking at is what Bob would say is very, very wooly and fuzzy still. But that doesn't mean just because it's hard, we shouldn't try or we shouldn't start thinking about it. And in a couple of our goals that's the case. It isn't as much the case on the healthy environment goals. There are data out there for indicators.

But I guess the area where we felt the indicators needed to be strengthened was in the area of issues like environmental justice and distribution over those environmental risks across the country. Are there segments of our society that either through cumulative environmental problems or site-specific environmental issues bear a bit more of the burden of environmental risks in our country?

Our second goal is economic prosperity. Probably the most noteworthy issue about that goal was the huge controversy and the long hours of discussion that took place around one word: Should you say "prosperity," or should you say "growth"? And I think if you were to poll the Council it would be a 50-50 division between those that feel very strongly about using the word growth and those who feel equally strongly that the word is prosperity and not growth. Both at the Council level and also in all of our public hearings, this duality between these two words created more emotional discussions and heated debates than perhaps any other issue that the Council faced. And we did, in our vision statement, we used the word growth; in our goals, we used the word prosperity. So I guess we straddled discussion. But underneath that symbolism that those two words took, I think finally the Council, at least, has come up with a vision of what they mean. Whether they use the word prosperity or use the word growth, what they viewed as a strong economy was more than just looking at a measurement of the GDP, how fast the GDP grows. It's got to be more complicated than that. It has to take into account how fast we are using up our environmental resources, and can we better measure some type of green GDP or an environmental account? It has to take into account better ways of looking at how income is distributed. The Council expressed its concern that we are seeing a polarization between the haves and the have-not's and for our economy in the long run in 25 years, that is not something that's going to be sustainable, economically or socially.

The third goal was our equity goal. And there we did not develop any indicators of progress. It is such a broad issue that to even attempt to narrow it down to four or five ways to measure equity was, at least for the Council, too hard to do. If anybody here has some ideas of where to get started on that, we'd welcome them. Because measuring equity, and I think in this case, it's equity of opportunity as opposed to equity of outcome, but I guess you'd consider more social distribution or income redistribution type areas.

The fourth goal is conservation of nature. Our look at conservation issues was really driven by a watershed approach. It was really looking at regional ecosystems, and one of the gaps that probably Bob Watson would comment on here is that we didn't look at some of the global threats as much as we perhaps should have or could have. Part of that was because one of the big ones, climate change, was a very controversial issue and we weren't able to come to a strong consensus on it. And part of it really was that we learned so much through our Natural Resource's task force on the site-specific watershed that it was really reflecting more of where our findings and our education came from.

The fifth goal is the stewardship goal. That one really sort of changed names from time to time. At one point it was called the personal responsibility goal; one time it was called, I think, an environmental stewardship goal; it ended up just stewardship. But that is really looking at, How much are we using? How well are we? And how efficiently are we using the resources that this nation has?

The sixth goal is sustainable communities. As Judith said, communities really were the heart of the PCSD. And we learned more listening and learning and watching what some of the communities were doing than perhaps from any other group that came to us or from any other experience that we had. There really are a thousand flowers blooming out there, and what our challenge will be at the Council level is how we try to start to find trends in what these communities have done. What commonalities are there between what Chattanooga has done and Seattle, Portland, or different places around the country—where what works in Chattanooga won't work in Seattle, Washington. Yet there are some areas of common ground where lessons are transferable, and I think the Council is hoping to be able to pick up from those transferable lessons. And that's what the sustainable communities goal is attempting to do.

Our seventh goal is another kind of vague one, and that's civic engagement. None of these concepts of sustainable development will work if you don't have an engaged and participating society. And it's essential; it's a hard thing to measure. We did, unlike equity, attempt to come up with some measurements of this—quantifying civic engagement. There's not much out there. There's a professor up at Harvard, Putnam, who is attempting to do it. He's starting to look at what makes some communities healthier from a civic point of view and what makes some fail in terms of having people involved in decision-making—not just government decision-making, but all sorts of decision-making. He calls it social capital. How do you build social capital? So that's what our indicators are attempting to get out there. There's a lot more work that's going to have to be done in that field.

The eighth goal is on population: stabilizing U.S. populations. The most important indicators—besides the more predictable measuring population growth—are probably the status of women and the role of women in society, particularly, the role of young women in society. And I think that our two co-chairs on the population task force, Tim Worth and Diane Dilinridgely, spent a number of days in roundtables talking about demographic trends and issues, and we try to capture that a little bit in our indicators.

The ninth goal is international responsibility: requiring the United States to take a leadership role on sustainable development. The indicators there could be a lot more extensive, but we focused primarily on the federal government. Treaties, international assistance, and that list could be far longer, but in our attempt to be focused we really looked at the federal government.

And the final one of the primary goals was on education. I think, perhaps along with communities, education was another theme that has to be woven, along with equity as well, into everything that we do—developing education that teaches some of these things, whether it's civic engagement or stewardship or personal responsibility. Again, none of that can be done if you don't have a well-educated and well-informed population. And there we looked at not just the formal K-12 education, but also non-formal education—lifelong learning issues; worker vocational retraining issues. There was a broad number of issues that both the task force and the Council looked at in the field of education.

And then what's still being discussed, and we'd love your advice on this, is will the Council have sectoral goals? You've got three of them in your document—one on energy, one on transportation, and one on agriculture. The problem is that once you start having sectoral goals, why isn't there one on manufacturing? This is a good question to ask. Each of you probably has a favorite that ought to be here and is not. And that's how you end up back with 50 or 200 or 500 goals again, when you start getting sector specific. But these were three that we had task forces that had done lot of work on, and it seemed to make some sense to try to deal with them. Certainly on energy and transportation, it did help us at least start addressing some issues like climate change through some of our behavior. So that's one of the advantages of having the sectoral-specific goals.

That kind of is the walk-through. I'm sure all of you have a lot more questions about both the overview of the PCSD that Judith gave and our specific goals.

Presentation

Gilbert S. Omenn

Dean, School of Public Health and Community Medicine, University of Washington

RISK ASSESSMENT AND RISK MANAGEMENT: REPORT ON THE PRESIDENTIAL/CONGRESSIONAL RISK COMMISSION

Thank you very much, Dr. Stever. It's a pleasure to be included in the program and to see many of you during your 10-day immersion in this important project.

The notion of identifying and building an agenda for national science and technology goals for as crucial and large an area as the environment is ambitious. I gather your task is to envision what the goals should be 25 years from now, and to build the research and technology base to prepare for achieving that vision.

Themes

Certain themes are well developed in the papers that have been prepared for this meeting. First, sustainable development. I think there's still a task to explain this widely used phrase, but basically it's the convergence of economic and environmental objectives for the long term.

The second theme, not often highlighted in environmental circles, is the crucial role of population numbers and consumption patterns. Human activities account for most of the environmental problems we're trying to redress or anticipate, though we should not neglect natural disasters and the opportunities to anticipate and mitigate their effects.

Third, we need a stewardship strategy that combines pollution prevention and aggressive cleanup. The most dramatic examples of lack of pollution prevention are the Department of Energy nuclear weapons production sites in this country and, even more so, the nasty situations in Eastern Europe and the former Soviet Union.

Finally, we must somehow engage everyone, through the media and through our communities, to translate interest and support for protection of health and environment into core values and personal responsibility. At an international level, we need to draw the United States and all other countries together to share objectives, to achieve harmonization of test methods and risk assessment for chemicals and other hazards, to recognize ways in which environmental issues may be used or misused in trade negotiations under new trade agreements, and to bring the environment into the center of discussions about international relations and economic development.

Risk is the coin of the realm in protecting health and the environment. Risk assessment, risk communication, and risk reduction strategies can help us determine priorities and help to persuade those who are paying, basically the taxpayers and consumers, that there is a decent return on our investment in risk management.

We are stretching beyond the limits of science to discuss risk. So it is not surprising that scientists disagree on risk estimates or on what should be done, if anything, to reduce those risks. Nevertheless, the public finds such disagreement disconcerting, and the cartoonists mock us!

In assessing health and ecological risks, we are stretching our knowledge of mechanisms and our capabilities below the range of exposures subject to the direct observations or experiments that are the domain of science. The extrapolation to low dose exposures reflects models, assumptions, speculation, and judgment. We need to explain better what is known and what is speculated.

Objectives of Risk Assessment

Risk assessment has been developed to address several different kinds of tasks. First, the laws covering pharmaceuticals and pesticides require that the responsible federal regulatory agencies (FDA and EPA) balance risks and benefits. Anticancer drugs, antimicrobial agents, and pesticides are designed to be toxic to living things. Balancing the benefits and risks depends on the margin of safety, patterns of use, and appropriate protections. Other laws, such as the Clean Air Act, do not explicitly authorize balancing benefits and risks.

A second risk management strategy is to set target levels of risk, usually as federal guidance to the states. Devolution to the states will be an increasing theme over the next few years, probably for the 25-year period you are addressing. There will be more and more responsibility laid on states and localities, which will have to come together to deal with the fact that environmental pollution does not respect geographic or political boundaries.

This strategy is used for food contaminants and water pollutants. Target levels are set, something that can be measured. If contaminant concentrations are above the target level, the food cannot be sold or the facilities discharging to water are out of compliance. If the contaminants are below the target level, the food or water is OK, "safe," presents negligible risk.

Third, everybody has to set priorities in the face of limited resources. Regulatory agencies, environmental groups, consumer groups, and manufacturers must decide where to invest their efforts, people, and dollars, including their effort to explain to the public what they're up against in trying to prevent hazards or develop products.

Finally, and most neglected, is the use of risk assessment to determine what we've accomplished. One source of skepticism about risk assessment and about our whole program of environmental stewardship and cleanup is that many people really don't understand how much has been accomplished in the last 25 years, despite well-documented gains in air quality, water quality, habitat protection, product safety, waste disposal, recycling, and pollution prevention. We should put ourselves to the test of predicting what can be done to reduce effects, or at least exposures, and then determine whether the steps that were taken, the investments that were made, really did accomplish those reductions.

For regulatory decision making, it is essential to have a guiding framework for risk management. We have such a framework for risk assessment, developed in the Carter Administration (Figure 1) and adapted in "The Red Book," the 1983 National Research Council report *Risk Assessment in the Federal Government: Managing the Process* (NRC 1983) (Figure 2). The first step is to determine whether a particular activity, chemical, microorganism, or radiation exposure could be hazardous to health or to the environment (Figure 1).

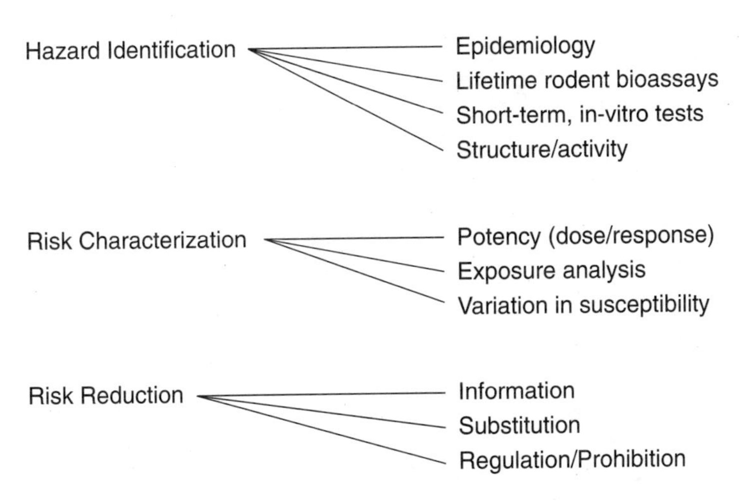

FIGURE 1 Framework for regulatory decision-making.

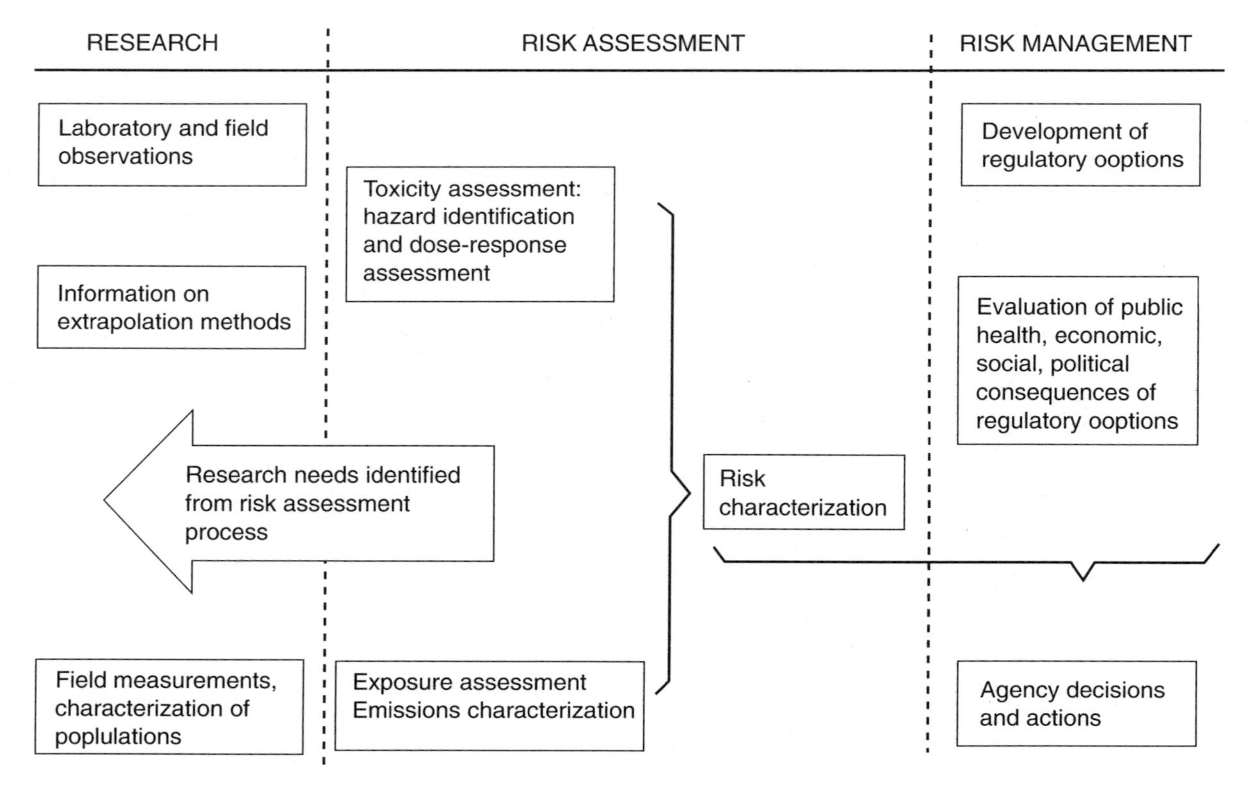
RESEARCH
RISK ASSESSMENT
RISK MANAGEMENT
Laboratory and field observations
Information on extrapolation methods
Research needs identified from risk assessment process
Field measurements, characterization of poplulations
Toxicity assessment: hazard identification and dose-response assessment
Exposure assessment Emissions characterization
Risk characterization
Development of regulatory ooptions
Evaluation of public health, economic, social, political consequences of regulatory ooptions
Agency decisions and actions

FIGURE 2

Whether there's a risk depends on there being an exposure, on potency of the chemical, and on individual variation in susceptibility due to genetics, nutrition, coexisting exposures, and protective actions. The characterization of risk in qualitative and quantitative terms serves as input to decisions and strategies for risk reduction. This framework can be applied quite generally. In fact, people use such thinking intuitively in their daily lives and in making judgments about any public policy of interest to them.

As I will tell you in a few moments, in the context of the Risk Commission, many people are uncomfortable with the term "risk." Probabilistic expressions and estimates of risk come out of science, engineering, and mathematical modeling; there is a perception that risk and risk analysis can be easily manipulated and is dominated by experts who see the world differently from ordinary people. We must take account of that discomfort and distrust if we hope to use these constructs broadly with all stakeholders.

The Presidential-Congressional Commission on Risk Assessment and Risk Management

I was asked to speak specifically about the Risk Commission. Table 1 lists the membership, all political appointees; six appointed by the Congress; three appointed by the Administration; the 10th member, Bernie Goldstein, appointed by the Academy.

It's a group with considerable scientific expertise; some of these people are well known to many of you. This Commission is the second part of a two-part mandate from the 1990 Clean Air Act Amendments, which first instructed EPA to fund a study at the National Academy of Sciences, in which Debbie Stine was a key staff person, on *Science and Judgment in Risk Assessment* (NRC 1994a).

TABLE 1 Presidential-Congressional Commission on Risk Assessment and Risk Management Membership

PETER Y. CHIU, MD, Kaiser Permanente, San Francisco
ALAN C. KESSLER, LLB, Buchanan Ingersoll, Philadelphia
SHEILA MCGUIRE, PhD, Iowa Health Research Inc, Boone, Iowa
NORMAN ANDERSON, American Lung Association, Augusta, Maine
DAVID P. RALL, MD, PhD, former director, NIEHS; Washington D.C.
JOHN DOULL, PhD, Toxicologist, Kansas University, Kansas City, Kansas
JOSHUA LEDERBERG, PhD, former president, Rockefeller University, New York City
GILBERT OMENN, MD, PhD, dean, University of Washington SPH, Seattle
VIRGINIA V. WELDON, PhD, vice president, Monsanto Company, St. Louis, Missouri
BERNARD GOLDSTEIN, MD, University of Medicine and Dentistry, Robert Wood Johnson Medical School, Piscataway, N.J.

Staff: Gail Charnley, PhD; Sharon Newsome; Joanna Foellmer

TABLE 2 Mandate

• Uses and limitations of risk assessment in decision making
• Appropriate exposure scenarios
• Uncertainty and risk communication
• Risk management policy issues
• Consistency across agencies

The title conveys the main theme, "science and judgment." That report is important input for our Commission.

Our mandate has five components (Table 2). I want to talk about each of these briefly and see if it stimulates discussion. First, the uses and limitations of risk assessment. I broadly hinted at one of the basic problems, that there are people in our society and well-organized groups who are skeptical about the intentions of risk assessment, who do not see risk assessment as a neutral framing mechanism, but as a tool of people who are expert in its use and employed by those who want to have risk assessments done. We must address that distrust and discomfort, and persuade folks that this is an approach to problem-solving that can be utilized by all.

The limitations are actually embedded in the science. Too much of what we've done in risk assessment, in my opinion and in the emerging opinion of the Commission as we prepare for recommendations to Congress next spring, has been focused on individual chemicals, one chemical at a time. Lay people do not associate exposures with one chemical at a time. They know that we have many, many exposures in our daily lives. There are hundreds of measured chemicals in contaminated water and contaminated air; there are thousands of chemicals, most of them already known to be carcinogenic, in cigarette smoke. So the fact that we talk about only one chemical at a time and present an elaborate risk assessment for one chemical stretches our credibility. We must find ways of thinking about mixtures and testing the effects of mixtures. In risk reduction we must indicate how much impact our actions could have in the broader contexts of all sources of air pollution or cancer rates or habitat protection.

The second limitation is that there has been an obsession with risk associated with chemicals, particularly from industrial and agricultural activities. For the most part, people have come to neglect the risks associated with microorganisms. But people in Milwaukee were reminded when an outbreak of cryptosporidiosis affected 400,000 people in 1993. In Seattle, and in several other cities and states, there was an outbreak of toxigenic *E. coli* with deaths of children and dozens of children now on long-term kidney dialysis as a result. Medical and industrial uses and, more dramatically, the Department of Energy nuclear weapons production sites remind us that we must think about mixtures of radiation and chemical exposures and understand the similarities and differences in their effects, their mechanisms, and their dose-response curves and take some lessons from one into the other field. For example, it's always surprised me that, while it is routine to do dose fractionation studies in radiobiology and in medical radiotherapy, we tend to ignore the rate of exposure in chemical toxicology and just average over a year or an eight-hour working period. Sometimes we have short-term exposure limits, but for the most part, we just average exposures, as if that covered the situation.

Those limitations in our science base limit the use of risk assessment as an analytical tool because the data are not there to address questions that state and local health officers and lay people logically expect us to address.

The second mandate involves exposure scenarios. Exposure assessment is the element of risk assessment that has been ridiculed the most—and for good reason. It became standard practice at EPA and in other agencies to postulate that the exposure of an individual or of a population could be characterized in a precautionary way by a "most-exposed individual," a hypothetical person with her or his nose in the fence line for 70 years breathing at a maximal breathing rate, with maximal assumptions compounded. To its credit, the EPA has moved away from this method in recent risk assessments toward a high-end, real-life individual or some subgroup of the upper end of the distribution. This is a promising development.

Another aspect of exposure assessment is recognition of different population groups. There's been quite a lot of attention lately, including an NRC report on *Pesticides in the Diets of Infants and Children* (NRC 1993) and environmental group advocacy, around the special protections required for infants and children, who have much smaller body mass and may have special dietary intakes, as in the case of fruit drinks and baby foods with pesticide residues.

We need to pay more attention to special exposures, differences in metabolism, and differences with age. We need to make them real and understandable to people who are the decision makers and ultimately to the public trying to understand what the exposure scenarios are and which populations most need to be protected.

Third, one of the best pieces of the NRC report *Science and Judgment in Risk Assessment*, in my opinion, was the distinction between variability and uncertainty. There is much variability in human populations, due to inherited differences in metabolism or in susceptibility at the site of action, nutrition, other exposures, effects of age early in life and late in life, coexisting diseases, and personal behaviors. All of that can be investigated and characterized.

But some aspects are not observable and rest on assumptions and models. This problem plagues the standard extrapolation from observable ranges, for example, with tumors in 10 to 100 percent of animals exposed, to try to make a judgment that the risk is less than one in a million for lifetime excess human cancer risk in a upper-bound, worst-case scenario. That enormous extrapolation of at least a factor of 100,000 covers uncertainty that cannot be addressed by real data. Modelers are undeterred, of course. Monte Carlo simulations and other methods are de rigueur in federal and state regulatory agencies, even at the state level, and in environmental consulting firms.

I think the Risk Commission is going to come out with a two-pronged recommendation here. We are quite keen to see probability distributions utilized for exposure parameters where you actually have data on adherence of chemicals to soils and release rates, on distribution of body weight in the population, on distribution of consumption of tap water or soil by children, and so forth. Exposure estimates can use real data, and the distributions can actually be validated for a particular site, if it's a site-specific problem.

We are not comfortable with probabilistic uncertainty analysis for the overall risk assessment or for the health effects when we don't even know whether the chemical can be properly classified as an agent causing cancer or birth defects, or not. We have been told over and over by non-technical specialists that decision makers don't want all that stuff. Qualitative judgments and full descriptions of the reasons for uncertainty are most welcome, however. What they really want from the technicians, the risk assessors, is guidance. Is it a problem or is it not a problem? I'm a busy person, I've got a full plate; is this something that requires my attention or not? That is not well captured with probability distributions and Monte Carlo simulations. There's a great risk that those folks just glaze over, walk away, and distrust all the risk assessment information.

We are aware that there's quite a debate about what can be called "bright lines," the notion that if you're above a certain value it's a problem, if you're below a certain value it's not a problem. Of course, there are enormous assumptions that go into a bright line. You may set, say, 10-' upper bound for lifetime cancer excess risk as the cut point for what's acceptable or not. Maybe, it should be 10-', one in 100,000. That bright line was adopted by the State of California and accepted by industry and the environmentalists in California for labeling of chemicals under Proposition 65. That compromise between 10-' preferred by industry and 10-6 preferred by environmentalists was crucial to the smooth work on Proposition 65 after all the strife at the beginning (Roe and Omenn 1995).

But let's be clear; there are numerous assumptions that go into that risk calculation, reflecting different future use scenarios, particularly. The Commission would like to encourage technical work to support choices of bright lines for exposures and emissions. As I have emphasized, risk cannot be measured, but exposure levels can be measured. We do this now for water pollutants and for food contaminants. We say you can sell peanuts and corn with aflatoxin B1 below a particular measured level, but not above. In this way, risk assessors and risk managers help bridge the gap between the people who want to have a practical basis for making a decision and taking action and explaining it and those who would like to deal with all the risk-related assumptions and details fresh every time.

The fourth area of the Commission's mandate is a set of risk management issues for risk reduction. The biggest items here have to do with economic analyses and comparisons of risks. It's no secret that the regulatory reform agenda in the Congress is heavily tied to the notion of benefit-cost analysis. It is shocking to us that those bills go on at length about the assumptions and uncertainties that must be described in detail for risk assessment, but assume the estimates of costs and of monetized benefits would be highly reliable, not requiring attention to assumptions, uncertainties, and peer review! We're amazed that numbers like the gross domestic product or the census or the unemployment rate are published with no uncertainty comment. We know there's plenty of uncertainty in those numbers. They are estimates. They are commonly revised, as you know. Maybe others would like to address this matter.

We have strongly urged the Congress to encourage cost-effectiveness analysis. We believe that costs should be considered and should be evaluated in making decisions about health and environmental protection. The last five Presidents, including the present President, have demanded that of their agency heads through executive order. Once the objective is determined to reduce exposures to certain levels to protect against cancers, birth defects, neurotoxicity, or immunological effects, or to protect habitat or achieve other environmental objects, it should be clear that we want to do so in the least costly manner. It should not be a burden to an agency head to stand before the public and explain that the benefits expected to be achieved through these regulatory actions justify the costs, by whatever metric that person feels capable of marshaling to build public support for the decision.

Finally, we were mandated to address consistency across federal agencies. It is a bone of contention with numerous parties that agencies, including programs within EPA, seem to take different approaches with similar data. Sometimes that's required by current statutes, or by the current interpretations of those statutes, at least. Other times, it's lack of coordination, lack of a common strategy, a failure to share data, or failure to compromise on assumptions like body weight extrapolation from rodents to humans. We believe there are many ways to improve consistency. This matter was of great interest to the Carnegie Commission Task Force on Regulatory and Judicial Decision Making.

There's quite a lot of work internationally on harmonization, testing protocols, risk assessment strategies, and standards. We expect harmonization to be increasingly important with globalization of trade and impacts of the World Trade Organization, NAFTA, and GATT. In fact, there's a cartoon character called GATTzilla, depicting the role of trade agreements on the environmental discussions.

Observations About Public Health and Environment

I was asked to make some comments about the relevant fields of public health sciences, especially epidemiology, biostatistics, and environmental health. Epidemiology is the core field of public health, investigating factors that cause or prevent diseases, deaths, injuries, or poor health. Epidemiology is undergoing a dramatic transformation with much more interest about underlying biological

TABLE 3 Characteristics of ATSDR Health Assessments and EPA Risk Assessments

ATSDR Public Health Assessment

- Qualitative, site specific; uses environmental contamination, health outcomes, and community health concerns data
- Medical and public health perspectives are weighted to assess health hazards
- Used to evaluate human health impacts and to identify public health interventions
- Is advisory
- May lead to pilot health effects studies, surveillance, epidemiologic studies, or exposure registry

EPA Risk Assessment

- Quantitative, compound oriented, not site specific
- Statistical and/or biologic models of dose-response and of exposure are used to calculate numerical estimates of health risk
- Used to facilitate remediations or other risk management actions
- Bears regulatory weight
- May lead to selection of particular remediation measures at a site

mechanisms and many more tools to relate animal studies to human studies. Biomarkers and mechanisms should be exciting scientific connections between ecology and human health.

Biostatistics is crucial to the design of studies and evaluation of data. Environmental health covers biological, chemical, and physical hazards; investigations of mechanisms; and assessments of risks, exposures, and ways to reduce exposures. Engineering plays a big role here, as do behavioral interventions.

I started off by saying that one of the problems with risk assessment is it tends to be chemical specific. Table 3 shows a contrast between EPA risk assessment and health assessment as normally practiced by local health departments, state health departments, and the Agency for Toxic Substances and Disease Registry, part of the U.S. Public Health Service, around Superfund sites or other places of possible contamination and exposure for local populations.

From a health point of view, people want to know more about what should be done. Qualitatively, is it a problem? Do I have to drink water from special bottled sources? Do we have to evacuate people? Do we have to take other kinds of precautions, or not? It relates to the site as a whole and not to particular individual chemicals.

Medical and public health perspectives are weighted and put in context. The response tends to be advisory. It doesn't bear the weight of regulation, although state and local health officers may have plenty of regulatory authority, as for quarantine. The health assessment may lead to further studies, whereas the risk assessment is intended to lead to remediation proposals at these sites.

One of the more interesting papers in recent years is one by Michael

McGinnis, just stepping down as long-time head of the Office for Disease Prevention and Health Promotion in the Department of Health and Human Services, and Bill Foege, former head of the Centers for Disease Control (McGinnis and Foege 1993) . This paper included Table 4, which shows the official 10 leading medical causes of death in the United States in 1990 and the lifestyles leading to half of these, 2.1 million deaths per year. As shown here, heart disease and cancers still account for a majority of all the deaths.

But when you ask what are the "real causes" of death (i.e., the lifestyles leading to these deaths), we get some guidance for public health attention. There's no question that the leading cause, by far, is tobacco. It's interesting that there's a whole new struggle developing between the President and FDA and the Congress over what, if anything, will be done to try to deal with the scourge of cigarette smoking in our country and increasingly around the world.

The second is a big number, but a lot less certain: diet and especially sedentary life-style contribute mightily to mortality rates in this country. Then comes alcohol. In fact, it's easy to say, in Johnny Carson's style, that the four "biggies" are cigarettes, alcohol and drugs, vehicles, and guns. Ask yourself if any of those

TABLE 4 The 10 Leading Medical Causes of Death . . .

Heart disease	720,000
Cancer	505,000
Cerebrovascular disease	144,000
Accidents	92,000
Chronic pulmonary disease	87,000
Pneumonia and influenza	80,000
Diabetes	48,000
Suicide	31,000
Liver disease, cirrhosis	26,000
AIDS	25,000

. . . and Lifestyle Factors Leading to Half of Them

Tobacco	400,000
Diet, sedentary life-style	300,000
Alcohol	100,000
Infections	90,000
Toxic agents	60,000
Firearms	35,000
Sexual behavior	30,000
Motor vehicles	25,000
Illicit drug use	20,000
Total	1,060,000

The nation's investment in prevention is estimated at less than 5 percent of the total annual health care cost.

are on the list of priority risks for this meeting or if any of those are regulated significantly by the EPA or other regulatory agencies or your state agencies. There are political reasons why they have been excluded, but we should have some perspective on what the major causes of death and disability really are in this country.

Toxic agents come in right in the middle of the list. Sixty thousand deaths, roughly estimated, is not an insignificant number. This estimate is fairly well justified in the original article and by many background articles, and it is certainly worthy of extensive reduction.

Before closing, let me mention one other item, the cleanup at DOE facilities, the legacy of 50 years of accumulation, decomposition, and migration of radioactive and chemical contamination. The polling paper in your meeting book describes how technology has been at the core of many human activities that have led to pollutants and contamination, as well as population growth and its attendant problems. At the same time, people have a great faith that technology can provide a fix to any problem.

After crucial technical advances that helped our Nation prevail in World War II and now in the Cold War, we are left with a task that will take at least 50 years for the cleanup. The estimated price tag of $250 billion, in 1990 dollars, represents promissory notes against the precious discretionary funds of the U.S. government, surely a source of tremendous future state and federal confrontation.

The Department of Energy has responded to the need for effective and continuing stakeholder involvement. There is considerable uncertainty about the technical assessments and priorities upon which those cleanup plans and associated promissory notes were built, and there has been a congressional demand for risk-based, integrated assessment of the present risks and the risks and cost-effectiveness of remediation options. The recommendations of the National Research Council report called *Building Consensus* (NRC 1994b), turned out in 60 days from the time DOE Assistant Secretary Tom Grumbly made the request to the Academy at a meeting in November 1993, are very compatible with how our Risk Commission feels about stakeholder involvement in comprehensive, iterative risk assessments.

The word "iterative" has a lot of baggage. *Science and Judgment in Risk Assessment* suggested a very conservative first analysis, a screening analysis to be followed by a more substantial analysis. Our Commission is anxious about that recommendation, because we believe all this work should be done in the open, disclosed to the public, involving the public. Once a very high risk estimate is generated under extremely conservative assumptions, we believe it is impossible to retain the public's confidence when a more careful assessment with more data is reported to justify a much lower risk estimate. It is very hard, we feel, to overcome the sense that somebody influential had a vested interest in reducing that risk estimate.

At the complex DOE sites, an iterative process as things are done and more is learned is very appropriate. Risk assessment(s) should compare potential outcomes and cost-effectiveness. *Building Consensus* recommended that risk assessments should involve the public, evaluate the risks of remediation, and involve an external organization. My colleagues and I have organized that new external entity, the Consortium for Risk Evaluation with Stakeholder Participation (CRESP). We are paying attention to potential grief to workers, ecosystems, and public health from the cleanup efforts themselves.

Finally, as we look 25 years ahead, we don't want to be in the position that we have picked up the problems, but not found the solutions. As a cartoonist put it, a driver of a tank truck is telling his hitchhiker, "Didn't you know, we just drive around. This is a mobile toxic waste dump!"

I look forward to comments from the panel and from all participants. My best wishes with your work.

References

Calkins D.R., R.L Dixon, C.R. Gerber, et al. "Identification, characterization, and control of potential human carcinogens: A framework for federal decision-making." *JNCI* 61:169-175 (1980).

McGinnis J.M., W.H. Foege. "Actual causes of death in the United States," *Journal of the American Medical Association* 270:2207-2212 (1993).

NRC (National Research Council), *Risk Assessment in the Federal Government: Managing the Process* (Washington, D.C.: National Academy Press, 1983).

NRC (National Research Council), *Pesticides in the Diets of Infants and Children* (Washington D.C.: National Academy Press, 1993).

NRC (National Research Council), *Science and Judgment in Risk Assessment* (Washington D.C.: National Academy Press, 1994a).

NRC (National Research Council), *Building Consensus Through Risk Assessment and Management of the Department of Energy's Environmental Remediation Program* (Washington, D.C.: National Academy Press, 1994b).

Omenn G.S., "Can Systematic, Integrated Risk Assessment with Full Stakeholder Participation Enhance Cleanup at DOE's Sites?" in (G.W. Gee and R. Wing, editors) The 1994 Herbert H. Parker Lecture presented at the Thirty-Third Hanford Symposium on Health and the Environment *In-situ Remediation: Scientific Basis for Current and Future Technologies* (Battelle Press, 1994, Part I). p. xv-xxx

Roe D., and G.S. Omenn, "California Has Successful Model of Regulatory Risk Assessment," (Oped), *Seattle Post-Intelligencer* p. A9 (July 25, 1995).

PART IV

APPENDIXES

APPENDIX A

Committee Member and Staff Biographical Information[1]

NATIONAL FORUM ON SCIENCE AND TECHNOLOGY GOALS—NO. 1: ENVIRONMENT

Co-Chairs

John F. Ahearne [NAE] is lecturer in public policy at Duke University and director of the Sigma Xi Center. Previously, he was executive director of Sigma Xi, The Scientific Research Society. He was deputy and principal deputy assistant secretary of defense, 1972-1977; White House Energy Office and deputy assistant secretary of energy, 1977-1978; and commissioner of the U.S. Nuclear Regulatory Commission, 1978-1983 (chairman, 1979-1981). From 1984 to 1989, he was vice president and senior fellow, Resources for the Future, where he is currently adjunct scholar. He received a bachelor of engineering physics and an MS from Cornell University and both an MA and a PhD in physics from Princeton University. Dr. Ahearne joined the Air Force in 1959 and resigned in 1970 with the rank of major. He then worked at the Air Force Weapons Center on nuclear weapons effects and taught at the Air Force Academy, Colorado College, and the University of Colorado (Colorado Springs).

H. Guyford Stever [NAS, NAE] is trustee at a variety of scientific agencies and a consultant on science issues. He was science and technology adviser to President Ford, 1976-1977. From 1972 to 1976, he was director of the National Science Foundation. He was president of Carnegie-Mellon University from 1965 to 1972, chief scientist of the Air Force from 1955 to 1965, and professor of aero-

[1][NAS] Member of the National Academy of Sciences / [NAE] Member of the National Academy of Engineering / [IOM] Member of the Institute of Medicine.

nautical engineering at Massachusetts Institute of Technology from 1946 to 1965. He received degrees from Colgate and California Institute of Technology. He was National Academy of Engineering foreign secretary from 1984 to 1988. In 1991, he was awarded the National Medal of Science.

Membership

Alvin L. Alm became assistant secretary for environmental management at the Department of Energy in May 1996. Before assuming that position, Mr. Alm was sector vice president and director responsible for the Environmental Business Area in the Science Applications International Corporation (SAIC). From 1985 to early 1987, Mr. Alm was chairman of the board and chief executive officer of Thermal Analytical Corporation. He stayed on the board until 1989. From 1987 to 1989, he was chief executive officer of Alliance Technologies Corporation and senior vice president of the parent company, TRC Companies, Inc. He became senior vice president and SAIC board member in June 1989. Mr. Alm received his BA in 1960 from the University of Denver and his MPA in 1961 from Syracuse University.

Barbara L. Bentley is professor of ecology at the State University of New York at Stony Brook, where she has been since 1973. While on the faculty, she was also dean of the Graduate School in 1983-1987. Her research interests include nitrogen fixation in tropical environments, plant and insect interactions, geographic variation in termite populations, and plant-herbivore interactions in *Lupinus*. She has been vice president for education for the Organization for Tropical Studies (1978-1985) and a member of the board of directors since 1974; vice president of the Ecological Society of America; director on the board of the Community Health Plan of Suffolk County, Long Island, New York; an appointee to the Environmental Conservation Board for the Village of Head of the Harbor, New York; a consultant to Brookhaven National Laboratory; a National Science Foundation visiting professor; a Fulbright fellow; and a member of the board of directors of the American Institute of Biological Sciences. Dr. Bentley received an MA from the University of California at Los Angeles and a PhD in ecology from the University of Kansas.

Jan E. Beyea is a consultant and former senior scientist with the National Audubon Society in New York. Before his current position, he was on the research staff at Princeton University from 1976 to 1980, an assistant professor of physics at Holy Cross College from 1970 to 1976, and a research associate in physics at Columbia University from 1968 to 1970. He is a member of the American Physical Society, the American Association for the Advancement of Science, and the Health Physics Society. Dr. Beyea received a BA from Amherst College in 1962 and a PhD in physics from Columbia University in 1968.

Harvey Brooks [NAS, NAE, IOM] is Benjamin Pierce Professor of Technology and Public Policy, emeritus, in the Kennedy School of Government, and Gordon McKay Professor of Applied Physics, Emeritus, in the Division of Applied Science at Harvard University. Dr. Brooks graduated from Yale University and did graduate physics at Cambridge University, England, and at Harvard University (PhD, 1940). He joined General Electric in 1946, where he served as associate head of the Knolls Atomic Power Laboratory. He returned to Harvard in 1950 as Gordon McKay Professor of Applied Physics. From 1957 to 1975, he served as dean of the Division of Engineering and Applied Physics at Harvard. Dr. Brooks's research has been in solid-state physics, nuclear engineering, underwater acoustics, and more recently, science and public policy. He has served on many committees related to science policy. He was a member of the National Science Board from 1962 through 1974. He has received six honorary DSc degrees from Kenyon College, Union College, Yale University, Harvard University, Brown University, and Ohio State University. In 1994, he was recipient of the Philip Hauge Abelson Prize of the American Association for the Advancement of Science.

Patricia A. Buffler [IOM] is the dean of the School of Public Health, University of California, Berkeley, where she also holds a faculty appointment as professor of public health and epidemiology. Before becoming dean at Berkeley in December 1991, she was a member of the faculty at the University of Texas-Houston Health Sciences Center, School of Public Health, where she has held numerous positions since 1979. From 1980 to 1984, she served as associate dean for research at the school, while holding positions as professor of epidemiology and director of the Epidemiologic Research Unit. In 1988, she was named by the Health Sciences Center as director of the Southwest Center for Occupational and Environmental Health. In 1989, Dr. Buffler was named by the University of Texas as the first Ashbel Smith Professor of Public Health. She has served as president of the American College of Epidemiology. She has served on many National Research Council committees. She holds an MPH and a PhD in epidemiology from the University of California, Berkeley.

John B. Carberry has been with DuPont since 1965, in professional and management assignments in research, technical assistance, operating supervision, and business development at eight company locations from New Jersey to Texas. Most of his assignments have involved the development of chemical processes or of new products. He is serving on the chemical-engineering advisory board at Cornell, on a National Research Council committee advising the Navy on technologies for dealing with shipboard wastes, and on the American Chemical Society Pollution Prevention Program Committee. He represents DuPont as the U.S. regional coordinating partner in the IMS Initiative for Cleaner Technologies. Mr. Carberry holds a BS (1963) and an MS in chemical engineering from Cornell University, and an MBA (1974) from the University of Delaware.

Emilio Q. Daddario [NAS, IOM] was a member of the U.S. Congress (1958-1972) and first director of Congress's Office of Technology Assessment (1973-1977). He was a professor at Massachusetts Institute of Technology (1970-1971); mayor of Middletown, Connecticut (1946-1948); and a municipal-court judge (1948-1950). He received the Ralph Coats Roe Award from the American Society of Mechanical Engineers in 1974, the honor award and medal from Stevens Institute of Technology in 1975, the Public Welfare award from the National Academy of Sciences in 1976, the Distinguished Service Award from the National Science Foundation in 1990, and the W.R. Grace Award from the American Chemical Society in 1992. Dr. Daddario holds a BA from Wesleyan University (1939), an LLB from the University of Connecticut (1942), a DSc from Wesleyan University (1967), and an LLD from Rensselaer Polytechnic Institute (1967).

Perry L. McCarty [NAE] is Silas H. Palmer Professor of Civil Engineering and director of the Western Region Hazardous Substance Center at Stanford University. He was also chairman of the Department of Civil Engineering from 1980 to 1985. Before then he was professor and associate professor of civil engineering (1962-1975). Before his career at Stanford, he was assistant professor of sanitary engineering at the Massachusetts Institute of Technology from 1958 to 1962. Dr. McCarty received a BS in civil engineering in 1953 from the Wayne State University and both an SM (1957) and an ScD (1959) in sanitary engineering from the Massachusetts Institute of Technology.

Rodney W. Nichols is president and chief executive officer of the New York Academy of Sciences. Before his current position, he was scholar-in-residence at the Carnegie Corporation from 1990 to 1992. He was vice president and executive vice president of the Rockefeller University from 1970 to 1990, having served as special assistant for research and technology in the office of the Secretary of Defense from 1966 to 1970. An applied physicist and systems analyst with industrial and international experience, he has written widely on science and technology policy and was a member of the Carnegie Commission's Executive Committee while serving on several of the commission's panels.

Paul R. Portney is president of Resources for the Future (RFF), an independent, nonpartisan research and educational organization concerning itself with natural resources and the environment. Immediately before becoming president of RFF, he held the position of vice president. Prior to that, he was director of its Center for Risk Management and its Quality of the Environment Division. From January 1979 to September 1980, he served as chief economist at the Council on Environmental Quality in the Executive Office of the President. From 1977 to 1979, he was a visiting professor at the Graduate School of Public Policy, University of California, Berkeley; in 1992, he was a visiting lecturer at Princeton University's Woodrow Wilson School of Public and International Affairs. He received his BA in economics and mathematics from Alma College and his PhD in economics from Northwestern University. He is the author or coauthor of a number of journal articles and books, the most recent of which is *Footing the Bill for Superfund Cleanups: Who Pays and How?*

F. Sherwood Rowland [NAS, IOM] has been Donald Bren Professor of Chemistry since 1989 at the University of California, Irvine, where he has taught since 1964 and was department chairman in 1964-1970. He was awarded the Nobel Prize for Chemistry in October 1995, which he shares with Paul Crutzen and Mario Molina. Before going to Irvine, Dr. Rowland taught at the University of Kansas (1956-1963) and at Princeton University (1952-1956). He has more than 300 scientific publications in atmospheric chemistry, radiochemistry, and chemical kinetics. Dr. Rowland is co-discoverer, with Dr. Molina, of depletion of the ozone layer of the stratosphere by chlorofluorocarbon gases. Dr. Rowland has been president and is a fellow of the American Association for the Advancement of Science. He holds an MS and a PhD from the University of Chicago and numerous honorary degrees.

Robert M. White [NAE] is president emeritus of the National Academy of Engineering and senior fellow of the University Corporation for Atmospheric Research at the American Meteorological Society. He holds a BA in geology from Harvard University and MS and ScD degrees in meteorology from the Massachusetts Institute of Technology. He served under five U.S. presidents from 1963 to 1977, first as chief of the U.S. Weather Bureau and finally as the first administrator of the National Oceanic and Atmospheric Administration. In those capacities, he brought about a revolution in the U.S. weather warning system with satellite and computer technology, helping to initiate new approaches to the balanced management of the country's coastal zones.

Staff Biographical Information

Lawrence E. McCray is director of the National Research Council's Policy Division and executive director of the Committee on Science, Engineering, and Public Policy. Dr. McCray held positions in the U.S. Environmental Protection Agency, the U.S. Regulatory Council, and the Office of Management and Budget before coming to the academies in 1981. He has directed academy studies in carcinogenic risk assessment, export controls, nuclear winter, and federal science budgeting. A Fulbright scholar in 1968, he received the Schattschneider Award in 1972 from the American Political Science Association for the best dissertation in American government and politics. In 1987, he received the National Research Council Staff Award.

Deborah D. Stine is study director and associate director of the Committee on Science, Engineering, and Public Policy (COSEPUP). Dr. Stine has been working on various projects throughout the academy complex since 1989. She received a National Research Council group award for her first study for COSEPUP on policy implications of greenhouse warming and a Commission on Life Sciences staff citation for her work in risk assessment and management. Other studies have addressed graduate education, environmental remediation, the national biological survey, and corporate environmental stewardship. She holds a bachelor's degree in mechanical and environmental engineering from the University of California, Irvine; a master's degree in business administration; and a PhD in public administration, specializing in policy analysis, from American University. Before coming to the academy, she was a mathematician for the Air Force, an air-pollution engineer for the State of Texas, and an air-issues manager for the Chemical Manufacturers Association.

Patrick P. Sevcik is the program assistant for the National Forum on Science and Technology Goals. Before his work at the National Research Council, Mr. Sevcik was an assistant program officer with the International Republican Institute (IRI) from 1990 to 1993 working primarily in Central and Eastern Europe. He has held positions at the White House in the Office of Political Affairs (1989-1990) and on Capitol Hill (1987-1988) in the office of Representative John DioGuardi (R-NY). During this time, Mr. Sevcik also held concurrent positions in several Slovak-American organizations. He holds a BA in international affairs, with an emphasis on Soviet and Eastern European studies, from the George Washington University. He has also studied Russian language and culture at the Leningrad Polytechnic Institute in former Leningrad, USSR.

APPENDIX
B

Forum Agenda

NATIONAL FORUM ON SCIENCE AND TECHNOLOGY GOALS: ENVIRONMENT

The Arnold and Mabel Beckman Center of
the National Academies of Sciences and Engineering
100 Academy Drive
Irvine, California 92715

August 21–24, 1995

Beckman Center Auditorium

Monday, August 21

9:00 Registration

9:30 Welcome and Forum Overview
Committee Co-Chairs: John F. Ahearne and H. Guyford Stever

9:40 Now that the country is 25 years from Earth Day, what should be the country's environmental goals 25 years from today?
Keynote Speakers:
D. James Baker, National Oceanic and Atmospheric Administration
Thomas Grumbly, U.S. Department of Energy (via video link)
Robert Watson, Office of Science and Technology Policy
Barry Gold, U.S. Department of the Interior (for Ronald Pulliam)
Harlan Watson, House Committee on Science
David Garman, Senate Committee on Energy and Natural Resources

12:00 Lunch

Review of Proposed National Environmental Goals

1:30 U.S. Environmental Protection Agency
John Wise, Deputy Regional Administrator, Region 9; and Peter Truitt, Senior Analyst, Office of Policy, Planning, and Evaluation, and Manager, National Environmental Goals Project, EPA

2:30 President's Council on Sustainable Development
Judith Espinosa, Former Secretary of Environment, New Mexico & Member, PCSD; and Peggy Duxbury, Coordinator, Principles, Goals, and Definitions Task Force, and Staff, PCSD

3:30 Committee on Environment and Natural Resources of the National Science and Technology Council
Robert Watson, Associate Director for Environment, Office of Science and Technology Policy

4:30 Commissioned Paper: Review of National Environmental Goals
Richard Morgenstern, Resident Consultant and Visiting Fellow, Resources for the Future
Discussant: Carl Mazza, Science Advisor, Office of U.S. Senator Daniel P. Moynihan

5:30 Adjourn for the Day

5:30 Reception—Terrace

Tuesday, August 22

8:00 Registration

Review of Current Environmental Status

9:00 Commissioned Paper: Review of Environmental Quality Status
N. Phillip Ross, Chief, Office of Regulatory Management and Evaluation, U.S. Environmental Protection Agency
Discussant: John Shanahan, Policy Analyst, Environment, The Heritage Foundation

10:00 Commissioned Paper: Review of Public Opinion on Environmental Quality
Karlyn Bowman, Resident Fellow, American Enterprise Institute for Public Policy Research
Discussant: Clinton Andrews, Professor, Woodrow Wilson School, Princeton University

11:30 Public Comment and Discussion Opportunity

12:30 Lunch

Other Environmental Goals

2:00 Commissioned Paper: Review of Other Countries' National Environmental Goals
Konrad von Moltke, Senior Fellow, The Institute on International Environmental Governance, Dartmouth College

Discussant: William Stewart, Senior Associate, Pacific Institute for Studies in Development, Environment, and Security

3:00 Commissioned Paper: Review of State, Local, and Community Environmental Goals
Richard Minard, Associate Director, Center for Competitive, Sustainable Economies, National Academy of Public Administration
Discussant: Carol Whiteside, Director, Office of Intergovernmental Affairs, State of California

4:00 Public Comment and Discussion Opportunity

5:00 Adjourn for the Day

Wednesday, August 23

8:00 Registration

Industry Goals

9:00 Risk Assessment and Management Commission
Gilbert Omenn, Dean, School of Public Health and Community Medicine, University of Washington

10:00 Commissioned Paper: Review of Industry Environmental Goals
John Ehrenfeld, Director, Program on Technology and Environment, Massachusetts Institute of Technology
Discussant: Dorothy Ellington, Environmental Compliance Manager, Research Oil Company

11:00 Public Comment and Discussion Opportunity

1:00 Lunch

Ecological Goals

2:00 Commissioned Paper: Review of Ecological Goals
Walter Reid, Vice President for Programs, World Resources Institute
Discussant: Rodney Fujita, Senior Scientist, Environmental Defense Fund

3:00 Public Comment and Discussion Opportunity

Funding and National Environmental Goals

3:30 Commissioned Paper: Review of Environmental Goals Relative to Funding Priorities
Albert Teich, Head, Directorate for Science and Policy Programs, American Association for the Advancement of Science
Discussant: Kwai-Cheung Chan, Director of Program Evaluation in the Physical Systems Area, General Accounting Office

5:00 Public Comment and Discussion Opportunity

5:30 Adjourn for the Day

Thursday, August 24

9:00 Breakout Discussion Groups

12:00 Lunch

1:00 Adjournment

APPENDIX
C
Forum Participants

Forum Attendees

George Alapas
Program Manager, ORD
U.S. Environmental Protection Agency
Annandale, Va.

Bill Alevizon
Aquatic Ecologist
The Bay Institute
San Rafael, Calif.

Clinton J. Andrews
Professor
Woodrow Wilson School
Princeton University
Princeton, N.J.

Jack Azar
Manager
Environmental Design and Resource Conservation
Xerox Corporation
Webster, N.Y.

The Honorable D. James Baker
Undersecretary for Oceans and Atmosphere
National Oceanic and Atmospheric Administration
U.S. Department of Commerce
Washington, D.C.

Ray Beebe
Consultant
Tucson, Ariz.

Matt Bonaiuto
Research Associate
Council of Governors' Policy Advisors
Hall of the States
Washington, D.C.

Lee Botts
Independent Environmental Consultant
Gary, Ind.

Karlyn Bowman
Resident Fellow
American Enterprise Institute for Public Policy Research
Washington, D.C.

William J. Carroll
Vice-Chairman
Montgomery-Watson, Inc.
Pasadena, Calif.

Kwai-Cheung Chan
Director
Program Evaluation in the Physical Systems Area (PEPSA)
U.S. General Accounting Office
Washington, D.C.

Edwin Clark
President
Clean Sites
Alexandria, Va.

William J. Cook
Chief, Research Grants Division
Office of Technology Development
Pennsylvania Department of Commerce
Harrisburg, Pa.

Robert Coppock
Board on Sustainable Development
National Academy of Sciences
Washington, D.C.

David H. Critchfield
Director
Recycling Programs
International Paper
Memphis, Tenn.

Robert T. Drew
Director
Health and Environmental Services Department
American Petroleum Institute
Washington, D.C.

Peggy Duxbury
President's Council on Sustainable Development
The White House
Washington, D.C.

Thomas J. Dwyer
Manager
Marketing Development/ Environmental Products Division
Corning, Inc.
Corning, N.Y.

Shaun S. Egan
Program Manager
Colorado Center for Environmental Management
Denver, Colo.

John Ehrenfeld
Director
MIT Program on Technology and Environment
Massachusetts Institute of Technology
Cambridge, Mass.

Dorothy M. Ellington
Environmental Compliance Manager
Research Oil Company
Cleveland, Ohio

Judith Espinosa
Former Secretary of Environment, New Mexico
Albuquerque, N.M.

Timothy R. Fennell
Chemical Industry Institute of Toxicology
Research Triangle Park, N.C.

Frederick W. Freeman
Environmental Adviser
U.S. Agency for International Development
Washington, D.C.

Rodney M. Fujita
Senior Scientist
Environmental Defense Fund
Oakland, Calif.

David Gardiner
Assistant Administrator
Office of Policy Planing and Evaluation
U.S. Environmental Protection Agency
Washington, D.C.

David K. Garman
Professional Staff Member
Subcommittee on Energy R&D
Senate Committee on Energy and Natural Resources
Washington, D.C.

Barry Gold
Scientific Planning and Coordination
National Biological Service
U.S. Department of the Interior
Washington, D.C.

Jonathan Greenberg
Director, Environmental Policy
Browning Ferris Industries
Washington, D.C.

Thomas Grumbly
[via video-link from Washington]
Assistant Secretary for Environmental Management
U.S. Department of Energy
Washington, D.C.

Karen C. Guevara
Office of Management and Budget
Washington, D.C.

Ellen Stern-Harris
Executive Director
Fund for the Environment
Beverly Hills, Calif.

Frank B. Kapper
Director
Advanced Government Programs
Corning, Inc.
Washington, D.C.

Douglas M. Kleine
Executive Vice President
Soil and Water Conservation Society
Ankeny, Iowa

Frank Kreith
American Society of Mechanical Engineers, Legislative Fellow
National Conference of State Legislators
Denver, Colo.

David A. Litvin
Vice President
HSEQ
Kennecott Corporation
Salt Lake City, Utah

Jerry L. May
Manager
Lockheed Idaho Technologies Co.
Idaho Falls, Idaho

Carl Mazza
Science Advisor
Office of Senator Moynihan
Washington, D.C.

Elizabeth McKay
National Chapter Coordinator
Student Pugwash USA
Washington, D.C.

Becky McKelvey
Texas Instruments
Lewisville, Tex.

Richard A. Minard, Jr.
Associate Director
Center for Competitive, Sustainable Economies
National Academy of Public Administration
Washington, D.C.

Konrad von Moltke
Institute on International Environmental Governance
Dartmouth College
Hanover, N.H.

Richard Morgenstern
Resources for the Future
Washington, D.C.

B. Kim Mortensen
Epidemiology and Environmental Health Services
Upper Arlington, Ohio

Rebecca G. Moser
Special Assistant to the Under Secretary for Oceans and Atmosphere
National Oceanic and Atmospheric Administration
U.S. Department of Commerce
Washington, D.C.

Robert O'Keefe
Director, Program Strategy
Health Effects Institute
Cambridge, Mass.

Gilbert S. Omenn
Dean
School of Public Health and Community Medicine
University of Washington
Seattle, Wash.

Stanley Paytiamo
Environmental Protection Specialist
Pueblo of Acoma
Acomita, N.M.

Dennis R. Poulsen with Bonnie L. Poulsen
Environmental Services Manager
California Steel Industries, Inc.
Fontana, Calif.

Alfredo E. Prelat
Senior Scientist/Texaco Honorary Fellow
Texaco Exploration and Production Department
Houston, Tex.

Walter V. Reid
Vice President for Programs
World Resources Institute
Washington, D.C.

David Rejeski
Office of Science and Technology Policy
The White House
Washington, D.C.

Deanna Richards
Senior Program Officer
National Academy of Engineering
Washington, D.C.

David Z. Robinson
Executive Director
Carnegie Commission on Science, Technology, and Government
New York, N.Y.

N. Phillip Ross
Office of Policy and Planning
Environmental Information and Statistical Division
U.S. Environmental Protection Agency
Washington, D.C.

Alex Sapre
Corporate Manager
ESP Technology
Hughes Electronics
Los Angeles, Calif.

C. Thomas Science
Sciance Consulting Services, Inc.
Hockessin, Del.

John Shanahan
Policy Analyst, Environment
The Heritage Foundation
Washington, D.C.

William Stewart
Senior Associate
Pacific Institute for Studies in Development, Environment, and Security
Oakland, Calif.

John C. Tao
Corporate Director
Technology Partnerships
Air Products and Chemicals, Inc.
Allentown, Pa.

Roy L. Taylor
Rancho Santa Ana Botanic Garden
Claremont, Calif.

Albert H. Teich
Directorate for Science and Policy Programs
American Association for the Advancement of Science
Washington, D.C.

Peter G. Truitt
Manager
National Environmental Goals Project
U.S. Environmental Protection Agency
Washington, D.C.

Harlan L. Watson
Staff Director
Subcommittee on Energy and Environment
House Committee on Science
Washington, D.C.

Robert Watson
Office of Science and Technology Policy
The White House
Washington, D.C.

Carol G. Whiteside
Director
Office of Intergovernmental Affairs
State of California
Sacramento, Calif.

William R. Wiley
Senior Vice President
Science and Technology
Battelle Memorial Institute
Richland, Wash.

John C. Wise
Deputy Regional Administrator, Region 9
U.S. Environmental Protection Agency
San Francisco, Calif.

Committee Members In Attendance

John F. Ahearne, *Co-chair*
Lecturer in Public Policy
Duke University
Research Triangle Park, N.C.

H. Guyford Stever, *Co-chair*
Science Consultant
Washington, D.C.

Alvin L. Alm
Director and Sector Vice President
Science Applications International Corporation
McLean, Va.

Barbara L. Bentley
Department of Ecology and Evolution
State University of New York, Stony Brook
Stony Brook, N.Y.

Jan E. Beyea
Senior Scientist
National Audubon Society
New York, N.Y.

Patricia A. Buffler
Dean
School of Public Health
University of California, Berkeley
Berkeley, Calif.

John B. Carberry
Director, Environmental Technology
DuPont Research and Development
Wilmington, Del.

Emilio Q. Daddario
Former Member of Congress
Washington, D.C.

Robert A. Frosch
Senior Research Fellow, CSIA
John F. Kennedy School of Government
Harvard University
Cambridge, Mass.

Perry L. McCarty
Silas H. Palmer Professor of Civil Engineering
Stanford University
Department of Civil Engineering
Stanford, Calif.

Rodney W. Nichols
Chief Executive Officer
New York Academy of Sciences
New York, N.Y.

Paul R. Portney
Vice President
Resources for the Future
Washington, D.C.

F. Sherwood Rowland
National Academy of Sciences
Foreign Secretary
Department of Chemistry
University of California at Irvine
Irvine, Calif.

Robert M. White
President Emeritus, National Academy of Engineering
Senior Fellow, University Corporation for Atmospheric Research
American Meteorological Society
Washington, D.C.

The Honorable Richard Thornburgh, *Co-chair designee, NFSTG—No. 2*
Kirkpatrick & Lockhart
Washington, D.C.

Staff

Lawrence E. McCray
Director, Policy Division
National Academy of Sciences
Washington, D.C.

Deborah D. Stine
Study Director, NFSTG
National Academy of Sciences
Washington, D.C.

Patrick P. Sevcik
Program Assistant, NFSTG
National Academy of Sciences
Washington, D.C.

APPENDIX D

Summary of Responses to Call for Comments

Provided below is a brief summary of the responses received to the call for comments:

Question 1: What can science and technology contribute to meeting current national environmental goals?

Cost-effectiveness was one of the most common elements mentioned throughout the survey and was obviously one of the key parts of the responses to all the questions.

- Form a defensible basis for policy and strategy formation focusing on cost-effective, socially acceptable, and phased solutions.
- Define problems better, including their scale and possible actions or solutions to resolve problems.
- Communicate answers and provide tools to members of the public so that they can understand problems and solutions.
- Conduct effective and reliable risk and cost-benefit analysis so that resources are allocated to solve real problems and priorities can be set.
- Demonstrate and define the effectiveness and efficiency of available environmental remediation techniques and make these techniques faster, safer, cheaper, more innovative, and more effective.

A smaller group of more technical responses followed some of the themes outlined above:

- Determine the fundamental mechanism causing a problem and the source of that cause, including spelling out the underlying physical processes involved, identifying potentially unsustainable uses of resources, and determining the significance of environmental stress.

- Develop new and improved industrial, transportation, and energy "green processes" that reduce levels of primary pollutants, use more recycled materials, produce more easily recycled products, reduce the use of hazardous waste, and prevent pollution.
- Develop more advanced measurement instrumentation and platforms that provide more accurate, detailed, timely, and cost-effective assessment of pollutant distributions, background biogeochemical states and processes, interactive monitoring, and real-time information.
- Better manage science and technology, including balancing facts and figures with human dimensions of cooperation, attitudes, and issues.
- Undertake rigorous analysis of available information relevant to goals.

Question 2: What do you believe should be the nation's environmental goals for the future?

As indicated by one participant:

> Environmental goals are, by nature, multi-faceted, and a detailed listing of all important issues is subject to preferences and priorities. To avoid these choices at this stage, we believe that stating the following overall goal is more productive and allows specifics to be developed later. The nation's environmental goal should be to achieve an economy built on the principles of Sustainable Development. . . . This can guide the creation of more specific goals and focused objectives.

After that one overarching response, there was no particular consensus. Other key responses were these:

- Preserve (or improve) all natural resources as they exist.
- Reform environmental legislation and regulation so that they are more cost-effective and flexible without reducing environmental quality.

Question 3: How can science and technology contribute to meeting these future goals?

- Provide framework for any future environmental goals.
- Create cost-effective technical opportunities.
- Develop new options.
- Communicate to public and politicians.
- Develop measurement tools.
- Understand long-term consequences of today's solutions.
- Reduce degree of uncertainty in problem solving.
- Understand complex systems.
- Develop sound scientific foundation.

Question 4: Provided below are the environmental goals for the United States developed by the Environmental Protection Agency. Please rank what you consider to be the top five goals by placing a number from 1 to 5 before those items:

Clean Air
Climate Change Risk Reduction
Stratospheric Ozone Layer Restoration
Clean Waters
Healthy Terrestrial Ecosystems
Safe Indoor Environments
Safe Drinking Water
Safe Food
Safe Workplaces
Preventing Spills and Accidents
Preventing Wastes and Toxic Products
Safe Waste Management
Restoration of Contaminated Sites

Please list other goals that you believe should be in the top five, and indicate what rank you would give them:

- Generally, clean air, water, food, and soil were the top concerns.
- Specific comments indicated that it was not appropriate to separate these topics on a medium basis and that economics should be included in the analysis.

Question 5: Provided below are a few of the environmental goals for the United States currently being developed by the Presidential Commission on Sustainable Development. What are your comments on these goals?

- *Human health and equity.* We envision an American society where healthy and economically secure people sustain—and are sustained by—a healthy environment. Every person breathes clean air; drinks clean water; eats safe food; and lives, works, and plays in clean, pleasant, and safe surroundings.
- *Ecosystems.* Ensure the health of ecosystems and natural processes, including protection of biological diversity and the quality of water, air, and soil. The health of ecosystems must be accomplished through efforts to restore damaged ecosystems and through management of the use and enjoyment of ecosystems.
- *Environmental quality.* Attain a safe and clean environment by making pollution prevention, waste reduction, and product stewardship standard practice.
- *Efficient production and resource utilization.* Achieve a constant and significant improvement in the efficiency of materials use and production of all stages of resource development—extraction, production, manufacturing, and end-

use (and make pollution prevention, waste reduction, and product stewardship standard practice), with corresponding reductions in resource use to sustainable levels in environmental risks.

- —It is difficult not to agree with these goals.
- —Is the presence of one of these contradictory to another (economic growth versus environmental protection)?
- —Key is to define terms (e.g., clean, safe, pleasant, quality).
- —The goals are mushy and weak.

Question 6: How can science and technology contribute to achieving your selected top environmental goals?

Generally, respondents either specifically indicated that there was no difference in the answer to this question and question 3, or they gave answers that were similar to those for question 3.

Question 7: What are the chief barriers to achieving your top five environmental goals?

These fell into five categories: (1) economic, (2) legislative and regulatory, (3) leadership and political, (4) knowledge, and (5) education.

Some specific responses included:

- Failure to understand relation between economic activity and environmental perturbations
- Lack of individual responsibility and understanding
- Lack of focus on the 20% of issues that are causing 80% of the problems
- Government regulation
- Political forces
- Lack of systems approach
- Capital and human resources
- Cost-competitive technologies
- Adversarial relationship among business, government, and envirocrats
- Effective market incentives.

Question 8: What would be necessary for you to conclude each goal was achieved?

Responses generally fell into the following categories:

- Goals will never be achieved.
- Metrics are necessary for defining a healthy environment.
- Net growth in environmental perturbations is 0.
- There should be full integration of environmental goals into business.

- Environmental monitoring is necessary.
- Economic incentives are present.
- There must be clear indicators of progress.
- Public understanding needs to be achieved.
- There must be an adequate number of success stories.

Question 9: What information regarding science and technology's ability to aid in meeting the nation's environmental goals do you wish you had now that is unavailable?

Responses included the following:

- Integration of science and technology with socioeconomics
- Response of sociopolitical system
- Science and technology electronic database (data, technology, programs and projects)
- Forums for idea sharing
- Clear, simply worded documents for public use

Question 10: What questions should we have asked, and what are your answers?

Some suggested questions follow:

- How do we take action on environmental goals?
  —Technology development
  —Relationship with other countries
  —Scientific research
  —Pollution prevention
  —Society communication
  —Policy and institutional design
  —Tools development
  —Balancing of cost and benefits
- What are possible, probable, and preferable outcomes in next 20-50 years if we don't achieve our goals?
- What are the limits as to what science and technology can do to achieve goals?

APPENDIX E

Respondents to Call for Comments

Tundi Agardy
World Wildlife Fund
Washington, D.C.

David P. Albright
President
Alliance for Transportation Research
Albuquerque, N.M.

George C. Allen, Jr.
Manager, Environmental Programs
Sandia National Laboratories
Albuquerque, N.M.

Brad Allenby
AT&T ERC
Princeton, N.J.

J. Allies
President
TNT Technology Company
Tempe, Ariz.

Victor Ashe
Mayor
City of Knoxville
Knoxville, Tenn.

Michael Barker, AICP
Executive Director
American Planning Association
Washington, D.C.

R.R. Beebe
Mining and Metallurgical Society of America
Tucson, Ariz.

Harvey M. Bernstein
President
Civil Engineering Research Foundation
Washington, D.C.

David Blaskovich
Director of Marketing
Government and Environment Markets
Field Marketing
CRAY Research Inc.
Washington, D.C.

David Bodansky
Chairman, Panel on Public Affairs
American Physical Society
Washington, D.C.

David L. Bodde
Vice President
Midwest Research Institute
Kansas City, Mo.

Lynn M. Bradley, FAIC
Environmental Health Project Director
Association of State and Territorial Health Officials
Washington, D.C.

Lewis M. Branscomb
Kennedy School of Government
Harvard University
Cambridge, Mass.

Dale E. Brooks
Managing Director
Government Relations
American Institute of Chemical Engineers
Washington, D.C.

Annice Brown
Environment and Natural Resources Division
Asia Technical Department
World Bank
Washington, D.C.

Robert D. Brown
President
National Association of University Fisheries and Wildlife Programs
Department of Wildlife and Fisheries Sciences
Texas A&M University
College Station, Tex.

Susan L. Brown
Director
Economic Development Department
City of Knoxville
Knoxville, Tenn.

Tom Buechler
Black & Veatch
Kansas City, Mo.

Gregory H. Canavan
Senior Scientist
Physics Division
Los Alamos National Laboratory
Los Alamos, N.M.

Bruce B. Canty
George Washington University
Ashburn, Va.

John B. Carberry
DuPont Science & Engineering
Wilmington, Del.

William J. Carroll
President
WFEO/FMOI
Pasadena, Calif.

Arthur H. Chappelka
Associate Professor
School of Forestry
Auburn University, Ala.

Kenneth W. Chilton
Director, Center for the Study of American Business
Washington University
St. Louis, Mo.

Richard A. Conway
Union Carbide
South Charleston, W.Va.

John K. Crum
Executive Director
American Chemical Society
Washington, D.C.

Kevin S. Curtis
Senior Associate
The Keystone Center
Washington, D.C.

I.H. Cushman
Oak Ridge National Laboratory
Oak Ridge, Tenn.

Robert D. Day
Executive Director
Renewable Natural Resources Foundation
Bethesda, Md.

Christopher F. D'Elia
Professor and Director
Maryland Sea Grant College
University of Maryland System
College Park, Md.

Dan Dessecker
Forest Ecologist
The Ruffed Grouse Society
Rice Lake, Wisc.

Thomas W. Devine
RMT, Inc.
Greenville, S.C.

Susan Eisenberg
Executive Director
National Association of Environmental Professionals
Washington, D.C.

Laurie A. Fathe
Assistant Professor
Department of Physics
Occidental College
Los Angeles, Calif.

Robert L. Ford
Center for Energy and Environmental Studies
Southern University Baton Rouge Campus
Baton Rouge, La.

Robert Fri
Resources for the Future
Washington, D.C.

Elizabeth V. Gardener
Conservation Officer
Denver Water
Denver, Colo.

Joan N. Gardner
President
Applied Geographics, Inc
Boston, Mass.

Edward Gerjuoy
Professor of Physics (emeritus)
University of Pittsburgh
Pittsburgh, Pa.

R. Jane Ginn
Ginn & Associates
Newcastle, Wash.

William T. Golden
Carnegie Commission on Science, Technology, and Government
New York, N.Y.

Andrew J. Goodpaster
Co-Chair
The Atlantic Council of the United States
Washington, D.C.

Larry Gordon
School of Public Administration
University of New Mexico
Albuquerque, N.M.

Debra Grabowski
Environmental Epidemiologist
Albuquerque Environmental Health Department
Albuquerque, N.M.

John Graham
Harvard Center for Risk Analysis
Boston, Mass.

Jon Greenberg
Director of Environmental Policy
Browning Ferris Industries
Washington, D.C.

Brad Grems
Waste Policy Institute
Blacksburg, Va.

Alan Gressel
President
Research Oil Company
Cleveland, Ohio

Kenneth D. Haddad
Department of Environmental Protection
St. Petersburg, Fla.

Jay D. Hair
President
National Wildlife Federation
Washington, D.C.

Philip W. Hamilton
Managing Director
Public Affairs
American Society of Mechanical Engineers
Washington, D.C.

William M. Haney III
President and CEO
Molten Metal Technology, Inc.
Waltham, Mass.

Robert E. Hegner
Battelle Pacific Northwest Laboratory
Washington, D.C.

Ronald H. Henson
Vice President
Environment, Health and Safety/ Engineering Assembly and Testing
Pratt & Whitney
United Technologies
East Hartford, Conn.

George M. Hidy
College of Engineering
Center for Environmental Research and Technology
University of California, Riverside
Riverside, Calif.

Thomas E. Hitchins, AIA
Architect
Hewitt, N.J.

K. Elaine Hoagland
Executive Director
Association of Systematics Collections
Washington, D.C.

Naomi U. Kaminsky
Scientific Affairs Program Manager
American Pharmaceutical Association
Washington, D.C.

Elizabeth A. Kay
Vice President
Ecological Solutions, Inc.
Boston, Mass.

David R. Kiesling
Technical Director
Fluor Daniel Environmental Services
Irvine, Calif.

C. Judson King
Vice Provost for Research
University of California
Oakland, Calif.

Richard L. Klimisch
Vice President
Engineering Affairs Division
American Automobile Manufacturers Association
Detroit, Mich.

Charles E. Kolb
President
Aerodyne Research, Inc.
Billerica, Mass.

Richard F. Kosobud
Department of Economics
University of Illinois at Chicago
Chicago, Ill.

A.B. Krewinghaus
Shell Development Company
Houston, Tex.

Floy Lilley
Program Manager
University of Texas at Austin
College of Engineering
Austin, Tex.

Morton Lippman
Professor
Nelson Institute of Environmental Medicine
New York University Medical Center
Tuxedo, N.Y.

A.P. Malinauskas
Director
Office of Environmental Technology Development
Lockheed Martin Energy Systems
Oak Ridge, Tenn.

Michael Markels, Jr.
Chairman Emeritus
Versar, Inc.
Springfield, Va.

Edward J. Martin
Executive Director
Hazardous Materials Control
Resources Institute
Rockville, Md.

Peter May
Department of Political Science
University of Washington
Seattle, Wash.

Linda S. McCoy
Deputy Director
Advanced Technologies and Planning Division
Department of Energy
Idaho Operations Office
Idaho Falls, Idaho

Alan Miller
Center for Global Change
University of Maryland
College Park, Md.

Dennis F. Miller
Manager, International Programs
Idaho National Engineering Laboratory
Washington, D.C.

James M. Murray
President
Timberlock Company
La Jolla, Calif.

Gerald Nehman
Director
Environmental Institute for Technology Transfer
University of Texas at Arlington
Arlington, Tex.

George B. Newton
Director
System Planning Corporation
Arlington, Va.

J. Patrick Nicholson
Chief Executive
N-VIRO International Corporation
Toledo, Ohio

Charles Noss
Water Environment Research Foundation
Alexandria, Va.

Michael Novacek
Vice President and Provost of Science
American Museum of Natural History
New York, N.Y.

Gilbert S. Omenn
Dean, Public Health and Community Medicine
University of Washington
Seattle, Wash.

James F. Pankow
Department of Environmental Science Engineering
Portland, Ore.

Rod Parrish
Society of Environmental Toxicology and Chemistry
Pensacola, Fla.

Elisabeth Pate-Cornell
Department of IE-EM
Stanford University
Stanford, Calif.

Marcus Peacock
House Committee on Transportation and Infrastructure
Congress of the United States
Washington, D.C.

Madison J. Post
National Oceanic and Atmospheric Administration
Environmental Technology Laboratory
Boulder, Colo.

Dennis R. Poulsen, CEP
Manager, Environmental Services
California Steel Industries, Inc.
Fontana, Calif.

Peter H. Raven
Director
Missouri Botanical Garden
St. Louis, Mo.

James A. Roberts
Global Environmental
Sacramento, Calif.

Jonas Salk
The Salk Institute for Biological Studies
San Diego, Calif.

Edward S. Sarachik
Atmospheric Sciences
University of Washington
Seattle, Wash.

Lynn Scarlett
Vice President, Research
Reason Foundation
Los Angeles, Calif.

Raymond C. Scheppach
Executive Director
National Governors' Association
Washington, D.C.

Wayne A. Schmidt
Research Manager
National Wildlife Federation
Great Lakes Natural Resource Center
Ann Arbor, Mich.

Edward D. Schroeder
Department of Civil and Environmental Engineering
University of California, Davis
Davis, Calif.

C. Thomas Sciance
Sciance Consulting Services, Inc.
Hockessin, Del.

Henry Shaw
New Jersey Institute of Technology
Newark, N.J.

Paul S. Sheng
University of California, Berkeley
Berkeley, Calif.

John T. Sigmon
School of the Environment
Duke University
Durham, N.C.

Scott Sklar
Executive Director
Solar Energy Industries Association
Washington, D.C.

Eugene B. Skolnikoff
Massachusetts Institute of Technology
Cambridge, Mass.

Marina Skumanich
Research Scientist
Battelle Seattle Research Center
Seattle, Wash.

Maggie Smith
Michael Baker Jr., Inc.
Virginia Beach, Va.

Robert Solow
Princeton University
Princeton, N.J.

Christine Stevens
President
Animal Welfare Institute
Washington, D.C.

Sue A. Tolin
President
American Phytopathological Society
St. Paul, Minn.

Michael Toman
Resources for the Future
Washington, D.C.

Linda K. Trocki
Program Director
Los Alamos National Laboratory
Energy Technology Programs Office
Los Alamos, N.M.

R. Rhodes Trussell
Montgomery Watson, Inc.
Pasadena, Calif.

Francis Y. Tsang
Global Technologies, Inc.
Idaho Falls, Idaho

Jane Hughes Turnbull
Electric Power Research Institute
Palo Alto, Calif.

James M. Vail
Senior Science Advisor
American Petroleum Institute
Washington, D.C.

William R. Waldrop
Tellico Village
Loudon, Tenn.

Philip G. Watanabe
Blaire, Wash.

Mike Way
Executive Director
Colorado Alliance for Environmental Education
Denver, Colo.

Stephen R. Weil
Bechtel Hanford, Inc.
Richland, Wash.

Conrad G. Welling
Ocean Minerals Co.
Menlo Park, Calif.

Iddo Wernick
Research Associate
Program for the Human Environment
The Rockefeller University
New York, N.Y.

Roy F. Weston
Roy F. Weston, Inc.
West Chester, Pa.

Grace Wever
Vice President
Council of Great Lakes Industries
Rochester, N.Y.

W.R. Wiley
Senior Vice President
Science and Technology Policy
Battelle Memorial Institute
Richland, Wash.

Jane Willeboordse
NGO/Government Liaison
International Solar Energy Society
Columbia, Md.

Julian Wolpert
Professor of Geography, Public Affairs and Urban Planning
Woodrow Wilson School of Public and International Affairs
Princeton University
Princeton, N.J.

APPENDIX

F

Summary of Breakout-Group Discussions

On the last day of the forum, participants divided into four breakout groups to discuss what they had heard from the authors of the commissioned papers and those who gave keynote speeches and to discuss the process itself. A summary of the discussion by each group is presented below.

GROUP I

Group I enjoyed the breakout session portion of the meeting and felt that this portion should be expanded in future years to provide more time for discussion. As to issues that were addressed in the forum, Group I felt that the following issues were not sufficiently discussed during the forum:

- Energy
- Water
- Transportation
- Population
- Land use

Its overall recommendation was that science and technology could help society by developing an infrastructure for a systems approach to monitoring that would provide an early warning system for such emerging areas as biotechnology; a decentralized monitoring system for data collected by local governments; and a centralized adaptive management approach to federal government activities such as that at the Advanced Research Projects Agency and the National Institutes of Health.

In addition, Group I felt that the committee should set milestones for the scientific and engineering community. The committee should indicate that environmental research and development should be part of the normal activities of a business—as opposed to the current separation. Also, science and engineering schools should teach environmental ethics and more information about the social constraints on engineering and scientific activities so as to raise overall awareness of scientists and engineers in these matters.

GROUP II

Group II indicated that one of the current dilemmas that makes it difficult for the scientific and engineering community to respond to societal needs is that everything has high priority. Other issues that need to be addressed are the level to which the public is engaged in risk-assessment and why the environmental effort at the federal level is separated into so many agencies, as opposed to a single institution.

Issues that Group II felt should be addressed were these:

- Monitoring of biological, physical, and chemical changes
- Development of a source of available, inexpensive, renewable, noncarbon energy, while keeping in mind that conservation is still the least expensive source of energy
- Understanding of complex ecological, human, and other dynamic systems
- Development of negentropic technologies for mixing and separating products

Such diverse subjects could be linked by a high-quality robust federal research and development system that focuses on the environment and is capable of coupling societal goals to science and technology.

GROUP III

Group III felt that the key issues were setting priorities and developing a knowledge and information base. Knowledge and information can be developed via a process that involves the broader scientific community, that adapts to new information, that takes action before damage occurs, and that takes into account the social context of environmental goals.

Some problems that need to be addressed include the following:

- Multigenerational effects
- Groundwater pollution
- Ocean pollution
- Nonpoint-source pollution
- Fish stocks
- Industrial ecology

- Clean water supply
- Nuclear power
- Pollution and the science and engineering infrastructure of developing countries
- Assimilative capacity of ecosystems
- Air quality trends
- Population growth and economic development
- Space and waste management
- Incremental pollution increase
- Triggers for pollution

GROUP IV

Group IV felt that some of the key issues society will wish that it had addressed 25 years in the future in terms of their impact on the environment are a decline in literacy and the fast pace of technological change (e.g., the life expectancy of computers is only two years).

Science and technology can contribute to society's environmental goals by conducting research in

- energy sciences, particularly in decarbonization and renewable energy;
- environmental monitoring and dissemination of its results to evaluate progress (or lack thereof) in protecting the environment;
- economic science—how to balance benefits and costs, the cost of environmental compliance, the willingness of the current generation to sacrifice for the future, and the appropriate role of government in fostering environmental science and technology particularly with respect to incentives to companies to invest in such technologies;
- materials science in high-strength fibers, composites, chemicals, and electronic chemicals; and
- institutional impediments in the ecological system, transportation, biotechnology, and industrial progress.

APPENDIX

G

Detecting Changes in Time and Space

BARBARA BENTLEY

Natural environments change: environmental "problems" are essentially unacceptable changes in the rates or direction of this change. Thus, understanding the environmental impacts of human activities requires measuring and understanding those changes. However, we often think of goals as fixed endpoints and the setting of goals as rates and directions of change. The latter can be far more important in terms of the ultimate environmental result.

NATIONAL ENVIRONMENTAL MONITORING NETWORK

A National Environmental Monitoring Network (NEMN) could be used to

- establish baseline data from which environmental change can be measured;
- detect trends (both slope and direction) of environmental measures;
- determine areas in which research, remediation, or regulation needs to be applied; and
- assess the effectiveness (and ultimately cost-effectiveness) of environmental programs in achieving progress in the directions and at the rates that have been tentatively established as goals.

Sites

NEMN sites could be set up to adequately cover, to the extent possible, all appropriate natural, rural, urban, and offshore systems. The relative density of the sites in any given area would be a function of the complexity of the area (e.g., more dense in complex coastal areas than in the short grass prairie). Site selection could take advantage of existing monitoring stations, such as those at local, regional, or national parks, airports, nature reserves, biological research stations, or other areas with appropriate monitoring facilities. The establishment of new sites is likely to be required in some areas, either because of the low density of existing facilities in the area or because of the unusually high complexity of the area.

Data Collected

Each site could be equipped with the appropriate instrumentation to collect and transmit basic physical and biological data including the following:

- Physical data and sampling
- Weather: temperature, precipitation, cloud cover, insolation,[1] wind speed, wind direction, and variability
- Atmospheric chemistry
- Soil chemistry and microorganisms
- Water chemistry and flow rates
- Biological data and sampling
- Species richness and diversity
- Population dynamics of target species
- Land-use maps

Standardizing Formats

The methodologies for data collection must be standardized among all sites in the network. These methodologies will be established by a Task Force on Data Standardization consisting of qualified representatives from the physical, biological, and statistical sciences. Although a majority of the data collected will be the same throughout the network, some site-specific data sets and protocols must be established to account for variation among the sites.

To ensure that the protocols continue to be the best available, the methodologies will be reviewed by an external ad hoc committee convened every five years. In addition, site sampling and pilot studies could be used to establish sampling strategies in various environments.

Data Entry, Storage, and Access

Data collected at each of the sites will be entered into a central database on a regular basis (daily, monthly, quarterly, or annually as appropriate). Data entry could be via electronic media, and current data could be available upon demand by scientists, policy makers, and the general public.

[1]Insolation is the amount of incoming radiation.

Scientists must have unencumbered access to the raw data. The only requirement to gain access to the data is that the scientist supply NEMN with reprints of publications resulting from use of the data.

It shall be the responsibility of NEMN to maintain the database, prepare and provide summary statistics of the data, and become the curator of physical samples accumulated in conjunction with the collection of data.

Data Interpretation

The NEMN will sponsor an ongoing program of solicited papers reviewing the trends and implications detected in selected data sets. These papers must be peer reviewed and could be disseminated widely. The target audiences for these papers may vary to include, at different times, scientists, policy makers, and the general public.

Prior to publication, all solicited papers must be subjected to peer review to ensure the highest quality of data presentation and interpretation.

The NEMN will also maintain a library of publications resulting from the use of data generated by NEMN sites. A bibliography of these publications will be kept current and made widely available.

Personnel

Data collection and collation require a major commitment of personnel:

- *On-site technicians*. Qualified technicians must be on-site to supervise data monitoring instrumentation and to forward data sets to the central database. Qualifications need not include a college degree but could include training in instrument maintenance and data entry.
- *Summer parataxonomists*. To collect the volumes of biological data necessary at each site, students, amateur naturalists, and other community volunteers could work with professional field biologists to assess changes in species richness and diversity. Because most biological systems are seasonal, this work will be done primarily during the summer. Although the design of collection protocol and identification of the organisms require professional training, actual field work can be done by enthusiastic volunteers.
- *Professional scientists*. Detailed analyses and interpretations of data collected throughout the network could be done by a cadre of scientists working in NEMN. The tasks of these scientists should also include monitoring the integrity of the data by developing protocols, running quality checks, and overseeing the educational materials and solicited papers or other publications coming from NEMN.

APPENDIX
H

Enabling the Future: Linking Science and Technology to Societal Goals

A Report of the Carnegie Commission on Science, Technology, and Government

CONTENTS

EXECUTIVE SUMMARY

As for the Future, your task is not to foresee, but to enable it.
—Antoine de Saint-Exupéry, *The Wisdom of the Sands*

The end of the Cold War, the rise of other economically and scientifically powerful nations, and competition in the international economy present great opportunities for the United States to address societal needs: policymakers may now focus more attention on social and economic concerns and less on potential military conflicts. In the next decade and those that follow, the United States will confront critical public policy issues that are intimately connected with advances in science and technology. Policy decision making will require the integration of numerous considerations, including accepted scientific knowledge, scientific uncertainty, and conflicting political, ethical, and economic values. Policy issues will not be resolved by citizens, scientists, business executives, or government officials working alone; addressing them effectively will require the concerted efforts of all sectors of society. As Vannevar Bush wrote in his 1945 report to the President, *Science: The Endless Frontier*:

> Science, by itself, provides no panacea for individual, social, and economic ills. It can be effective in the national welfare only as a member of a team,

> whether the conditions be peace or war. But without scientific progress no amount of achievement in other directions can insure our health, prosperity, and security as a nation.[1]

The task force recognizes that many sectors of society contribute to the setting and achievement of long-term science and technology (S&T) goals, particularly the state governments and the industrial sector. Many policy areas with which state governments have had decades of experience, such as transportation, education, and agriculture, have come to the top of the national policy agenda. Nearly every state has a science and technology policy advisor or economic development program centered on science and technology, and it is through the states that many of our national S&T policies are implemented.[2] Even though the private sector is largely influenced by shorter term economic forces, it still employs the majority of scientists and engineers in the country and performs most of the nation's R&D. As a consequence, industry plays an important role in establishing and achieving long-term S&T goals.

Furthermore, we feel that it is important to recognize the role of international cooperation and development in government decision making in S&T. As discussed in a recent report by the Carnegie Commission, the distinction between "domestic" and "foreign" goals for science and technology is obsolete in the face of the explosion of global technology, information, capital, and people. If they are to be forward-thinking, our policies must now integrate national and international views.[3]

With this consideration in mind, our report focuses primarily on the role of the federal government in establishing and achieving long-term S&T goals. It also suggests some ways in which current problems can be managed and future issues can be identified and addressed. We discuss opportunities for opening the science policy process to a broader spectrum of society by creating and institutionalizing a forum for exchanging ideas. We also present mechanisms through which society and public officials can deal with the inevitable and continuing conflicts in goal-setting.

VOYAGES OF DISCOVERY

Basic scientific research is a voyage of discovery, sometimes reaching the expected objective, but often revealing unanticipated new information that leads, in turn, to new voyages. Some might say that setting long-range goals may harm basic researchers by overcentralizing and removing flexibility from the system. Long-range S&T goal-setting certainly should not hamper, but rather encourage, this freedom to discover. Furthermore, goal-setting should be a pluralistic, decentralized process.

The federal government is largely responsible for setting major goals and broad budget priorities between and among major disciplines (for example, biology and physics). It also plays a major role in setting priorities within disciplines (for example, particle and solid state physics), and must encourage the symbiotic combinations of differing fields (for example, biology and chemistry with respect to biotechnology products).

The relationships between scientific and technological advancement and government support are complex, and the stakes in these decisions are high, not just for scientists and engineers, but for society as a whole. Consequently, a better understanding of the process of articulating goals, both within and outside science, is vital.

THE CHOICE FOR AMERICA

We believe that America faces a clear choice. For too long, our science and technology policies, apart from support of basic research, have emphasized short-term solutions while neglecting longer-term objectives. If this emphasis continues, the problems we have encountered in recent years, such as erosion of the nation's industrial competitiveness and the difficulties of meeting increasingly challenging standards of environmental quality, could overwhelm promising opportunities for progress. However, we believe there is an alternative. The United States could base its S&T policies more firmly on long-range considerations and link these policies to societal goals through more comprehensive assessment of opportunities, costs, and benefits.

We emphasize the necessity for choice because there is nothing inevitable about the shape of the future: the policy decisions we make today will determine whether historic opportunities will be seized or squandered. American science could repeat its past successes: in the past three decades, American S&T has helped eradicate diseases, reverse the pollution of many of our rivers and lakes, reach the moon, launch the computer age, and spread the Green Revolution around the world. We may be able to achieve a new age of vitality and leadership in the world community. Or the problems of recent years—such as the loss of technological and commercial advantage to other nations, or our continuing dependence on foreign energy supplies—could prove irreversible. In short, the future is limited only by our ingenuity. As Frank Press, President of the National Academy of Sciences, said recently, "Without a vision of the future, there is no basis for choosing policies for science and technology that will be appropriate for the years ahead."[4]

This report seeks ways to improve the knowledge, understanding, and information available to the federal government on the long-term nature of the S&T enterprise as it relates to societal goals. As the government goes about the complex annual process of setting budget priorities and developing program plans for the S&T enterprise, it could use this knowledge, understanding, and information to ensure that both long- and short-term objectives are taken into account.

The report focuses on an interconnected set of ideas that, if implemented, would help accomplish this aim. The underlying theme of the set of recommendations is an effort to improve the capacity of the federal government to establish and achieve long-term S&T goals. At the core of our report is the recognition that there are significant efforts already under way within the federal government, but departments and agencies must be encouraged to direct more attention to long-term thinking. We describe the activities of several units of both the executive and legislative branches of government, recommend ways to strengthen their capabilities, and suggest mechanisms through which long-range, strategic planning can help federal departments and agencies fulfill their missions.

In addition to our recommendations directed to established governmental units, we have proposed the creation of a National Forum on Science and Technology Goals that would bring representatives of the science and technology community together with others from a broad set of fields who are interested in societal activities that have major S&T components. The Forum would work to identify ways in which science and technology can contribute to the definition and refinement of societal objectives and to their realization. Ultimately, it would try to articulate S&T goals, monitor efforts to achieve them, and maintain sustained support for particular objectives. The Forum would also define and develop criteria in support of dynamic goals such as the future needs of the several components of the science and technology base—basic research, generic technology, education and training, research facilities, and information dissemination, to name a few—in an effort to ensure their long-term health.

Several key considerations underlie our recommendation for a National Forum. The first is that a private forum must have long-term continuity in order to become an important contributor to federal policies. There are inherently long lead times associated not only with goals but also with the dynamics of major technological change. It is the mismatch between these realities and more immediate economic and political concerns that must be wrestled with. The second key consideration is the recognition that many organizations exist, both within and outside government, that do some long-term strategic planning. The Forum should make maximal use of these worthwhile efforts.

Furthermore, if the products of the Forum are to be useful, it must have strong linkages to the executive and legislative branches of the federal

government as well as the state governments. Finally, a balanced and effective interaction is needed between the scientific and engineering communities and those representing a broad range of other societal interests.

Our report does not address the issue of setting specific societal goals, because we believe this is primarily a political process. We do list a broad set of societal goals to indicate the general directions toward which S&T should be applied. Most of our report is devoted to the *process* of establishing S&T goals; however, we do present some examples of S&T goals for illustrative purposes.

RECOMMENDATIONS

Although this report touches on a number of goal-related themes, our recommendations focus on a few key issues: improving our national capacity to define and revise long-term S&T goals; linking S&T programs and goals more closely and clearly to broader societal goals; and building more effective linkages between governments (especially the federal government) and other sectors of society in debating, articulating, and pursuing these goals while assessing progress toward their achievement. To this end, we present a set of interconnected recommendations. We believe that each recommendation, in itself, is useful and should be implemented; however, the recommendations have been designed to support and strengthen each other and should be viewed as a whole.

In developing recommendations in this report, we sought to identify mechanisms to bring the major sectors of society—government, industry, academia, nongovernmental organizations, and the public—together to examine ways in which science and technology can be focused on achieving the nation's long-term objectives. Centralization of planning is not the answer, as the failures of command economies have demonstrated. However, we badly need a focusing of national attention and resolve. We also need to ensure that we are taking full advantage of the knowledge resulting from our national research and development efforts as we work to achieve societal objectives. Bridging the gap between research and policymaking is essential, and the assessment process is an effective bridging mechanism that must be used more frequently in the future as policymakers work to devise strategies for achieving long-range goals.

Throughout our work, we have been mindful of the great diversity of processes that help define the direction of national policy. There is no simple way to promote systematic long-term thinking about policy directions. For this reason we devote our recommendations to a variety of mechanisms within and outside government to foster discussion and debate about potential long-term S&T goals and the means of achieving them.

▪ **A nongovernmental National Forum on Science and Technology Goals should be established to facilitate the process of defining, debating, focusing, and articulating science and technology goals in the context of federal, national, and international goals, and to monitor the development and implementation of policies to achieve them.** The National Forum, as we envision it, would be responsible for undertaking several key activities (see Box 6 on p. 50). The Forum would convene individuals from industry, academia, nongovernmental organizations, and the interested public to explore and seek consensus on long-term S&T goals and the potential contribution of scientific and engineering advances to the achievement of societal goals.

The importance of the long-term goal-setting task is matched by the difficulty of carrying it out. For example, great diligence, fair-mindedness, and imagination would be needed to ensure that the Forum did not become either a vehicle for self-promotion by scientists and engineers or a venue for lodging grievances arising from technological change. The goal-setting process must involve individuals who have exhibited the ability to take a broad statesman-like view of complex issues.

We suggest two options for administering the Forum: the National Academies complex or a new, independent, nongovernmental organization. Regardless of the option chosen, we believe that the activities of the National Forum should be overseen by a Board of Directors responsible for selecting the members of the Forum's Council. The Council should be made up of representatives of a broad spectrum of our society who are appointed for fixed length, rotating terms. The Board should ensure that the Council is provided with the necessary institutional facilities, financial management, personnel, and other administrative backing to carry on the Forum's mission. We envision the Council as the leadership organization for the Forum.

▪ **Congress should devote more explicit attention to long-term S&T goals in its budget, authorization, appropriation, and oversight procedures.** Congressional support is key to the long-term productivity of science and technology. Budget, authorization, appropriation, and oversight procedures are complex and highly decentralized, and there are opportunities to improve the ways in which Congress addresses S&T issues. We have not, however, focused too closely on these opportunities. The Committee on Science, Technology, and Congress of the Carnegie Commission will address these issues in an upcoming report.[5]

We believe that one of the most effective ways for Congress to consider S&T issues in the longer term would be for the House Committee on Science, Space, and Technology, which has responsibility for cross-cutting science policy considerations, to hold a series of hearings, on an annual or biennial basis, on long-term goals for science and technology. The purpose of these hearings would be to step back from the budget process and near-term political considerations and consider science and technology from the long-term perspective. However, we also believe that each legislative committee in the House and Senate that has jurisdiction over major segments of federal S&T activities should periodically, perhaps biennially, devote formal attention to more specific questions regarding long-term S&T goals in its area of responsibility.

Congressional committees could ask the appropriate federal agencies and a full spectrum of responsible nongovernmental interests for their views on long-term S&T goals, hold hearings, and issue reports embodying the committees' conclusions. As the proposed Forum matures and gains public confidence, the leadership of the Senate and the House of Representatives may wish to develop mechanisms to use the Forum's output throughout congressional S&T policymaking activities.

- **In order to provide Congress with the information, analysis, and advice necessary to make policy decisions in this area, the Office of Technology Assessment and other congressional support agencies should evaluate national efforts to establish and achieve long-term science and technology goals in the context of societal goals.** The support agencies should work with congressional committees to consider what kinds of analyses of long-term S&T goals would help inform their legislative agendas. OTA, in particular, should apply its well-tested assessment process to analyzing long-term S&T goals and the procedures by which federal agencies articulate and work toward their achievement. The Congressional Budget Office (CBO), although it has limited responsibilities for S&T policy, has considerable expertise in economic analysis, which is an essential component of the goal-setting process. CBO should put its expertise to use in evaluating economic considerations with respect to long-range science and technology policy.

More specifically, we believe that CBO and OTA should establish an ongoing coordinated activity designed to combine their strengths in analyzing economics and science and technology in order to evaluate goals and budget priorities for science and technology. Furthermore, because we believe that interactive linkages are the key to solving complicated problems, we suggest that OTA, with the cooperation of the other congressional support agencies, assist congressional committees and the congressional leadership in reviewing and evaluating the products of the Forum.

■ **The Office of Science and Technology Policy (OSTP) and the Office of Management and Budget (OMB) within the Executive Office of the President should actively contribute to the establishment of federal science and technology goals and should monitor the progress of departments and agencies in attaining these goals.** Establishing long-term goals and communicating them to the federal agencies is a process that must be conducted separately from the annual budget process. With specific goals in mind, the agencies can create a budget that balances their vision of the future with the realities and constraints of the present.

OSTP and OMB should communicate long-term S&T goals to departments and agencies before the beginning of the budget cycle each year. In addition, both OSTP and OMB should work with these departments and agencies throughout their budget-planning processes to assure that long-term S&T goals are considered and advanced in their internal policy-planning activities.

OSTP should also monitor, critically evaluate, and report to the President and Congress on the progress of federal programs in achieving long-term S&T goals. In particular, OSTP should function as one liaison point between the National Forum and the Executive Branch. With OSTP leadership, the Federal Coordinating Council on Science, Engineering, and Technology (FCCSET) should extend its promising efforts in shaping long-term S&T goals involving more than one federal agency and emphasize the articulation of specific long-term goals through a more explicit planning process. Furthermore, the President's Council of Advisors for Science and Technology (PCAST) should play a more extensive role in guiding the goal-setting process within the Executive Office.

■ **Federal departments and agencies should enhance their policymaking efforts, integrating considerations of long-term science and technology goals into annual budgeting and planning efforts.** Federal agencies should enhance their strategic planning capabilities and develop explicit long-term S&T goals in the context of broader national goals established by Congress and the President. In order to do this, open communication and cooperation among the senior R&D administrators of departments and agencies should be encouraged. These individuals should meet periodically to discuss longer-term objectives and ways in which their work might contribute to or compete with broader goals and stated policies. If this approach proved effective, it could become a more formal step in the policymaking process. Furthermore, federal agencies should be required to present publicly each year an analysis of how their planned activities relate to their long-range S&T programs. Resource requirements to support the achievement of these goals should be incorporated into annual budget plans.

In addition, we recommend that federal agencies support extramural policy studies that can aid in developing and evaluating long-term S&T goals. The National Science Foundation (NSF) should develop and monitor indicators of the health and productivity of the science and technology enterprise and its contributions to societal goals. NSF should expand its competitive grants program in science and technology policymaking and work to involve scientists and engineers in the S&T goal-setting process. NSF, in conjunction with OSTP and other federal agencies, should establish continuing programs to develop the information base necessary to monitor progress in achieving long-term S&T goals. Furthermore, the National Science Board should assume greater responsibility for devising approaches to setting long-term goals with respect to the S&T base.

INDEX

D

E

F

J

K

L

M

Q

R

V

W

Z